METHODS IN MOLECULAR BIOLOGY™

Series Editor
John M. Walker
School of Life Sciences
University of Hertfordshire
Hatfield, Hertfordshire, AL10 9AB, UK

For further volumes:
http://www.springer.com/series/7651

Next-Generation MicroRNA Expression Profiling Technology

Methods and Protocols

Edited by

Jian-Bing Fan

Illumina, Inc., San Diego, CA, USA

Humana Press

Editor
Jian-Bing Fan
Illumina, Inc.
San Diego, CA, USA
jfan@illumina.com

ISSN 1064-3745 e-ISSN 1940-6029
ISBN 978-1-4939-6217-4 ISBN 978-1-61779-427-8(eBook)
DOI 10.1007/978-1-61779-427-8
Springer New York Dordrecht Heidelberg London

Softcover re-print of the Hardcover 1st edition 2012

Printed on acid-free paper

Humana Press is part of Springer Science+Business Media (www.springer.com)

Preface

The rapid pace of microRNA (miRNA) research continues to drive the advances of techniques for miRNA expression profiling. However, several unique attributes of miRNAs, including their small size, which limits the choice of probe selection, huge dynamic range of expression level, lack of polyadenylated tails, significant sequence homology among family members, and tendency to cross-hybridize to their mRNA targets with imperfect sequence homology, have made them challenging to quantify. To meet all these challenges, innovative technologies that are more sensitive, specific, quantitative, and that are compatible with a wide range of biospecimens have been developed during the past few years.

In this volume of the *Methods in Molecular Biology* series, we assembled a broad spectrum of methods and protocols that cover most of the next-generation miRNA expression profiling technologies. Two introductory chapters serve as references for the context of the more specialized chapters that follow. Chapter 1 (Gommans and Berezikov) provides a general overview to the miRNA field, such as miRNA biogenesis and regulation in human development and diseases, and Chapter 2 (Aldridge and Hadfield) provides a summary of most of the widely used miRNA assay technologies as well as insightful cross-platform comparisons.

We also include comprehensive coverage of methodologies that have been developed for miRNA profiling, including (1) quantitative PCR using either a stem-loop RT-PCR method (Chapter 3: Hurley et al.) or a Poly(T) adaptor RT-PCR approach (Chapter 4: Shi et al.), (2) in situ hybridization (Chapter 5: Nielsen), and (3) microarray analysis with a variety of microarray platforms, such as Agilent (Chapter 6: Andrade and Fulmer-Smentek), Illumina (Chapter 7: Chen et al.), Affymetrix (Chapter 8: Dee and Getts), Luminex (Chapter 9: Sorensen), Febit (Chapter 10: Beier et al.), and LC Sciences (Chapter 11: Zhou et al.). All these techniques allow high-throughput measurement of well-annotated miRNAs.

The past few years have witnessed the rapid development of next-generation sequencing technology and fast adoption of the technology for miRNA research. This area is covered by several excellent chapters that involve the Illumina miRNA-Seq platform (Chapter 12: Luo), Roche 454 GS FLX technology (Chapter 13: Soares et al.), Life Technologies SOLiD platform (Chapter 14: Linsen and Cuppen), and Helicos single-molecule sequencing technology (Chapter 15: Kapranov et al.). The sequencing approach provides several technical advantages over microarrays. It offers (1) more comprehensive coverage and *de novo* discovery potential, (2) single base specificity, and (3) better detection sensitivity and dynamic range.

In addition to hands-on "wet-lab" protocols, this volume also covers the use of miRNA databases, which deal with the annotation and discovery of miRNAs and other noncoding RNAs (Chapter 16: Yang and Qu), and the function of miRNAs in human diseases and biological processes (Chapter 17: Ruepp et al.), as well as data normalization methods (Chapter 18: D'haene et al.), next-generation sequencing data analysis (Chapter 19: Gunaratne et al.), and integrated miRNA expression analysis and target prediction (Chapter 20: Ritchie and Rasko).

Furthermore, specialized applications, such as cancer studies (Chapter 21: Zhong et al.) and miRNA-based noninvasive biomarker development (Chapter 22: Debey-Pascher et al.), are also addressed.

The content within this book is intended for students, researchers, and scientists in the field, at both the beginner and advanced levels, and contains sufficiently detailed protocols, particularly pointers, that will assist with troubleshooting. While each of the methods has some technical limitations, many of them have been used successfully in broad scientific researches. The choice of the method mainly depends on the users' desired application.

Finally, I would like to thank all the authors for their outstanding contributions to this timely developed protocol book. I would also like to thank Dr. Craig April for his assistance with the preparation of the chapters, as well as Professor John Walker, the *Methods in Molecular Biology* series editor, and David Casey at Humana Press. I truly hope this book will help accelerate the expression analysis of miRNA and expand our understanding of the biological functions of miRNA in different species and human diseases.

San Diego, CA, USA ***Jian-Bing Fan***

Contents

Contributors

SARAH ALDRIDGE • *CRUK Cambridge Research Institute, Cambridge, UK*
CRAIG S. APRIL • *Illumina, Inc., San Diego, CA, USA*
MARKUS BEIER • *febit biomed gmbh, Heidelberg, Germany*
EUGENE BEREZIKOV • *Hubrecht Institute, Utrecht, The Netherlands*
VALESCA BOISGUÉRIN • *febit biomed gmbh, Heidelberg, Germany*
ANDREW BOND • *Life Technologies Corporation, Foster City, CA, USA*
CAIFU CHEN • *Life Technologies Corporation, Foster City, CA, USA*
JING CHEN • *Illumina, Inc., San Diego, CA, USA*
VINCENT L. CHIANG • *Forest Biotechnology Group, Department of Forestry and Environmental Resources, North Carolina State University, Raleigh, NC, USA*
CRISTIAN COARFA • *Bioinformatics Research Laboratory, Baylor College of Medicine, Houston, TX, USA*
GEORGE COUKOS • *Department of Obstetrics and Gynecology, Center for Research on Early Detection and Cure of Ovarian Cancer, University of Pennsylvania, Philadelphia, PA, USA*
EDWIN CUPPEN • *Hubrecht Institute, Utrecht, The Netherlands*
PETULA N. D'ANDRADE • *Agilent Technologies, Santa Clara, CA, USA*
BARBARA D'HAENE • *Biogazelle, Zwijnaarde, Belgium*
SVENJA DEBEY-PASCHER • *Life and Medical Sciences (LIMES), Genomics and Immunoregulation, University of Bonn, Bonn, Germany*
SUZANNE DEE • *Affymetrix, Santa Clara, CA, USA*
CHRISTOPH EICKEN • *LC Sciences, LLC, Houston, TX, USA*
JIAN-BING FAN • *Illumina, Inc., San Diego, CA, USA*
STEPHANIE FULMER-SMENTEK • *Agilent Technologies, Santa Clara, CA, USA*
XIAOLIAN GAO • *Department of Biology and Biochemistry, University of Houston, Houston, TX, USA*
ROBERT C. GETTS • *Genisphere LLC, Hatfield, PA, USA*
WILLEMIJN M. GOMMANS • *Hubrecht Institute, Utrecht, The Netherlands*
PREETHI H. GUNARATNE • *Department of Biology and Biochemistry, University of Houston, Houston, TX, USA*
JAMES HADFIELD • *CRUK Cambridge Research Institute, Cambridge, UK*
JAN HELLEMANS • *Biogazelle, Zwijnaarde, Belgium*
JAMES HURLEY • *Life Technologies Corporation, Foster City, CA, USA*
PHILIPP KAPRANOV • *Helicos BioSciences Corporation, Cambridge, MA, USA*
DAVID KEYS • *Life Technologies Corporation, Foster City, CA, USA*
ANDREAS KOWARSCH • *Institute for Bioinformatics and Systems Biology (MIPS), Helmholtz Zentrum München, German Research Center for Environmental Health, Neuherberg, Germany*

Sam E.V. Linsen • *Hubrecht Institute, Utrecht, The Netherlands*
Shujun Luo • *Illumina, Inc., Hayward, CA, USA*
Pieter Mestdagh • *Center for Medical Genetics, Ghent University, Ghent, Belgium*
Patrice M. Milos • *Helicos BioSciences Corporation, Cambridge, MA, USA*
Boye Schnack Nielsen • *Exiqon A/S, Diagnostic Product Development, Vedbæk, Denmark*
Fatih Ozsolak • *Helicos BioSciences Corporation, Cambridge, MA, USA*
Patrícia M. Pereira • *RNA Biology Laboratory, Department of Biology and CESAM, University of Aveiro, Aveiro, Portugal*
Liang-Hu Qu • *Key Laboratory of Gene Engineering of the Ministry of Education, State Key Laboratory for Biocontrol, Sun Yat-sen University, Guangzhou, P.R. China*
John E.J. Rasko • *Cell and Molecular Therapies, Royal Prince Alfred Hospital, Camperdown, NSW, Australia;Gene and Stem Cell Therapy Program , Centenary Institute, University of Sydney , Sydney , NSW , Australia*
William Ritchie • *Gene and Stem Cell Therapy Program, Centenary Institute, University of Sydney, Sydney, NSW, Australia*
Doug Roberts • *Life Technologies Corporation, Foster City, CA, USA*
Andreas Ruepp • *Institute for Bioinformatics and Systems Biology (MIPS), Helmholtz Zentrum München, German Research Center for Environmental Health, Neuherberg, Germany*
Manuel A.S. Santos • *RNA Biology Laboratory, Department of Biology and CESAM, University of Aveiro, Aveiro, Portugal*
Nijing Sheng • *Atactic Technologies Inc., Houston, TX, USA*
Rui Shi • *Forest Biotechnology Group, Department of Forestry and Environmental Resources, North Carolina State University, Raleigh, NC, USA*
Ana Raquel Soares • *RNA Biology Laboratory, Department of Biology and CESAM, University of Aveiro, Aveiro, Portugal*
Benjamin Soibam • *Department of Physics, University of Houston, Houston, TX, USA*
Keld Sorensen • *Siemens Healthcare Diagnostics, Flanders, NJ, USA*
Andrea Staratschek-Jox • *Life and Medical Sciences (LIMES), Genomics and Immunoregulation, University of Bonn, Bonn, Germany*
Ying-Husan Sun • *Forest Biotechnology Group, Department of Forestry and Environmental Resources, North Carolina State University, Raleigh, NC, USA*
Arpit Tandon • *Bioinformatics Research Laboratory, Baylor College of Medicine, Houston, TX, USA*
Fabian Theis • *Institute for Bioinformatics and Systems Biology (MIPS), Helmholtz Zentrum München, German Research Center for Environmental Health, Neuherberg, Germany*
Jo Vandesompele • *Center for Medical Genetics, Ghent University, Ghent, Belgium*
Thorsten Voss • *QIAGEN GmbH, Diagnostic Sample Preparation and Stabilization, R&D Department, QIAGEN, Hilden, Germany*
Jian-Hua Yang • *Key Laboratory of Gene Engineering of the Ministry of Education, State Key Laboratory for Biocontrol, Sun Yat-sen University, Guangzhou, P.R. China*
Litao Yang • *Department of Biology and Biochemistry, University of Houston, Houston, TX, USA*

LIN ZHANG • *Department of Obstetrics and Gynecology, Center for Research on Early Detection and Cure of Ovarian Cancer, University of Pennsylvania, Philadelphia, PA, USA*
XIAOLIN ZHANG • *Atactic Technologies Inc., Houston, TX, USA*
XING-HAI ZHANG • *Department of Biological Sciences, Florida Atlantic University, Boca Raton, FL, USA*
XIAOMIN ZHONG • *Department of Obstetrics and Gynecology, Center for Research on Early Detection and Cure of Ovarian Cancer, University of Pennsylvania, Philadelphia, PA, USA*
XIAOCHUAN ZHOU • *LC Sciences, LLC, Houston, TX, USA*
QI ZHU • *LC Sciences, LLC, Houston, TX, USA*

Chapter 1

Controlling miRNA Regulation in Disease

Willemijn M. Gommans and Eugene Berezikov

Abstract

Our understanding of the importance of noncoding RNA molecules is steadily growing. One such important class of RNA molecules are microRNAs (miRNAs). These tiny RNAs fulfill important functions in cellular behavior by influencing the protein output levels of a high variety of genes through the regulation of target messenger RNAs. Moreover, miRNAs have been implicated in a wide range of diseases. In pathological conditions, the miRNA expression levels can be altered due to changes in the transcriptional or posttranscriptional regulation of miRNA expression. On the other side, mRNA molecules might be able to escape the regulation by miRNAs. In this review, we give an overview on how miRNA biogenesis can be altered in disease as well as how mRNAs can avoid the regulation by miRNAs. The interplay between these two processes defines the final protein output in a cell, and thus the normal or pathological cellular phenotype.

Key words: microRNA, microRNA biogenesis, microRNA expression regulation, microRNA regulation in disease, microRNA expression in disease

1. Introduction

The discovery that the small RNA lin-4 could regulate the expression of lin-14 in the nematode *Caenorhabditis elegans* (1) nearly 20 years ago was the beginning of a major shift in biological research. It contributed to our current awareness that RNA molecules are not just simple intermediates between DNA and proteins, but also can play important roles in gene regulation. These regulatory functions of RNA molecules can explain, at least in part, how such complex organisms can arise despite that they harbor only a limited set of protein-encoding genes. Although a tremendous progress in the understanding of small RNA-mediated gene regulation has been achieved in recent years, there is still a whole layer of relatively unexplored RNA-mediated mechanisms to be discovered that play a role in fine-tuning cellular behavior.

Jian-Bing Fan (ed.), *Next-Generation MicroRNA Expression Profiling Technology: Methods and Protocols*, Methods in Molecular Biology, vol. 822, DOI 10.1007/978-1-61779-427-8_1, © Springer Science+Business Media, LLC 2012

After the initial discovery of lin-4, the different classes and number of small regulatory RNAs has expanded dramatically. One important class of such small RNA molecules are microRNAs (miRNAs), of which there are currently several thousand annotated in various species in the publically available database miRBase (2). miRNAs are 20–23 nucleotides long, and are estimated to regulate the expression of as much as one-third of genes in mammals (3). miRNAs can downregulate gene expression by binding to complementary regions within the 3′-untranslated region of mRNAs, thereby effecting the translation or stability of the mRNA. In this chapter, we intend to give an overview on the current knowledge on miRNA biogenesis, their mode of action, and factors that influence miRNA-mediated regulation in normal and disease development.

2. The Biogenesis of an miRNA

In the human genome, miRNAs exist as either individual genes or in clusters. A substantial fraction of miRNA genes are clustered together, which might be related to their evolutionary mechanism for spreading throughout the genome (4). These relatively closely located miRNAs can be transcribed as one polycistronic transcript, giving rise to multiple co-expressed mature miRNAs. Nonclustered miRNA genes, on the other hand, are expressed as individual transcripts and have their own promoters (Fig. 1). Such transcriptional organization of miRNA genes provides for numerous combinations of miRNAs that are individually or simultaneously expressed in a spatiotemporal manner.

Also the genomic context of miRNAs can differ between the individual miRNAs. The genes encoding for miRNAs can be embedded in the intronic region of a protein-coding gene, as well as located in an intergenic region. Those miRNAs that are imbedded within other genes can have two forms of transcriptional regulation. They are either co-transcribed and processed together with the protein-coding gene, or have their own intronic promoters (5) (Fig. 1). Thus, the expression levels of intronic miRNAs do not always have to correlate with the expression levels of the host gene. Also, the independent transcriptional control of miRNAs underscores their role as independent regulators of gene transcription in eukaryotic biology.

MiRNAs are transcribed as long precursor molecules by the RNA polymerase II, and these so-called primary (pri-)miRNAs can be up to several kilobases long. During and after transcription, pri-miRNAs fold back into imperfect double-stranded RNA molecules that are the stepping stones for subsequent processing into the mature miRNAs. The maturation of pri-miRNA transcripts

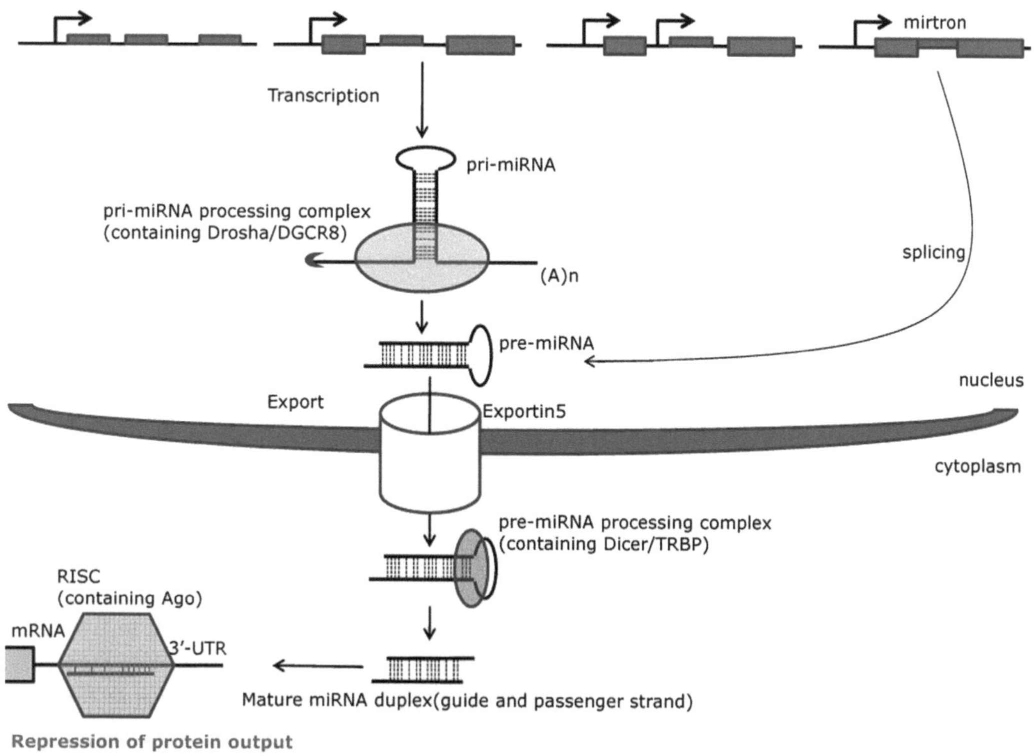

Fig. 1. The miRNA biogenesis pathway. MicroRNAs (*thin boxes*) can be located in intergenic regions or are embedded in the intronic region of a protein-encoding gene (exonic regions are shown here as *thick boxes*). In addition, microRNAs can be co-expressed with the genes in which they are implanted, or they have their own promoter sequence (*arrows*). After transcription, the pri-miRNA sequence is processed by the Drosha complex, the resulting pre-miRNA sequence is exported to the cytoplasm, and subsequently cleaved by the Dicer complex into a mature miRNA sequence. On the other hand, mirtrons are not cleaved by Drosha, but the pre-miRNA sequence is generated through splicing, after which it follows the biogenesis pathway of canonical miRNAs. The mature miRNA sequence is loaded into the RISC complex, which results in a reduction of protein output after binding to the target mRNA. See the main text for more details.

into mature miRNAs involves several consecutive protein-mediated processing steps (6) (Fig. 1). In mammalian systems, the first processing step of the pri-miRNA to precursor pre-miRNA occurs in the nucleus by the microprocessor complex that consists of at least 20 distinct proteins, including the RNAse III enzyme Drosha and the double-stranded RNA-binding enzyme DGCR8. A single pri-miRNA contains a typical stem structure of around 33 bp, and Drosha is guided by DGCR8 to cleave about 11 bp away from this stem-ssRNA junction (7). This cleavage step results in pre-miRNAs that are around 60–80 nt in length and contain a typical 2 nt 3′-overhang due to the Drosha cleavage.

The following processing step takes place in the cytoplasm. Thus, in order to become mature miRNAs, the pre-miRNAs are first transported from the nucleus into the cytoplasm. This transport is mediated by nuclear pore complex exportin5 and RanGTP (8–10), and the length of the stem loop structure as well as the presence of the 3′-overhang is important for recognition of the pre-miRNA by exportin-5 (9, 11). In the cytoplasm, the pre-miRNAs are cleaved into mature miRNA molecules by the RNAse III enzyme Dicer. This generates the final double-stranded miRNA products of 21–23 nucleotides, containing 2 nucleotide 3′-overhangs on both sides of the molecule. The Dicer protein alone is sufficient for generating the mature miRNAs from pre-miRNAs, but it does interact with a series of other proteins, such as TRBP (12).

In order for an miRNA to be functional in the regulation of gene expression, it has to be incorporated in the multiprotein complex (miRNA-induced silencing complex (miRISC). TRBP can form a bridge between the miRNA processing by Dicer and the ultimate functionality of the miRNA, by forming a complex with Dicer and recruiting Argonaute (Ago), an essential member of the miRISC complex (12). In general, the miRNA strand with the thermodynamically less stable 5′-end is incorporated in the miRISC complex (6) and functions as a template for scanning mRNA 3′-UTRs for sequences complementary to the miRNA. The other strand, referred to as the star or passenger strand is in most cases rapidly degraded.

Strikingly, the miRNA does not have to possess a completely perfect complementary sequence to the 3′-UTR for mediating gene repression. More precisely, the number of perfect matches needed between the miRNA and the target mRNA seems to differ from miRNA to miRNA to mRNA. The only critical sequences for targeting seem to be in the 5′-end of the miRNA, the so-called seed sequence, which comprises nucleotides 2–8, although weaker seed sequences might be compensated through pairing of the 3′-end of the miRNA to the mRNA. This lack of black-and-white well-defined miRNA target recognition rules thus makes it extremely difficult to identify bona fide miRNA targets in silico (13).

3. Post-transcriptional Factors That Modulate miRNA Biogenesis

The biogenesis of miRNAs is subjected to complex regulatory mechanisms. The regulation of miRNA production can provide the cell with a fast-acting response to environmental changes. Also, this effect will be wide-spread, as miRNAs act on many different mRNAs simultaneously. Influencing the efficiency of miRNA biogenesis can therefore be a quick and powerful approach to anticipate to external and internal cellular fluctuations.

As mentioned above, the canonical processing of pri-miRNAs into mature miRNAs occurs in two consecutive steps (Fig. 1). The transcribed pri-miRNAs are cleaved by the Drosha complex into pre-miRNAs, of which subsequently mature miRNAs are generated by the Dicer-processing complex. The processing efficiency of subsets of pri-miRNAs into the final mature miRNAs is influenced by a battery of other proteins, generating a complex network of miRNA expression regulation, which we are only just beginning to understand. We discuss several examples below.

During differentiation, there is a steady increase in mature Let-7 g levels. However, the pri-miRNA levels of Let-7 g remain fairly constant (14). This indicates that there is some kind of Drosha processing block for pri-miRNA Let-7 g processing, either due to posttranscriptional modification of Drosha, or due to the influence of other regulatory proteins. One of these regulatory proteins is the tumor suppressor p53, as p53 can enhance the processing efficiency of several miRNAs by interacting with a protein in the Drosha complex (15). Intriguingly, of the miRNAs that were tested in this study, the miRNAs that were influenced by p53 had growth suppressive functions. How p53 selects between different miRNAs is yet unknown, but it directly couples the tumor suppressive network to the process of miRNA biogenesis. Another family of proteins that acts on the level of Drosha cleavage are Smad proteins (16). These proteins recognize and bind to a specific motif present within the stem loop region of certain miRNAs. Binding of the Smad proteins subsequently enhances the docking and cleavage by Drosha.

Thus, both Smad proteins as well as p53 enhance the processing step of pri-miRNAs into pre-miRNAs. On the other side, there are also proteins that can inhibit the processing of pri-miRNAs. Upon binding of the estrogen receptor α to estrogen, the receptor associates with the Drosha complex and inhibits the maturation of a set of miRNAs (17). In addition, depletion of the nuclear factor (NF) NF90 and NF45 results in an accumulation of pri-miRNAs in the cells (18).

A subgroup of miRNAs that skips the Drosha processing step altogether are mirtrons (19–21). These miRNAs are encoded within small intronic regions of transcribed genes. Upon splicing, the intronic region folds into a pre-miRNA hairpin structure. These pre-miRNA hairpin mimics subsequently follow the canonical miRNA biogenesis pathway to mature miRNAs.

The step following Drosha cleavage is the transport of the pre-miRNA into the cytoplasm. However, to what extent miRNA biogenesis is controlled by the efficiency of the pre-miRNA export is still a matter of debate (22). In the cytoplasm, the processing of miRNAs can be controlled by factors that influence the efficiency of Dicer. The protein TRBP is phosphorylated through the MAPK signaling pathway, and phosphorylated TRBP does stabilize the Dicer-processing complex. This stabilization results in an increased miRNA production and miRNA-mediated target silencing (23). Thus, crucial signaling pathways, such as the tumor suppressive p53 pathway and the oncogenic MAPK/Erk pathway, are involved in miRNA biogenesis. This suggests that these cell growth-regulating pathways can also partly exert their effects through regulating miRNA expression, and that misregulation of miRNA biogenesis can be involved in oncogenesis, as is discussed in more detail below.

There are also proteins that can interfere with both the Drosha- as well as the Dicer-mediated cleavage step. The protein Lin28 binds to a specific sequence present in the loop structure of the miRNA let-7. This binding inhibits both the Drosha and Dicer cleavage step (24–27). Moreover, Lin28 recruits a terminal uridylyl transferase that uridylates pre-let-7, guiding pre-let-7 into a degradation pathway (28). However, protein binding can also have an opposite effect, resulting in an enhanced Drosha and Dicer processing. In the case of the miRNA mir-18a, the protein hnRNP A1 binds both the terminal loop structure as well as the stem loop of the pri-miRNA (29), and knockdown of hnRNP A1 reduces the levels of both pri- and pre-mir-18a (30). Also the protein KSRP can promote the processing of several miRNAs. KSRP is a component of both the Drosha- and Dicer-multiprotein complex, and it binds to the terminal loop region of pri- and pre-miRNAs (31). Binding of this protein can subsequently result in the recruitment or correct positioning of the Drosha- or Dicer-complex.

The above-described proteins directly interact with proteins of the miRNA-biogenesis machinery. Another family of proteins that has a completely different mode of action is the ADARs. These proteins can alter the primary sequence of double-stranded RNA molecules, such as pri-miRNAs (32–36), through the hydrolytic deamination of adenosines. This results in the conversion of an adenosine into an inosine, and the inosine behaves similar to guanosine. This A-to-I editing of pri-miRNAs can alter the processing efficiency of Drosha and/or Dicer and is expected to occur in roughly 16% of all pri-miRNA molecules (37). Most interestingly, editing also occurs in the seed region of miRNAs. The editing at these positions can alter the target recognition of the miRNA, with the edited version targeting another set of genes than the unedited version (34). Although difficult to assess (38), the editing of the mature miRNA sequence seems to be limited and the main effect of miRNA editing therefore appears to be the alteration of miRNA biogenesis.

The final generated mature miRNA is incorporated into the miRISC complex and its ultimate half-life is dependent upon several factors (39). The miRNA is protected against decay through incorporation into the Argonautes, and ectopic expression of the Ago proteins results in increased miRNA levels (39, 40). However, for example in neurons there is still a rapid turnover of miRNAs with a half life of approximately 1 h (41). In the nematode *C. elegans*, the degradation of the exposed mature single-stranded miRNAs is mediated by the 5′-3′ exonuclease XRN-2 (42). This protein seems to have an intriguing double function, as it can also mediate the release of the miRNA from the Ago-complex if the miRNA is not interacting with its target molecule. This regulatory mechanism might be important for the rapid transitions seen in miRNA expression profiles.

Also posttranscriptional modification of the miRNA could increase its half-life. However, this seems to be a much more common phenomenon in plants rather than in animals. Although deep-sequencing studies indicated that mammalian miRNAs frequently contain untemplated adenine or uracil additions (43), a functional effect has so far only been demonstrated for miR-122, where the addition of an adenine at the 3′-end enhances its stability (44).

There is thus a whole array of regulatory proteins that influence the homeostasis of mature miRNAs, and more proteins and regulatory mechanisms are likely yet to be discovered (22, 39). The interplay between all these proteins will generate a broad spectrum of miRNA expression levels, which cannot be deducted solely from their initial transcription levels. Another pending question is how precisely miRNAs subsequently regulate the repression of gene expression. As is discussed in more detail below, also here there might be more (regulatory) roads that are leading to Rome.

4. miRNA-Mediated Regulation of Gene Repression

In animal systems, miRNAs can regulate gene expression through either the inhibition of protein translation or initiation of mRNA degradation (6). This is a major difference with the plant kingdom, where miRNAs form highly complementary Watson-Crick base pairing with the target mRNA and subsequently solely induce mRNA cleavage (45). Nonetheless, in animals there seems to be a correlation with the extent of base-pairing between the miRNA and its target, and subsequent cleavage or translational repression (46).

In mammals, there are four members of the Argonaute family (AGO1-4) (47), of which only AGO2 has mRNA cleavage ("slicer") activity and is also the only AGO protein that functions in the RNA interference (RNAi) pathway by cleaving the mRNA in the center of the mRNA-siRNA duplex (47, 48). Interestingly, there is also an miRNA mir-451, whose maturation is directly regulated by the slicer activity of AGO2. Instead of Dicer, this pre-miRNA is cleaved by AGO2 and likely subsequently polyuridylated and trimmed into the mature miR-451 sequence (49). There is thus a close interplay between miRNA production and activity. Besides the slicer activity of AGO2, miRISCs containing either AGO2 or other AGO proteins can cause a reduction in the protein output without directly diminishing the mRNA levels. In these instances, reduction in protein expression is caused by interfering with the translational machinery.

There are currently several working models that describe the repression of translation after binding of the miRISC complex to the mRNA molecule. Broadly, these mechanisms can be divided into inhibition before or after the initiation of translation. Inhibition of translation at the initiation step can occur through interference with the protein assembly on the mRNA molecule, prior to translation. In brief, proteins that bind to either the 5′- cap or the 3′-polyA tail of the same RNA molecule interact with each other, bringing the two ends of the RNA molecule together. This circularization stimulates the initiation of translation. In the case of miRNA-mediated repression, AGO2 might interact with the 5′-cap of the RNA molecule, competing with the binding of the regular translation initiation factors (6, 50). In addition, miRNA-mediated repression can result in the shortening of the 3′-polyA tail through deadenylation (51–54), which might cause the disruption of the circularization of the RNA molecule. Thus, inhibiting the translation by miRNAs can be caused by targeting either of both the 5′- or 3′-ends of the RNA molecule.

As with many new rapidly evolving research fields, there is not yet one solid answer to the question of how miRNAs regulate the inhibition of protein translation; besides inhibiting the translational machinery at the initiation step, there is also evidence that miRNAs can repress gene expression during the elongation of translation. An intriguing example is the study from Petersen et al. (55) that showed that short RNA-induced repression can also occur in the downstream open reading frame of a bi-cistronic reporter containing and IRES. The translation of this downstream RNA molecule is thus independent of the 5′-cap structure. Moreover, other data indicated that translating ribosomes do assemble but might prematurely drop-off again. It is very well feasible that miRNAs can act as inhibitors of translation on both the level of translation initiation as well as translation elongation. However, the importance of either mechanism is still fuel for an ongoing scientific debate.

Besides mRNA cleavage or the inhibition of the translational machinery, other mechanisms that reduce the protein output can play a role as well. miRNAs can induce the deadenylation of the mRNA polyA tail, leading to de-capping and subsequent degradation of the RNA molecule through 5′ → 3′ exonucleolytic activity (52–54, 56). So there seem to be many different mechanisms by which a miRNA can cause a reduction in the protein output. At this moment, it is not yet clear what determines whether the RNA is destabilized or if there will be inhibition of translation. Also, it is unclear whether RNA deadenylation results from translational repression, or whether this is a primary event. A recent study from Guo et al. (57) demonstrated that over 80% of the decreased protein production was assigned to a decrease in mRNA levels, thus supporting the mRNA destabilization model.

5. miRNA Regulation in Disease

miRNAs play an important regulatory role in many crucial cellular pathways, such as development, differentiation, and apoptosis. However, knockdown of specific miRNAs in flies and worms in many cases did not reveal crucial viable or developmental defects (58–60). But when environmental conditions changed, the defects due to knockdown of certain miRNAs became apparent (61). This indicates that miRNAs can play a role in maintaining homeostasis in a cell, and there is a loss of fine-tuning when conditions are altered. It is therefore not surprising that an altered miRNA-mediated regulation can contribute to pathological conditions, such as cancer, cardiovascular disease, and many others. This altered regulation of the miRNA target can be the result of changed miRNA levels, but also because the mRNA target might simply escape miRNA-controlled regulation (Fig. 2).

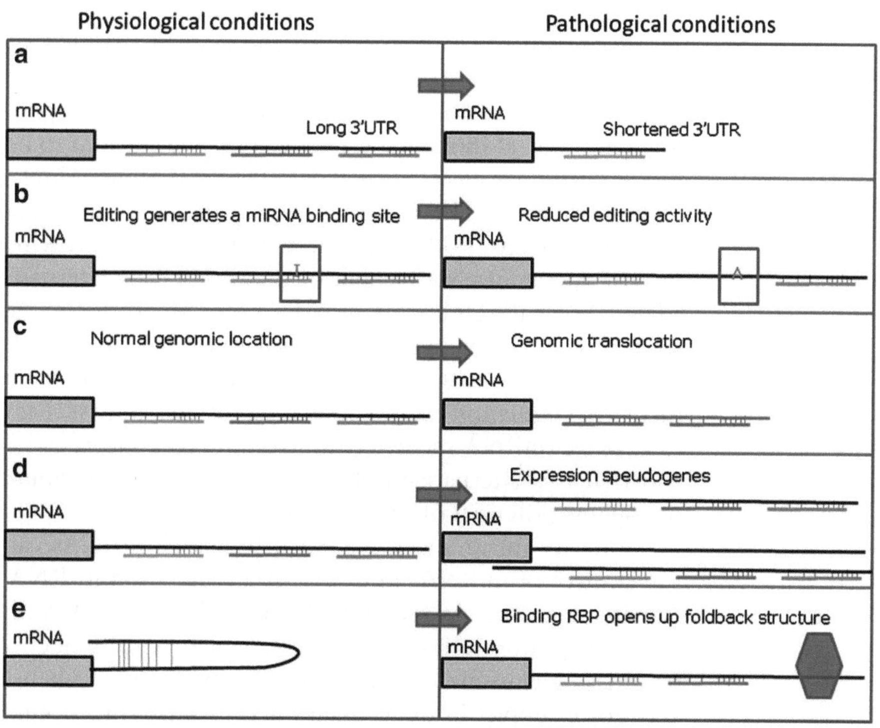

Fig. 2. Preventing the regulation by miRNAs (**a**) Shortening of the 3′-UTR results in less binding sites for miRNAs. (**b**) Decreased A-to-I RNA editing activity causes an increased prevalence of an adenosine over an inosine, which subsequently causes the loss of an miRNA binding site. (**c**) Genomic translocation leads to another 3′-UTR sequence and consequently other miRNA-binding sites. (**d**) The expression of a pseudogene captures the miRNAs away from the original mRNA target sequence. (**e**) Binding of an RNA-binding protein opens up the structure of the 3′-UTR and makes it more accessible for the miRNA to target the mRNA.

6. Escaping miRNA-Mediated Regulation

The interaction between the miRNA and the target mRNA occurs predominantly in the 3′-UTRs of the mRNA molecule. This results either in inhibition of translation or RNA degradation, as outlined above. Thus, a way for mRNAs to escape the regulation by miRNAs is to shorten the length of the 3′-UTR by alternative polyadenylation (Fig. 2a). This intriguing possibility was first investigated by Sandberg and coworkers (62), who demonstrated the widespread shortening of 3′-UTRs upon activating/proliferating stimuli in a variety of immune cells. To further extrapolate their results, they also investigated a wide variety of cell lines. The length of the 3′-UTR was consistently lower in cell lines in comparison to the tissues from which they originated, further indicating that the length of the 3′-UTR was associated with the cellular proliferation potential. Using a luciferase-based system, a longer 3′-UTR was shown to decrease protein output in comparison to the short 3′-UTR sequence, likely due to the inhibition of translation or RNA stability. Finally, they suggest a model in which the shortening of the 3′-UTR region results in reducing the potential for targeting by antiproliferative factors, such as certain families of miRNAs (62).

In another key study, the shortening of the 3′-UTRs is linked to oncogenic transformation rather than proliferation per se (63). Mayr and Bartel show that cancer cell lines compared to nontransformed cell lines have shortened 3′-UTRs, despite a similar proliferation potential. These shorter mRNA forms are more stable and characteristically produce about ten times more protein. In the case of oncogenes, shortening of the 3′-UTR by alternative polyadenylation can result in an increased protein production and consequent cellular oncogenic transformation. The increased protein output could at least in part be explained by escaping miRNA-mediated regulation (63). These observations suggest that mRNAs can escape miRNA-mediated control due to a shortened 3′-UTR, and that this alternative polyadenylation could contribute to the oncogenic potential of tumor cells.

Another intriguing mechanism by which mRNAs can escape the control of miRNAs in disease is an alteration in RNA editing activity (Fig. 2b). As mentioned above, A-to-I RNA editing converts an adenosine into an inosine, which is subsequently interpreted as guanosine. RNA editing is a widespread phenomenon, and, for example, in brain tissues it is estimated to occur once in every 17,000 nt (64). Most editing events take place in the untranslated regions, including 3′-UTRs. A study investigating the interplay between editing and miRNA regulation demonstrated that mRNA editing seems to avoid miRNA seed sequences (65). However, several sites were found where editing abolished an miRNA binding site, or generated a new miRNA recognition site.

The editing efficiency at an individual position can range from <2% to 90% (66), and its activity is altered in cancer and a variety of other diseases (67). This generates a complex interplay between two important posttranscriptional mechanisms to fine-tune gene regulation in normal and diseased conditions.

A more permanent methodology to eliminate miRNA-binding sites is the acquisition of genomic mutations in the 3′-UTR during tumor development (Fig. 2c). One such example is the oncogene Hmga2 (68), in which a chromosomal translocation causes disruption of the 3′-UTR. The Hmga2 3′-UTR contains several binding sites for the tumor-suppressive miRNA let-7, and disruption of these binding sites can lead to higher levels of Hmga2 and consequent oncogenic transformation. Thus, not only mutations in the open reading frame of specific proteins can result in pathological conditions, but also genomic alterations in the 3′-UTR that disrupt or create miRNA-binding sites, can play a role in disease development.

More subtle genomic alterations can occur through single nucleotide polymorphisms (SNPs). When located in the 3′-UTR of a gene, these SNPs can either disrupt or create miRNA binding sites. SNPs are allele-specific, and measuring the differences in allele-specific mRNA expression levels indeed revealed that genes carrying an SNP-generated miRNA-binding site were repressed in those tissues where the miRNA was expressed as well (69). As two unrelated humans have a distinct set of SNPs, many genes will be expressed in an individual-specific manner and can subsequently also underline differences in disease risk and clinical outcome.

The final protein levels from a certain transcript depend both on the number of miRNA molecules and the number of mRNA molecules present in a certain cell. However, if the miRNA binds to other RNA molecules, the original target mRNA is thus derepressed and translation can take place. This function can be fulfilled by pseudogenes, such as PTENP1 and KRASP1, which can modulate the protein production of respectively the tumor-suppressor PTEN and the oncogene KRAS, by functioning as a decoy for miRNA binding (Fig. 2d). Intriguingly, in certain cancer types, the pseudogene PTENP1 has indeed a loss of genomic copy number, whereas the KRAS1P locus is amplified in a variety of human tumors, which is consistent with their tumor-suppressive and oncogenic roles (70). As pseudogenes are nearly as numerous as protein-coding genes in the transcriptome, many more examples might exist, attributing a novel role for pseudogenes in regulating the expression of coding genes.

Finally, another way for an mRNA molecule to escape the regulation by miRNAs is to hide the miRNA-binding site for the binding of the miRNA (71) (Fig. 2e). The miRNAs mir-221 and mir-222 are required in cancer to inhibit the expression of p27 and subsequently stimulate proliferation. Intriguingly, the levels of these miRNAs and p27 mRNA are not significantly altered in quiescent and cycling cells. However, in cycling cells the RNA-binding proteins PUM1 and PUM2 are activated and bind to the 3′-UTR of p27. It is this binding that opens up the secondary structure of the p27 3′-UTR and subsequently causes the miRNAs to bind and repress p27 protein expression (71). There is thus an extensive interplay between the control of protein expression by miRNAs and the many factors that influence the overall availability of miRNA-binding sites. The balance between this availability, as well as mRNA and miRNA levels ultimately define the protein output of a certain transcript, and if this balance shifts it might contribute to the development of disease. Besides the alteration of factors influencing the availability of miRNA-binding sites in pathological conditions, the expression levels of miRNAs themselves are also often altered in disease states. This can be due to differences in expression levels and also due to factors influencing the processing of miRNAs, as discussed below.

7. Altered miRNA Expression Levels in Pathological Conditions

At present, there are hundreds of research papers describing a role for miRNAs in pathological conditions, and the number of miRNA publications is steadily growing each year (72). This is not surprising, as miRNAs can influence the expression levels of many proteins that are involved in cell development, homeostasis, and disease. The strength of a single miRNA lies predominantly in the fact that it can influence the expression of an array of proteins. Thus, often the inhibitory effect of one miRNA on one protein is only marginal, but it is the cumulative effect of one miRNA on many proteins that results in the final cellular behavioral response. Changes in the expression level of a single miRNA can thus be an important factor in severe pathological conditions.

Among the first miRNAs that were suggested to play a direct role in tumor biology were mir-15a and mir-16-1. They were originally found in a genomic area that is often deleted in chronic lymphocytic leukemia (CLL), and these miRNAs were shown to be lost in about 70% of developing CLL (73). Strikingly, many other miRNA genes are located in cancer-associated genomic regions or fragile areas as well (74). Besides mir-15-1 and mir-16-1, a variety of other miRNAs have now been shown to play a direct role in tumor biology, functioning either as a tumor-suppressors or oncogenes (75).

A striking example of how one miRNA can drastically alter a cellular phenotype is mir-21 (76). This miRNA is overexpressed in a great number of tumor types and induction of mir-21 expression in vivo results in a pre-B malignant lymphoid-like phenotype.

Upon subsequent inactivation of mir-21 expression, this phenotype completely regresses again within a few days, indicating that tumors can become addicted to oncogenic miRNAs and demonstrating the strength of a single miRNA to regulate cellular behavior.

mir-21 was artificially introduced in these cell types, which resulted in the oncogenic transformation. In cancer, however, the expression levels of a wide variety of miRNAs was shown to be reduced (77) to such an extent that clustering of miRNA expression profiles causes tumor tissues to group together, instead of clustering to the tissue from which they derive (77). These miRNA changes can be the consequence of altered miRNA gene transcription levels. Among others, widespread repression of miRNA gene expression can occur via the c-Myc oncogene (78) or hypermethylation of CpG island promoters (79–81). But global changes in the mature miRNA levels can also occur posttranscriptionally, as often the altered levels in mature miRNAs do not correlate with changes in the primary miRNA transcript (14). What is the cause of these altered mature miRNA levels is still relatively unknown, although alterations in mechanisms that control the steady-state levels of miRNAs could at least partly explain these observed wide-spread changes in miRNA expression profiles.

As mentioned above, the tumor-suppressor gene p53 influences the processing of a group of miRNAs. As the protein levels of p53 are reduced in a high number of tumor types, this might alter the expression of these miRNAs as well. The same can hold for other proteins that influence the processing of pri- or pre-miRNAs, such as SMADs, NF90, or ADARs. Differential expression of each of these proteins in diseased tissues can cause an alteration in the expression levels of subsets of miRNAs.

Yet another protein involved in the differential expression of miRNAs in disease is the key miRNA-processing enzyme Dicer. In a variety of cancers, the expression of Dicer was shown to be reduced, including lung cancer, hepatocellular carcinoma, and ovarian cancer (82–85). This reduced expression was in general correlated with a poor clinical outcome. In addition, knock-down of Dicer caused an enhanced proliferation and migration rate of tumor cells, and in mice decreased Dicer expression resulted in a reduced survival rate due to an increased tumor burden (86). Intriguingly, this reduced survival was correlated to a hemizygous instead of a homozygous loss of Dicer, indicating that Dicer functions as a haploinsufficient tumor suppression gene. Also, hemizygous deletion of Dicer in cell lines indeed resulted in reduced steady-state miRNA levels (87). These results indicate that the observed reduced Dicer expression in cancer at least partly contributes to the global reduction in miRNA expression levels that are typical for cancerous tissues (77, 82, 87).

No loss-of-function genetic mutations have been reported for the Dicer gene in tumor tissues, even though it acts as a tumor suppressor gene and shows reduced protein levels in a variety of tumor types. Reduced Dicer expression might occur due to changes in the transcriptional regulation of Dicer, mediated for example by TAp63 in metastasis (88). However, reduced Dicer expression can also at least partly arise from mutations in the gene encoding for TRBP (89). As discussed above, TRBP is involved in miRNA processing through direct interaction with the endonuclease Dicer. Heterozygous mutations in the TRBP gene are found in several types of cancer, where it causes a reduction in the TRBP protein levels. Reduced TRBP protein levels results in the destabilization of the Dicer protein and thus subsequently reduces Dicer protein levels. Moreover, mutations in TRBP also lead to reduced miRNA levels and this deregulation of miRNA processing caused an enhanced tumor growth (89). Thus, the global inhibition in miRNA expression as seen in cancer does not seem to be a mere consequence of a reduced differentiation state of the cell. Instead, a reduction in the general miRNA processing efficiency seems to have a direct effect on the oncogenic potential of tumor cells, which is also in agreement with the finding that Dicer can act as a tumor suppressor gene.

In general, proteins that are involved in the homeostasis of miRNAs could also influence the miRNA expression levels in diseased conditions, either by inducing or reducing the processing efficiency of subsets of miRNAs or globally altering the miRNA expression levels. These altered miRNA levels can subsequently have an influence on the cellular phenotype, such as an increase in the oncogenic potential of a cell. Pinpointing and influencing the underlying mechanisms that regulate miRNA homeostasis can therefore be relevant for finding novel therapeutic strategies.

8. Concluding Remarks

The biogenesis of an miRNA is still an intensive area of investigation for many research groups. Besides the pri- and pre-miRNA processing complexes there are also many other factors involved that fine-tune the regulation of miRNA expression after transcription. This fine-tuning is highly important, as miRNAs can play important roles in disease development and thus their expression should be tightly controlled. We are still only at the beginning of understanding the regulatory mechanisms that control the biogenesis of an miRNA, and it is expected that other regulatory proteins are yet to be discovered. This enhances our understanding on how miRNA expression can be deregulated in disease and might provide novel therapeutic avenues for controlling the miRNA balance within a cell.

In addition to miRNA expression alterations, some mRNA targets seem to directly escape miRNA-mediated regulation in diseased conditions. For example, the global shortening of 3′-UTRs as well as the genomic translocation of protein-encoding genes might directly result in increased protein levels due to a loss of miRNA control. For now, it remains to be investigated if this loss of miRNA-mediated control of gene expression directly results in an altered cellular behavior, or whether this is a secondary effect.

Overall, the balance between the miRNA levels as well as the availability of the mRNA target defines the final protein output. It is this balance that is disturbed in many diseases, and might play a more significant role than has been recognized during these last decades. Our understanding of the consequences that small molecules play in normal development and disease is steadily growing and will ultimately change our perspective on cellular behavior. In this respect, the availability of deep-sequencing methods to find novel small RNA species in addition to determining their expression levels and modifications as described in the following chapters, are of great importance for elucidating the effect of small RNAs on the final cellular phenotype.

References

1. Wightman, B., Ha, I., and Ruvkun, G. (1993) Posttranscriptional regulation of the heterochronic gene lin-14 by lin-4 mediates temporal pattern formation in C. elegans, *Cell* **75**, 855–862.
2. Kozomara, A., and Griffiths-Jones, S. (2011) miRBase: integrating microRNA annotation and deep-sequencing data, *Nucleic Acids Res* 39, D152–7.
3. Selbach, M., Schwanhausser, B., Thierfelder, N., Fang, Z., Khanin, R., and Rajewsky, N. (2008) Widespread changes in protein synthesis induced by microRNAs, *Nature* **455**, 58–63.
4. Altuvia, Y., Landgraf, P., Lithwick, G., Elefant, N., Pfeffer, S., Aravin, A., et al. (2005) Clustering and conservation patterns of human microRNAs, *Nucleic Acids Res* **33**, 2697–706.
5. Martinez, N. J., Ow, M. C., Reece-Hoyes, J. S., Barrasa, M. I., Ambros, V. R., and Walhout, A. J. (2008) Genome-scale spatiotemporal analysis of Caenorhabditis elegans microRNA promoter activity, *Genome Res* **18**, 2005–15.
6. Filipowicz, W., Bhattacharyya, S. N., and Sonenberg, N. (2008) Mechanisms of post-transcriptional regulation by microRNAs: are the answers in sight? *Nat Rev Genet* **9**, 102–14.
7. Han, J., Lee, Y., Yeom, K. H., Nam, J. W., Heo, I., Rhee, J. K., et al. (2006) Molecular basis for the recognition of primary microRNAs by the Drosha-DGCR8 complex, *Cell* **125**, 887–901.
8. Bohnsack, M. T., Czaplinski, K., and Gorlich, D. (2004) Exportin 5 is a RanGTP-dependent dsRNA-binding protein that mediates nuclear export of pre-miRNAs, *RNA* **10**, 185–91.
9. Lund, E., Guttinger, S., Calado, A., Dahlberg, J. E., and Kutay, U. (2004) Nuclear export of microRNA precursors, *Science* **303**, 95–8.
10. Yi, R., Qin, Y., Macara, I. G., and Cullen, B. R. (2003) Exportin-5 mediates the nuclear export of pre-microRNAs and short hairpin RNAs, *Genes Dev* **17**, 3011–16.
11. Zeng, Y., and Cullen, B. R. (2004) Structural requirements for pre-microRNA binding and nuclear export by Exportin 5, *Nucleic Acids Res* **32**, 4776–85.
12. Chendrimada, T. P., Gregory, R. I., Kumaraswamy, E., Norman, J., Cooch, N., Nishikura, K., et al. (2005) TRBP recruits the Dicer complex to Ago2 for microRNA processing and gene silencing, *Nature* **436,** 740–4.
13. Brodersen, P., and Voinnet, O. (2009) Revisiting the principles of microRNA target recognition and mode of action, *Nat Rev Mol Cell Biol* **10**, 141–8.
14. Thomson, J. M., Newman, M., Parker, J. S., Morin-Kensicki, E. M., Wright, T., and Hammond, S. M. (2006) Extensive post-transcriptional regulation of microRNAs and its implications for cancer, *Genes Dev* **20**, 2202–7.
15. Suzuki, H. I., Yamagata, K., Sugimoto, K., Iwamoto, T., Kato, S., and Miyazono, K. (2009)

Modulation of microRNA processing by p53, *Nature* **460**, 529–33.

16. Davis, B. N., Hilyard, A. C., Nguyen, P. H., Lagna, G., and Hata, A. Smad proteins bind a conserved RNA sequence to promote microRNA maturation by Drosha, *Mol Cell* **39**, 373–84.
17. Yamagata, K., Fujiyama, S., Ito, S., Ueda, T., Murata, T., Naitou, M., et al. (2009) Maturation of microRNA is hormonally regulated by a nuclear receptor, *Mol Cell* **36**, 340–7.
18. Sakamoto, S., Aoki, K., Higuchi, T., Todaka, H., Morisawa, K., Tamaki, N., et al. (2009) The NF90-NF45 complex functions as a negative regulator in the microRNA processing pathway, *Mol Cell Biol* **29**, 3754–69.
19. Berezikov, E., Chung, W. J., Willis, J., Cuppen, E., and Lai, E. C. (2007) Mammalian mirtron genes, *Mol Cell* **28**, 328–36.
20. Okamura, K., Hagen, J. W., Duan, H., Tyler, D. M., and Lai, E. C. (2007) The mirtron pathway generates microRNA-class regulatory RNAs in Drosophila, *Cell* **130**, 89–100.
21. Ruby, J. G., Jan, C. H., and Bartel, D. P. (2007) Intronic microRNA precursors that bypass Drosha processing, *Nature* **448**, 83–6.
22. Slezak-Prochazka, I., Durmus, S., Kroesen, B. J., and van den Berg, A. MicroRNAs, macrocontrol: regulation of miRNA processing, *RNA* **16**, 1087–95.
23. Paroo, Z., Ye, X., Chen, S., and Liu, Q. (2009) Phosphorylation of the human microRNA-generating complex mediates MAPK/Erk signaling, *Cell* **139**, 112–22.
24. Newman, M. A., Thomson, J. M., and Hammond, S. M. (2008) Lin-28 interaction with the Let-7 precursor loop mediates regulated microRNA processing, *RNA* **14**, 1539–49.
25. Piskounova, E., Viswanathan, S. R., Janas, M., LaPierre, R. J., Daley, G. Q., Sliz, P., et al. (2008) Determinants of microRNA processing inhibition by the developmentally regulated RNA-binding protein Lin28, *J Biol Chem* **283**, 21310–4.
26. Rybak, A., Fuchs, H., Smirnova, L., Brandt, C., Pohl, E. E., Nitsch, R., et al. (2008) A feedback loop comprising lin-28 and let-7 controls pre-let-7 maturation during neural stem-cell commitment, *Nat Cell Biol* **10**, 987–93.
27. Viswanathan, S. R., Daley, G. Q., and Gregory, R. I. (2008) Selective blockade of microRNA processing by Lin28, *Science* **320**, 97–100.
28. Heo, I., Joo, C., Cho, J., Ha, M., Han, J., and Kim, V. N. (2008) Lin28 mediates the terminal uridylation of let-7 precursor MicroRNA, *Mol Cell* **32**, 276–84.
29. Michlewski, G., Guil, S., Semple, C. A., and Caceres, J. F. (2008) Posttranscriptional regulation of miRNAs harboring conserved terminal loops, *Mol Cell* **32**, 383–93.
30. Guil, S., and Caceres, J. F. (2007) The multifunctional RNA-binding protein hnRNP A1 is required for processing of miR-18a, *Nat Struct Mol Biol* **14**, 591–6.
31. Trabucchi, M., Briata, P., Garcia-Mayoral, M., Haase, A. D., Filipowicz, W., Ramos, et al. (2009) The RNA-binding protein KSRP promotes the biogenesis of a subset of microRNAs, *Nature* **459**, 1010–4.
32. Blow, M. J., Grocock, R. J., van Dongen, S., Enright, A. J., Dicks, E., Futreal, P. A., et al. (2006) RNA editing of human microRNAs, *Genome Biol* **7**, R27.
33. Kawahara, Y., Zinshteyn, B., Chendrimada, T. P., Shiekhattar, R., and Nishikura, K. (2007) RNA editing of the microRNA-151 precursor blocks cleavage by the Dicer-TRBP complex, *EMBO Rep* **8**, 763–9.
34. Kawahara, Y., Zinshteyn, B., Sethupathy, P., Iizasa, H., Hatzigeorgiou, A. G., and Nishikura, K. (2007) Redirection of silencing targets by adenosine-to-inosine editing of miRNAs, *Science* **315**, 1137–40.
35. Luciano, D. J., Mirsky, H., Vendetti, N. J., and Maas, S. (2004) RNA editing of a miRNA precursor, *RNA* **10**, 1174–7.
36. Yang, W., Chendrimada, T. P., Wang, Q., Higuchi, M., Seeburg, P. H., Shiekhattar, R., et al. (2006) Modulation of microRNA processing and expression through RNA editing by ADAR deaminases, *Nat Struct Mol Biol* **13**, 13–21.
37. Kawahara, Y., Megraw, M., Kreider, E., Iizasa, H., Valente, L., Hatzigeorgiou, A. G., et al. (2008) Frequency and fate of microRNA editing in human brain, *Nucleic Acids Res* **36**, 5270–80.
38. de Hoon, M. J., Taft, R. J., Hashimoto, T., Kanamori-Katayama, M., Kawaji, H., Kawano, M., et al. (2010). Cross-mapping and the identification of editing sites in mature microRNAs in high-throughput sequencing libraries, *Genome Res* **20**, 257–64.
39. Kai, Z. S., and Pasquinelli, A. E. MicroRNA assassins: factors that regulate the disappearance of miRNAs, *Nat Struct Mol Biol* **17**, 5–10.
40. Diederichs, S., and Haber, D. A. (2007) Dual role for argonautes in microRNA processing and posttranscriptional regulation of microRNA expression, *Cell* **131**, 1097–108.
41. Krol, J., Busskamp, V., Markiewicz, I., Stadler, M. B., Ribi, S., Richter, J., et al. (2010). Characterizing light-regulated retinal microRNAs reveals rapid turnover as a common property of neuronal microRNAs, *Cell* **141**, 618–31.

42. Chatterjee, S., and Grosshans, H. (2009) Active turnover modulates mature microRNA activity in Caenorhabditis elegans, *Nature* **461**, 546–9.
43. Chiang, H. R., Schoenfeld, L. W., Ruby, J. G., Auyeung, V. C., Spies, N., Baek, D., et al. (2010) Mammalian microRNAs: experimental evaluation of novel and previously annotated genes, *Genes Dev* **24**, 992–1009.
44. Katoh, T., Sakaguchi, Y., Miyauchi, K., Suzuki, T., Kashiwabara, S., and Baba, T. (2009) Selective stabilization of mammalian microRNAs by 3' adenylation mediated by the cytoplasmic poly(A) polymerase GLD-2, *Genes Dev* **23**, 433–8.
45. Jones-Rhoades, M. W., Bartel, D. P., and Bartel, B. (2006) MicroRNAS and their regulatory roles in plants, *Annu Rev Plant Biol* **57**, 19–53.
46. Bartel, D. P. (2009) MicroRNAs: target recognition and regulatory functions, *Cell* **136**, 215–33.
47. Peters, L., and Meister, G. (2007) Argonaute proteins: mediators of RNA silencing, *Mol Cell* **26**, 611–23.
48. Tolia, N. H., and Joshua-Tor, L. (2007) Slicer and the argonautes, *Nat Chem Biol* **3**, 36–43.
49. Cifuentes, D., Xue, H., Taylor, D. W., Patnode, H., Mishima, Y., Cheloufi, S., et al. (2010) A novel miRNA processing pathway independent of Dicer requires Argonaute2 catalytic activity, *Science* **328**, 1694–8.
50. Kiriakidou, M., Tan, G. S., Lamprinaki, S., De Planell-Saguer, M., Nelson, P. T., and Mourelatos, Z. (2007) An mRNA m7G cap binding-like motif within human Ago2 represses translation, *Cell* **129,** 1141–51.
51. Bagga, S., Bracht, J., Hunter, S., Massirer, K., Holtz, J., Eachus, R.,et al. (2005) Regulation by let-7 and lin-4 miRNAs results in target mRNA degradation, *Cell* **122**, 553–63.
52. Behm-Ansmant, I., Rehwinkel, J., Doerks, T., Stark, A., Bork, P., and Izaurralde, E. (2006) mRNA degradation by miRNAs and GW182 requires both CCR4:NOT deadenylase and DCP1:DCP2 decapping complexes, *Genes Dev* **20**, 1885–98.
53. Giraldez, A. J., Mishima, Y., Rihel, J., Grocock, R. J., Van Dongen, S., Inoue, K.,et al. (2006) Zebrafish MiR-430 promotes deadenylation and clearance of maternal mRNAs, *Science* **312**, 75–9.
54. Wu, L., Fan, J., and Belasco, J. G. (2006) MicroRNAs direct rapid deadenylation of mRNA, *Proc Natl Acad Sci USA* **103**, 4034–9.
55. Petersen, C. P., Bordeleau, M. E., Pelletier, J., and Sharp, P. A. (2006) Short RNAs repress translation after initiation in mammalian cells, *Mol Cell* **21**, 533–42.
56. Eulalio, A., Huntzinger, E., Nishihara, T., Rehwinkel, J., Fauser, M., and Izaurralde, E. (2009) Deadenylation is a widespread effect of miRNA regulation, *RNA* **15**, 21–32.
57. Guo, H., Ingolia, N. T., Weissman, J. S., and Bartel, D. P. Mammalian microRNAs predominantly act to decrease target mRNA levels, *Nature* **466,** 835–40.
58. Bushati, N., and Cohen, S. M. (2007) microRNA functions, *Annu Rev Cell Dev Biol* **23**, 175–205.
59. Leaman, D., Chen, P. Y., Fak, J., Yalcin, A., Pearce, M., Unnerstall, U., et al. (2005) Antisense-mediated depletion reveals essential and specific functions of microRNAs in Drosophila development, *Cell* **121**, 1097–108.
60. Miska, E. A., Alvarez-Saavedra, E., Abbott, A. L., Lau, N. C., Hellman, A. B., McGonagle, S. M., et al. (2007) Most Caenorhabditis elegans microRNAs are individually not essential for development or viability, *PLoS Genet* **3**, e215.
61. Leung, A. K., and Sharp, P. A. MicroRNA functions in stress responses, *Mol Cell* **40**, 205–15.
62. Sandberg, R., Neilson, J. R., Sarma, A., Sharp, P. A., and Burge, C. B. (2008) Proliferating cells express mRNAs with shortened 3' untranslated regions and fewer microRNA target sites, *Science* **320**, 1643–7.
63. Mayr, C., and Bartel, D. P. (2009) Widespread shortening of 3'UTRs by alternative cleavage and polyadenylation activates oncogenes in cancer cells, *Cell* **138**, 673–84.
64. Paul, M. S., and Bass, B. L. (1998) Inosine exists in mRNA at tissue-specific levels and is most abundant in brain mRNA, *EMBO J* **17**, 1120–7.
65. Liang, H., and Landweber, L. F. (2007) Hypothesis: RNA editing of microRNA target sites in humans? *RNA* **13**, 463–7.
66. Athanasiadis, A., Rich, A., and Maas, S. (2004) Widespread A-to-I RNA editing of Alu-containing mRNAs in the human transcriptome, *PLoS Biol* **2**, e391.
67. Farajollahi, S., and Maas, S. Molecular diversity through RNA editing: a balancing act, *Trends Genet* **26**, 221–30.
68. Mayr, C., Hemann, M. T., and Bartel, D. P. (2007) Disrupting the pairing between let-7 and Hmga2 enhances oncogenic transformation, *Science* **315**, 1576–9.
69. Kim, J., and Bartel, D. P. (2009) Allelic imbalance sequencing reveals that single-nucleotide polymorphisms frequently alter microRNA-directed repression, *Nat Biotechnol* **27**, 472–7.

70. Poliseno, L., Salmena, L., Zhang, J., Carver, B., Haveman, W. J., and Pandolfi, P. P. A coding-independent function of gene and pseudogene mRNAs regulates tumour biology, *Nature* **465**, 1033–8.
71. Kedde, M., van Kouwenhove, M., Zwart, W., Oude Vrielink, J. A., Elkon, R., and Agami, R. A Pumilio-induced RNA structure switch in p27-3' UTR controls miR-221 and miR-222 accessibility, *Nat Cell Biol* **12**, 1014–20.
72. Hsu, S. D., Lin, F. M., Wu, W. Y., Liang, C., Huang, W. C., Chan, W. L., et al. (2011) miRTarBase: a database curates experimentally validated microRNA-target interactions, *Nucleic Acids Res* **39**, D163-9.
73. Calin, G. A., Dumitru, C. D., Shimizu, M., Bichi, R., Zupo, S., Noch, E., et al. (2002) Frequent deletions and down-regulation of micro- RNA genes miR15 and miR16 at 13q14 in chronic lymphocytic leukemia, *Proc Natl Acad Sci USA* **99**, 15524–9.
74. Calin, G. A., Sevignani, C., Dumitru, C. D., Hyslop, T., Noch, E., Yendamuri, S., et al. (2004) Human microRNA genes are frequently located at fragile sites and genomic regions involved in cancers, *Proc Natl Acad Sci USA* **101**, 2999–3004.
75. Ventura, A., and Jacks, T. (2009) MicroRNAs and cancer: short RNAs go a long way, *Cell* **136**, 586–91.
76. Medina, P. P., Nolde, M., and Slack, F. J. OncomiR addiction in an in vivo model of microRNA-21-induced pre-B-cell lymphoma, *Nature* **467**, 86–90.
77. Lu, J., Getz, G., Miska, E. A., Alvarez-Saavedra, E., Lamb, J., Peck, D., et al. (2005) MicroRNA expression profiles classify human cancers, *Nature* **435**, 834–8.
78. Chang, T. C., Yu, D., Lee, Y. S., Wentzel, E. A., Arking, D. E., West, K. M., et al. (2008) Widespread microRNA repression by Myc contributes to tumorigenesis, *Nat Genet* **40**, 43–50.
79. Lujambio, A., Calin, G. A., Villanueva, A., Ropero, S., Sanchez-Cespedes, M., Blanco, D., et al. (2008) A microRNA DNA methylation signature for human cancer metastasis, *Proc Natl Acad Sci USA* **105**, 13556–61.
80. Lujambio, A., Ropero, S., Ballestar, E., Fraga, M. F., Cerrato, C., Setien, F., et al. (2007) Genetic unmasking of an epigenetically silenced microRNA in human cancer cells, *Cancer Res* **67**, 1424–9.
81. Saito, Y., Liang, G., Egger, G., Friedman, J. M., Chuang, J. C., Coetzee, G. A., et al. (2006) Specific activation of microRNA-127 with downregulation of the proto-oncogene BCL6 by chromatin-modifying drugs in human cancer cells, *Cancer Cell* **9**, 435–43.
82. Faggad, A., Budczies, J., Tchernitsa, O., Darb-Esfahani, S., Sehouli, J., Muller, B. M., et al. (2010) Prognostic significance of Dicer expression in ovarian cancer-link to global microRNA changes and oestrogen receptor expression, *J Pathol* **220**, 382–91.
83. Karube, Y., Tanaka, H., Osada, H., Tomida, S., Tatematsu, Y., Yanagisawa, K., et al. (2005) Reduced expression of Dicer associated with poor prognosis in lung cancer patients, *Cancer Sci* **96**, 111–5.
84. Merritt, W. M., Lin, Y. G., Han, L. Y., Kamat, A. A., Spannuth, W. A., Schmandt, R., et al. (2008) Dicer, Drosha, and outcomes in patients with ovarian cancer, *N Engl J Med* **359**, 2641–50.
85. Wu, J. F., Shen, W., Liu, N. Z., Zeng, G. L., Yang, M., Zuo, G. Q., et al (2010) Down-regulation of Dicer in hepatocellular carcinoma, *Med Oncol.* Apr 20. doi: 10.1007/s12032-010-9520-5.
86. Kumar, M. S., Lu, J., Mercer, K. L., Golub, T. R., and Jacks, T. (2007) Impaired microRNA processing enhances cellular transformation and tumorigenesis, *Nat Genet* **39**, 673–7.
87. Kumar, M. S., Pester, R. E., Chen, C. Y., Lane, K., Chin, C., Lu, J., et al. (2009) Dicer1 functions as a haploinsufficient tumor suppressor, *Genes Dev* **23**, 2700–4.
88. Su, X., Chakravarti, D., Cho, M. S., Liu, L., Gi, Y. J., Lin, Y. L., et al. (2010) TAp63 suppresses metastasis through coordinate regulation of Dicer and miRNAs, *Nature* **467**, 986–90.
89. Melo, S. A., Ropero, S., Moutinho, C., Aaltonen, L. A., Yamamoto, H., Calin, G. A., et al. (2009) A TARBP2 mutation in human cancer impairs microRNA processing and DICER1 function, *Nat Genet* **41**, 365–70.

Chapter 2

Introduction to miRNA Profiling Technologies and Cross-Platform Comparison

Sarah Aldridge and James Hadfield

Abstract

MicroRNA analysis has been widely adopted for basic and applied science. The tools and technologies available for quantifying and analysing miRNAs are still maturing. Here, we give an introductory overview of the main tools and the challenges in their use. We also discuss the importance of basic experimental design, sample handling and analysis methods as the impact of these can be as profound as the choice of miRNA analysis platform. Whether the reader is interested in a gene-by-gene or genome-wide approach choosing the platform to use is not trivial. Careful thought given before starting an experiment will make the execution much easier.

Key words: MicroRNA, Microarray, Sequencing, Reverse transcription quantitative PCR, In situ hybridisation, Comparison

1. Introduction

MicroRNA (miRNA) analysis has rapidly gained a foothold in many labs and is quickly becoming a routine research tool being used in large cohort studies on clinical samples (1), is showing promise in cancer research (2) and has been reported as useful in tumour classification (3) (also reviewed in see ref. 4). MiRNAs hold particular appeal in clinical setting as they have been shown to be very stable in both plasma and serum (5–7). The tools used to measure and detect miRNAs have been largely borrowed from mRNA expression analysis and array-based comparative genomic hybridisation (aCGH) to interrogate DNA copy-number state. The use of microarrays in both mRNA and copy-number is now routine and gradually moving into clinical use; in fact, both techniques are

Jian-Bing Fan (ed.), *Next-Generation MicroRNA Expression Profiling Technology: Methods and Protocols*, Methods in Molecular Biology, vol. 822, DOI 10.1007/978-1-61779-427-8_2, © Springer Science+Business Media, LLC 2012

already being challenged by next-generation sequencing (NGS) methods as the emerging standard method (8, 9). And NGS analysis of miRNA has recently been reported as useful in prediction of clinical outcome (10). Analysis and comparison of different sequencing methods is only just starting to be published (11).

Array-based methods for mRNA and aCGH have been extensively compared (12–14) and the technologies have matured such that they are now routine experimental tools. The different platforms all produce high-quality data and the decisions on choice are often subjectively made. No single platform achieves gold-standard status. Developments in mRNA and aCGH arrays are now primarily an increase in the amount of data generated; i.e., total numbers of probes interrogated. MiRNA analysis techniques have not yet reached the same level of maturity, but the articles in this publication demonstrate how far we have come in a short time and discuss some of the issues yet to be resolved.

2. Problems with miRNA Detection and Quantitation

MiRNAs and other nucleic acids are detected, quantified, and otherwise analysed by three primary methods: hybridisation, PCR, and sequencing. The hybridisation-based Southern (15) and Northern (16) blotting techniques introduced in the 1970s ultimately led to the development of microarrays (17). The polymerase chain reaction (18) was further developed and reverse-transcription quantitative real-time PCR (RT-qPCR) has become the gold-standard technique for nucleic acid quantitation. Sanger sequencing (19) has developed much since its introduction in 1977 and is still the fundamental approach underlying NGS platforms (20–22), even though these systems rely on quite different methods to generate sequence data.

Compared to other nucleic acids however, analysis of miRNA is significantly complicated by several factors: miRNA length, discrimination between pre-, pri-, and mature miRNAs, variable Tm of primers or probes, RNA ligase sequence bias, high degrees of homology in miRNA families and high rates of miRNA discovery. Combinations of these issues impact the different methods for miRNA detection and quantitation and must be considered when designing miRNA experiments. A further complicating factor is that not all the miRNAs present in the central miRNA repository "miRBase" (23) are necessarily real. Resequencing experiments conducted by the Bartel lab (24) found that about 10% of miRBase miRNAs were not present in their dataset and may have been artefactual in other datasets.

A recent review of the major issues with miRNA detection and quantification (25) explains all of the issues listed above and more.

3. miRNA Analysis Technologies

3.1. Extraction Methods

The method chosen for preparation of RNAs, including small RNA from cells or tissues is an often overlooked but important aspect of any RNA analysis. The extraction method chosen for RNA, total RNA including the small RNA species or small RNAs alone, can have a downstream effect on the results obtained from a study. Techniques for assessing the quality of small RNA is still an area for which the best practise is still undecided. The impact of methods papers comparing nucleic acid extraction methods is under-valued and this information is often consigned to supplementary methods and commonly never makes it into a formal publication, although there are studies looking into the impact of extraction methodology (26–31). All of these give the very clear and easily understood message that any study should use standardised protocols throughout and employ a single RNA isolation method to avoid the small but potentially significant affects on gene expression analysis due to the RNA preparation method. This is further emphasised in the 2008 review by (25).

Debey et al. and Kim et al. both focussed on the impact of pre-analytical variables, including RNA extraction on gene expression profiling from blood. Debey et al. noted that none of the methods tested outperformed the others. Kim et al. reviewed several studies that tested RNA extraction methods but came to similar conclusions about standardisation. Campo Dell'Orto and Ach et al. both compared three methods of RNA extraction and compared gene expression measurements on arrays and with qRT-PCR. Ach et al. compared: TRIzol (Invitrogen) coupled with isopropanol precipitation, miRNeasy (Qiagen) and mirVana (Life Technologies). They used Agilent miRNA microarrays and real-time PCR to show that very few miRNA gene expression levels were affected by extraction method. Campo Dell'Orto et al. compared: miRNeasy, TRIzol, and TRIzol followed by RNeasy (Qiagen) cleanup. They used Affymetrix HG-U133 Plus 2.0 microarrays to show that the extraction method used does have an impact on gene expression experiments. They suggested the use of a single method but went further in recommending other pre-analytical variables be optimised before gene level analysis. Debey et al. compared: extraction of PBMC cells with TRIzol followed by RNeasy cleanup to whole blood PAXgene (Qiagen) or QIAamp (Qiagen) RNA extraction. They used Affymetrix HGU133A arrays to demonstrate the impact of extraction method on gene expression experiments. Git et al. investigated the best methods for RNA extraction and QC and have used extraction methods based on the Qiagen miRNeasy protocol in their 2010 comparison study. They also performed subsequent yield and quality assessment on Agilent Bioanalyser small RNA series II chips, spectrophotometric analysis and by urea/polyacrylamide gel electrophoresis.

Many of the microarray and sequence-based analysis methods now use total RNA as an input rather than fractionated miRNA. We would very strongly recommend the advice given in all the papers discussed above. A single sample handling and RNA extraction methods should be used in a study. Considering the relatively high cost, and time committed to performing miRNA expression studies efforts should be put into these upstream pre-analytical variables and ideally these should be performed by experienced or practised operators. Users should never compare samples that have been extracted using different methods.

3.2. Reverse Transcription Quantitative PCR

Reverse transcription Quantitative PCR (RT-qPCR) protocols are varied, but essentially rely on conversion of RNA to cDNA and subsequent locus-specific quantification by comparison to a standard reference gene or sample. It is the most sensitive assay technology currently available although re-sequencing may ultimately have equal single copy sensitivity (32). Recently, guidelines for reporting qPCR experiments have been published (33) further strengthening the reliability and intra-lab accessibility of such data.

Methods are available which use either TaqMan probes or SYBR Green. The TaqMan probe-based method (34) starts with a reverse transcription step using gene-specific stem loop primers, which will reverse transcribe both precursor and mature miRNA (35, 36). The alternative SYBR Green-based method (37) uses tagged and anchored oligo- dT primers for reverse transcription of polyadenylated small RNAs for mature miRNAs (38) followed by SYBR Green-based detection. Platforms for medium to high-throughput analysis of miRNA have been an area of intense development of recent years. These have taken the form of assay plates that can assess tens or hundreds of miRNAs across multiple samples in a highly parallel format (39, 40). If large numbers of samples are available for analysis, then RT-qPCR is hard to beat. However, unlike for mRNA, miRNA RT-qPCR is constrained by the detection limitations mentioned above. At least two studies have thus questioned the use of RT-qPCR as a "gold standard" for miRNA quantification (25, 29).

3.3. In Situ Hybridisation

The use of in situ hybridisation (ISH) allows miRNA analysis to be performed directly in tissues of interest and facilitates identification of miRNA expression in specific cell types in complex organs or heterogeneous tumours. Although not a high-throughput tool, ISH can be a very important validation technique once genome-wide miRNA analysis has been conducted. There are several published methods for miRNA ISH, largely using locked nucleic acids (LNA) probes. Probes with LNAs included in the design show increased hybridisation affinities for RNA and miRNA targets over standard probes (41–43). Incorporation of LNAs increases the thermal stability of the probe/RNA complex (44). This is important as probes for miRNAs need to be short, but the ISH conditions must be stringent to allow for accessibility and hybridisation of relatively short probes (45). Exiqon offers a commercial design service for miRNA ISH probes, which include a proportion of LNAs in the probe sequence. Low signal strength is one downside of LNA probes, but this can be improved by the use of 3′ and 5′ labelled probes (45).

3.4. Microarrays

Microarrays can be produced in-house (46) or purchased commercially (see Table 1 for a non-exhaustive list). Any array-based method is subject to the same problems of probe design and hybridisation artefacts as described previously. Another problem, discussed in some of the comparison studies, is that not all manufacturers are willing to freely distribute probe sequence information. This data is required for a thorough analysis of miRNA probe characteristics. There is also a risk of obsolescence with microarrays; the Illumina BeadArray and Ambion miRNA platforms were withdrawn in early 2010 both of which performed well in miRNA comparison studies. Users of these products have little control over decisions like this yet comparing results from datasets generated on different platforms is very complex.

The choice of microarray platform is not easy to make. Agilent has almost complete flexibility in array design, Ambion included

Table 1
Platforms analysed in different comparison studies

	RT-qPCR		Microarray platforms										Sequencing	
	SYBR	TaqMan	ABI LDA	Af	Ag	Am	C	E	Il	In	L	T	Illumina	SOLiD
Ach		•			•									
Baldwin			•	•	•			•	•				•	•
Chen		•									•			
Dreher		•		•				•		•				
Git	•	•			•	•	•	•	•	•			•	
Pradervand			•	•	•				•				•	
Sah				•	•	•		•	•					
Sato	•				•	•		•		•		•		
Yauk			•		•			•		•	•			

Af Affymetrix, *Ag* Agilent, *Am* Ambion, *C* Combimatrix, *E* Exiqon, *Il* Illumina, *In* Invitrogen, *L* LC sciences, *T* Toray

putative miRNAs not present in miRBase, Exiqon use LNA to increase specificity, Combimatrix support reuse of arrays, Affymetrix offer a single array containing miRNAs from five species, and other platforms all offer something unique. The technical differences in the available platforms include: printing and surface technology, slide format, labelling, hybridisation, one- or two-colour detection chemistries, probe design, and cost. The input RNA sample requirements also differ widely, from 100 ng of total RNA to 1 μg of small RNA fraction. Replicate spots are useful in downstream analysis and range from 1 to more than 300, with mean spot replicate numbers being from 2 to 5. Surprisingly, the number of replicates is not necessarily constant within a single array platform.

Microarrays are the most obvious choice for users with tens to hundreds of samples for which they wish to perform high-throughput miRNA analysis.

3.5. Next-Generation Sequencing

NGS technology was first used to profile small RNA sequences in *C. elegans* on the 454 platform (47). This study identified several small RNA species and demonstrated that NGS had great potential as a platform for small RNA analysis. Libraries are prepared for NGS using methods based on traditional small RNA cloning techniques. Adapters are ligated to the ends of the small RNA molecules and these are then used as templates for sequencing (48). Small RNA cloning methods for NGS have proved to be technically challenging and time consuming although protocols are improving and alternative methods are becoming available. It is known that biases can be introduced during library production and the implications this has for downstream NGS sequencing has been explored (49, 50). Several steps of the small RNA cloning protocol are noted as hotspots for bias introduction, including adapter ligation, PCR amplification, reverse transcription, and gel isolation.

NGS is particularly well suited to the discovery of novel small RNA species, as the technique is not constrained by the use of hybridisation probes for which prior knowledge of sequence is required. Advances in sequencing technology have accelerated both the discovery rate of new miRNAs and modifications to existing miRNA entries, reflecting subtle variations in mature miRNA sequences (e.g., post-transcriptional editing or terminal residue addition) (51). With the advent of next-generation sequencers with increased capacity for data generation, coupled with advancement in small RNA library preparation methods, many researchers are making use of methods for indexing and multiplexing pools of small RNA libraries to maximise data return. There is a lack of consensus over the best methods for data normalisation, a downside that this platform shares with other methods for small RNA analysis. In addition, associated tools for computational analysis are in their infancy.

4. Microarrays vs. Sequencing

Caveat lector; the discussion below will almost certainly be outdated by the time you read this. We would prefer readers to use this section as a springboard for discussions in their own labs. The rate of change in sequencing technologies is far too great to keep up with in written form. At the time of writing, for instance, Illumina had just announced a 1.14 Tb run on their HiSeq 2000 platform that would allow over 200 exomes in a single run.

There has been much discussion on when, not if, sequencing will supplant microarrays as the analysis method of choice. This discussion is happening almost everywhere that users are running microarrays and is particularly evident on forums like SEQanswers (http://www.seqanswers.com). In many cases, the quality of data obtainable from a sequence-based analysis is superior to microarray.

For gene expression analysis, the same levels of detail can be obtained from 10 M sequence reads vs. a standard 3′ gene expression array (8, 9). Montgomery et al. showed that this relatively small number of reads produced a similar dynamic range to microarrays but with improved ability to detect and quantify alternatively spliced and very abundant transcripts. Bashir et al. observed that 90% of observed transcripts in a 35 M read dataset can be detected with just 1 M sampled reads, which compares well with the Montgomery et al. analysis. They also noted that an initial sampling run, using highly multiplexed libraries, for instance, could be used as an experimental design tool for transcript sequencing projects. The analysis of alternative splicing has exploded with the advent of NGS. A recent comparison (52) of SOLiD sequencing and Affymetrix exon arrays looked specifically at expression of individual exons, and transcription outside currently annotated loci. They showed that over 80% of exons were detected on both platforms but that RNA-Seq appeared to have a lower background error rate. RNA-Seq was also more sensitive in detecting differentially expressed exons and they could find thousands of novel transcripts with previously unreported exon–exon junctions. Lastly, discovery of new transcripts (mRNAs, miRNAs, LINC RNAs, etc.) is simply not possible using a microarray.

For structural variation analysis, the same levels of detail can also be obtained from about 10 M reads (8). However, as much SV analysis is being done using genotyping intensities from microarrays and the SNP calls bring additional information on LOH that can be used in many studies, there is not a clear choice between the platforms. To obtain the same depth of SNP coverage as an array may require 10–30-fold sequencing of a genome. Sequencing will allow breakpoints and CNV junctions to be mapped to single-nucleotide resolution. Bashir et al. showed that they could resolve

90% of breakpoints using a mix of 200 bp and 2 kb insert size libraries. There is a trade-off between detection and resolution; for a given number of reads increasing library insert size increases the probability of structural variation detection; however, this decreases ultimate resolution of breakpoints. As the number of reads increases in datasets, this issue is reduced. However, many researchers will aim to perform structural variation analysis of tens, possibly hundreds of individuals in a single sequencing run in the near future. An important observation they made was that detection of small structural variations requires the use of libraries with a low insert size distribution and that the distribution must be smaller than the size of the structural variant itself. But even though technologies and methods are improving, long insert library preparation still requires large amounts of nucleic acid. In the case of clinical samples, this can be a major obstacle and the experimental design should balance "sample-cost" vs. structural variation detection and/or resolution.

It is likely that the choice between microarrays and sequencing will be made on secondary factors, such as the platforms locally and easily available.

5. Data Analysis

The use of microarrays for differential miRNA expression led to the adoption of the same or similar tools for their analysis. However, there is an assumption in many mRNA analysis tools that mean mRNA levels are relatively stable and that only a subset of mRNAs might be truly differentially expressed. This is certainly not the case for miRNA analysis, where the number of miRNAs expressed is quite low and there can be stark differences between samples when looking for differentially expressed miRNAs. The methods for processing data can have a similar impact on final results as the technology used in a study (53). Novel methods for miRNA analysis are, and will continue to be developed. Git et al. implemented a novel algorithm to get around the need to choose a reference technology or "gold standard" in their 2010 study.

Understanding the inherent biases in the technique being analysed is important if sensible and biologically meaningful results are to be obtained. The use of spike-in control miRNAs does not necessarily make analysis simpler. However, good experimental design where all variables are considered and a controlled randomised design is used with a single analytical technique will allow useful comparisons to be made. Different biases present in the varied technologies may make certain effects impossible to detect.

6. Comparison Studies

MiRNA analysis methods have only recently been systematically compared (26, 29, 53–59). While microarrays from nine array suppliers were used in these studies none has been used across all those discussed here, one study used six, two used five and four more used three or four array platforms each (Table 1). Since not all studies compared the same arrays, it is somewhat unfair to try and suggest which array platform performs "best"; of course, this is exactly what most readers of these papers want to find out!

Comparisons, and choices, are complicated by the debate about the merits of microarray vs. sequencing vs. real-time PCR as the method of choice. So while there is a large choice of microarray platforms, there is almost as much choice from the non-microarray-based systems (see Table 1). Table 1 shows which platforms were used in the comparison studies we compared. The different platforms generally showed good within- and between-platform reproducibility and correlated well with qPCR, as reported in each study.

Git et al. carried out the most extensive comparison, which encompassed six microarray platforms, real-time PCR using either SYBR Green following reverse transcription with a tagged and anchored oligo-dT primer or TaqMan-based assays with reverse transcription using a pool of gene-specific primers and NGS on the Illumina GAIIx platform.

7. Pitfalls of Comparisons

All comparison studies published have the same flaw; they are outdated as soon as they are available in print. The protocols for sample handling, microarray design or next-gen sequencing technologies improve at a rate far outstripping the ability of authors to produce and analyse comparison datasets. However, these studies are useful to others in deciding which platform to use in a project. Any comparison is likely to reveal shortcomings in the assumptions made about samples, platforms and analysis methods at the start of the process. These may not necessarily be resolvable once the study is complete.

The biological samples chosen and the methods used to extract, quantify, and quality assess them before any biological analysis is made can have a profound effect on the outcome of comparison studies. While many groups have suggested the use of standard samples for use as controls in biological studies, these can only have an impact if these standards are used in the majority of published experiments, which they are not. The samples used in the comparison studies addressed here varied significantly: Ach et al.

used Ambion normal human tissue RNA, HeLa, and ZR-75-1 cell lines; Baldwin et al. used two commercially available RNA samples; Chen et al. used mouse myoblast RNA; Dreher et al. used an HPV-transfected human cell line; Git et al. used an RNA pool from normal breast tissue, and two breast cancer cell lines that were representative of samples used in cancer research; Pradervand et al. used human heart and brain total RNA from Stratagene; Sah et al. used human placenta total RNA spiked with seven synthetic miRNAs in complex pools; Sato et al. used two human RNAs from Ambion; Yauk et al. used two pools of mouse tissue RNA. The majority of these used commercially available RNAs or cell lines that would be relatively easy for others to acquire if they wanted to repeat any aspects of these studies or use them as controls in other work.

A comparison study needs to consider the real-world application of any methods being compared. Protocols for microarray, RT-qPCR, and next-gen sequencing vary from lab to lab. Authors of comparison studies need to decide whether to use manufacturers recommended protocols and starting materials or use their own experience. Both significantly affect the performance of platforms for measuring miRNAs.

The challenges of probe design for miRNAs and the rapidly evolving miRBase database mean that it is important to only compare probes targeting the same miRNA sequence. Several of the comparison papers specifically compared probe sequences and Git et al. commented on the availability, or not, of probe sequence information from the companies compared.

8. Conclusions

Nearly, all technologies used in the comparison studies above performed acceptably in the different measures of performance discussed in each paper. As there are such large differences between and within the available technologies and between platforms, it is important to consider the choice for a particular experiment, and understand that each experimental factor will have an impact on the final results. Comparison studies allow us to quickly assay the performance of a wide variety of systems to measure miRNAs and are of very real benefit to individual scientists. Unfortunately, they do not carry the gravitas of primary scientific publications focussing on biological insights. It would help if these papers were referenced more frequently if the comparison paper aided the choice of platform.

The choice of platform for miRNA analysis needs to balance time, precision, accuracy, cost, and sample type. RT-qPCR is likely to yield the highest sensitivity, use minimal sample and cost least, but is not necessarily practical for profiling hundreds of miRNAs.

Microarrays allow the profiling of tens or hundreds of samples across the known miRNA'ome is shortened from miRNA transcriptome, but are limited by probe design. If it is important to discover new miRNAs, distinguish between isoforms or analyse RNA editing then sequencing is the only method to consider.

Acknowledgements

We thank Stefan Graf, Heidi Dvinge, Claudia Kutter, and Anna Git for their helpful comments on the manuscript.

References

1. Volinia, S., Galasso, M., Costinean, S., Tagliavini, L., Gamberoni, G., Drusco, A., et al. (2010). Reprogramming of miRNA networks in cancer and leukemia. *Genome Res* **20**, 589–99.
2. Lin, P-Y., Yu, S-L., and Yang, P-C. (2010). MicroRNA in lung cancer. *Br J Cancer* **103**, 1144–8.
3. Rosenfeld, N., Aharonov, R., Meiri, E., Rosenwald, S., Spector, Y., Zepeniuk, M., et al. (2008). MicroRNAs accurately identify cancer tissue origin. *Nat Biotechnol* **26**, 462–9.
4. Le Quesne, J., and Caldas, C. (2010). MicroRNAs and breast cancer. *Mol Oncol* **4**, 230–41.
5. Mitchell, P. S., Parkin, R. K., Kroh, E. M., Fritz, B. R., Wyman, S. K., Pogosova-Agadjanyan, E. L., et al. (2008). Circulating microRNAs as stable blood-based markers for cancer detection. *Proc Natl Acad Sci USA* **105**, 10513–18.
6. Kroh, E. M., Parkin, R. K., Mitchell, P. S., and Tewari, M. (2010). Analysis of circulating microRNA biomarkers in plasma and serum using quantitative reverse transcription-PCR (qRT-PCR). *Methods* **50**, 298–301.
7. Liu, R., Zhang, C., Hu, Z., Li, G., Wang, C., Yang, C., et al. (2010). A five-microRNA signature identified from genome-wide serum microRNA expression profiling serves as a fingerprint for gastric cancer diagnosis. *Eur J Cancer* **47**, 784–91.
8. Bashir, A. (2010). Designing deep sequencing experiments: detecting structural variation and estimating transcript abundance. *BMC Genomics* **11**, 385–99.
9. Montgomery, S. B, Sammeth, M., Gutierrez-Arcelus, M., Lach, R. P., Ingle, C., Nisbett, J., et al. (2010). Transcriptome genetics using second generation sequencing in a Caucasian population. *Nature* **464**, 773–7.
10. Hu, Z., Chen, X., Zhao, Y., Tian, T., Jin, G., Shu, Y., et al. (2010). Serum MicroRNA Signatures Identified in a Genome-Wide Serum MicroRNA Expression Profiling Predict Survival of Non-Small-Cell Lung Cancer. *J Clin Oncol* **28**, 1721–6.
11. Levin, J. Z., Yassour, M., Adiconis, X., Nusbaum, C., Thompson, D. A., Friedman, N., et al. (2010). Comprehensive comparative analysis of strand-specific RNA sequencing methods. *Nat Methods* **7**, 709–15.
12. Baumbusch, L. O., Aarøe, J., Johansen, F. E., Hicks, J., Sun, H., Bruhn, L., et al. (2008). Comparison of the Agilent, ROMA/NimbleGen and Illumina platforms for classification of copy number alterations in human breast tumors. *BMC Genomics* **9**, 379–401.
13. Curtis, C., Lynch, A. G., Dunning, M. J., Spiteri, I., Marioni, J. C., Hadfield, J., et al. (2009). The pitfalls of platform comparison: DNA copy number array technologies assessed. *BMC Genomics* **10**, 588–601.
14. MAQC. (2006). The MicroArray Quality Control (MAQC) project shows inter- and intraplatform reproducibility of gene expression measurements. *Nat Biotechnol* **24**, 1151–61.
15. Southern, E. M. (1975). Detection of specific sequences among DNA fragments separated by gel electrophoresis. *J Mol Biol* **98**, 503–17.
16. Alwine, J. C., Kemp, D. J., and Stark, G. R. (1977). Method for detection of specific RNAs in agarose gels by transfer to diazobenzyloxymethyl-paper and hybridization with DNA probes. *Proc Natl Acad Sci USA* **74**, 5350–4.
17. Schena, M., Shalon, D., Davis, R. W., and Brown, P. (1995). Quantitative Monitoring of

Gene Expression Patterns with a Complementary DNA Microarray *Science* **270**, 467–70.

18. Mullis, K. B. (1990). Target amplification for DNA analysis by the polymerase chain reaction. *Ann Biol Clin (Paris)* **48**, 579–82.
19. Sanger, F., and Coulson, A. R. (1975). A rapid method for determining sequences in DNA by primed synthesis with DNA polymerase. *J Mol Biol* **94**, 441–8.
20. Bentley, D. R., Balasubramanian, S., Swerdlow, H., Smith, G. P., Milton, J. (2008). Accurate whole human genome sequencing using reversible terminator chemistry. *Nature* **456**, 53–9.
21. Margulies, M., Egholm, M., Altman, W. E., Attiya, S., Bader, J. S., Bemben, L. A. et al. (2005). Genome sequencing in microfabricated high-density picolitre reactors. *Nature* **437**, 376–80.
22. McKernan, K. J., Peckham, H. E., Costa, G. L., McLaughlin, S. F., Fu, Y., Tsung, E. F., et al. (2009). Sequence and structural variation in a human genome uncovered by short-read, massively parallel ligation sequencing using two-base encoding. *Genome Res* **19**, 1527–41.
23. Griffiths-Jones, S., Saini, H. K., van Dongen, S., and Enright, A. J. (2008). miRBase: tools for microRNA genomics. *Nucleic Acids Res* **36**, D154-8.
24. Chiang, H. R., Schoenfeld, L. W., Ruby, J. G., Auyeung, V. C., Spies, N., Baek, D., et al. (2010). Mammalian microRNAs: experimental evaluation of novel and previously annotated genes. *Genes Dev* **24**, 992–1009.
25. Nelson, P.T., Wang, W-X., Wilfred, B. R., and Tang, G. (2009). Technical variables in high-throughput miRNA expression profiling: much work remains to be done. *Biochim Biophys Acta* **1779**, 758–65.
26. Ach, R. A, Wang, H., and Curry, B. (2008). Measuring microRNAs: comparisons of microarray and quantitative PCR measurements, and of different total RNA prep methods. *BMC Biotechnol* **8**, 69–85.
27. Campo Dell'Orto, M., Zangrando, A., Trentin, L., Li, R., Liu, W. M., te Kronnie, G., et al. (2007). New data on robustness of gene expression signatures in leukemia: comparison of three distinct total RNA preparation procedures. *BMC Genomics* **8**, 188–203.
28. Debey, S., Schoenbeck, U., and Hellmich, M. (2004). Comparison of different isolation techniques prior gene expression profiling of blood derived cells: impact on physiological responses , on overall expression and the role of different cell types. *Pharmacogenomics* **4**, 193–207.
29. Git, A., Dvinge, H., Salmon-Divon, M., Osborne, M., Kutter, C., Hadfield, J., et al. (2010). Systematic comparison of microarray profiling, real-time PCR, and next-generation sequencing technologies for measuring differential microRNA expression. *RNA* **16**, 991–1006.
30. Git, A., Spiteri, I., Blenkiron, C., Dunning, M. J., Pole, J. C., Chin, S. F., et al. (2008). PMC42, a breast progenitor cancer cell line, has normal-like mRNA and microRNA transcriptomes. *Breast Cancer Res* **10**, R54.
31. Kim, S. J., Dix, D. J., Thompson, K. E., Murrell, R. N., Schmid, J. E., Gallagher, J. E., et al. (2007). Effects of Storage, RNA Extraction, Genechip Type, and Donor Sex on Gene Expression Profiling of Human Whole Blood. *Clin Chem* **53**, 1038–45.
32. Fan, H. C., Blumenfeld, Y. J., Chitkara, U., Hudgins, L., and Quake, S. R. (2008). Noninvasive diagnosis of fetal aneuploidy by shotgun sequencing DNA from maternal blood. *Proc Natl Acad Sci USA* **105**, 16266–71.
33. Bustin, S. A., Benes, V., Garson, J. A., Hellemans, J., Huggett, J., Kubista, M., et al. (2009). The MIQE Guidelines: Minimum Information for Publication of Quantitative Real-Time PCR Experiments. *Clin Chem* **55**, 611–22.
34. Heid, C. A., Stevens, J., Livak, K. J., Williams, P. M. (1996). Real time quantitative PCR. *Genome Res* **6**, 986–94.
35. Chen, C., Ridzon, D. A., Broomer, A J., Zhou, Z., Lee, D. H., Nguyen, J. T., et al. (2005). Real-time quantification of microRNAs by stem-loop RT-PCR. *Nucleic Acids Res* **33**, e179-88.
36. Schmittgen, T. D., Lee, E. J., Jiang, J., Sarkar, A., Yang, L., Elton, T. S., et al. (2008). Real-time PCR quantification of precursor and mature microRNA. *Methods* **44**, 31–8.
37. Schneeberger, C., Speiser, P., Kury, F., and Zeillinger, R. (1995). Quantitative detection of reverse transcriptase-PCR products by means of a novel and sensitive DNA stain. *PCR Methods Appl* **4**, 234–8.
38. Shi, R., and Chiang, V. (2005). Facile means for quantifying microRNA expression by real-time PCR. *Biotechniques* **39**, 519–25.
39. Morrison, T., Hurley, J., Garcia, J., Yoder, K., Katz, A., Roberts, D., et al. (2006). Nanoliter high throughput quantitative PCR. *Nucleic Acids Res* **34**, e123–31.
40. Spurgeon, S. L., Jones, R. C., and Ramakrishnan, R. (2008). High throughput gene expression measurement with real time PCR in a microfluidic dynamic array. *PLoS One* **3**, 1662–9.
41. Kloosterman, W. P., Wienholds, E., Bruijn, E. D., Kauppinen, S., Plasterk, R. H. A. (2006).

In situ detection of miRNAs in animal embryos using LNA-modified oligonucleotide probes. *Nat Methods* **3**, 2005–7.
42. Válóczi, A., Hornyik, C., Varga, N., Burgyán, J., Kauppinen, S., Havelda, Z. (2004). Sensitive and specific detection of microRNAs by northern blot analysis using LNA-modified oligonucleotide probes. *Nucleic Acids Res* **32**, e175–82.
43. Wienholds, E., Kloosterman, W. P., Miska, E., Alvarez-Saavedra, E., Berezikov, E., de Bruijn, E., et al. (2005). MicroRNA expression in zebrafish embryonic development. *Science* **309**, 310–1.
44. Thomsen, R., Nielsen, P. S., Jensen, T. H. (2005). Dramatically improved RNA in situ hybridization signals using LNA-modified probes. *RNA* **11**, 1745–8.
45. Obernosterer, G., Martinez, J., Alenius, M. (2007). Locked nucleic acid-based in situ detection of microRNAs in mouse tissue sections. *Nat Protoc* **2**, 1508–14.
46. Liu, C. G., Calin, G. A., Meloon, B., Gamliel, N., Sevignani, C., and Ferracin, M., et al. (2004). An oligonucleotide microchip for genome-wide microRNA profiling in human and mouse tissues. *Proc Natl Acad Sci USA* **101**, 9740–4.
47. Ruby, J. G., Jan, C., Player, C., Axtell, M. J., Lee, W., and Nusbaum, C., et al. (2006). Large-scale sequencing reveals 21U-RNAs and additional microRNAs and endogenous siRNAs in C. elegans. *Cell* **127**, 1193–207.
48. Hafner, M., Landgraf, P., Ludwig, J., Rice, A., Ojo, T., Lin, C., et al. (2008). Identification of microRNAs and other small regulatory RNAs using cDNA library sequencing. *Methods* **44**, 3–12.
49. Linsen, S. E., de Wit, E., Janssens, G., Heater, S., Chapman, L., Parkin, R. K., et al. (2009). Limitations and possibilities of small RNA digital gene expression profiling. *Nat Methods* **6**, 474–6.
50. Tian, G., Yin, X., Luo, H., Xu, X., Bolund, L., and Zhang, X. (2010). Sequencing bias : comparison of different protocols of MicroRNA library construction. *BMC Biotechnol* **10**, 64–73.
51. Landgraf, P., Rusu, M., Sheridan, R., Sewer, A., Iovino, N., and Aravin, A., et al. (2007). A Mammalian microRNA Expression Atlas Based on Small RNA Library Sequencing. *Cell* **129**, 1401–14.
52. Bradford, J. R., Hey, Y., Yates, T., Li, Y., Pepper, S. D., and Miller, C. J. (2010). A comparison of massively parallel nucleotide sequencing with oligonucleotide microarrays for global transcription profiling. *BMC Genomics* **11**, 282–302.
53. Sah, S., McCall, M., Eveleigh, D., Wilson, M., and Irizarry, R. (2010). Performance evaluation of commercial miRNA expression array platforms. *BMC Res Notes* **3**, 80–6.
54. Baldwin, D. (2009). ABRF microRNA Profiling: Platform Comparison. www.abrf.org/ResearchGroups/Microarray/Activities/R7_Baldwin.pdf.
55. Chen, Y., Gelfond, J. A., McManus, L. M., and Shireman PK (2009). Reproducibility of quantitative RT-PCR array in miRNA expression profiling and comparison with microarray analysis. *BMC Genomics* **10**, 407–17.
56. Dreher, A., Rossing, M., Kaczkowski, B., Nielsen, F. C., and Norrild, B. (2010). Differential expression of cellular microRNAs in HPV-11 transfected cells. An analysis by three different array platforms and qRT-PCR. *Biochem Biophys Res Commun* **403**, 357–62.
57. Pradervand, S., Weber, J., Lemoine, F., Consales, F., Paillusson, A., and Dupasquier, M., et al. (2010). Concordance among digital gene expression, microarrays, and qPCR when measuring differential expression of microRNAs. *Biotechniques* **48**, 219–22.
58. Sato, F., Tsuchiya, S., Terasawa K., and Tsujimoto G. (2009). Intra-platform repeatability and inter-platform comparability of microRNA microarray technology. *PLoS One* **4**, 5540–52.
59. Yauk, C., Rowan-Carroll, A., Stead, J., and Williams, A. (2010). Cross-platform analysis of global microRNA expression technologies. *BMC Genomics* **11**, 330–57.

Chapter 3

Stem-Loop RT-qPCR for MicroRNA Expression Profiling

James Hurley, Doug Roberts, Andrew Bond, David Keys, and Caifu Chen

Abstract

Quantification of the microRNAs (miRNAs) in cells or tissues is a crucial step in understanding their biological functions. Development of the stem-loop reverse transcription procedure and TaqMan® miRNA assays enables accurate detection of miRNA expression levels by quantitative PCR. Increased experimental throughput permits the expression screening of larger number of miRNAs with small amounts of sample. Here, we demonstrate the use of both TaqMan® Array Card and OpenArray® platforms to accurately determine the level of miRNA gene expression in biological samples.

Key words: MicroRNA, Stem-loop RT, qPCR, Expression profiling, TaqMan® Array Cards, OpenArray® Plates

1. Introduction

MicroRNAs (miRNAs) are short, noncoding RNA molecules which play important roles in modulating gene expression in animals and plants (1, 2). To date, there are a total of 1,048 miRNAs in human according to the miRBase Release 16 (http://www.mirbase.org/cgi-bin/browse.pl). Accurate quantification of miRNAs is a crucial step toward understanding their biological functions and potentially using them as biomarkers for cancers and many other human diseases (3–6). Quantification of such small RNA molecules is challenging, mainly due to small size, sequence homology, end polymorphism, and large dynamic range in gene expression. Since 2004, a number of miRNA quantification methods, including TaqMan® MicroRNA Assays or Arrays (7–10), miRNA microarrays (11–14), SYBR®-based miRNA reverse transcription (RT)-quantitative PCR (qPCR) assays (15–17), BeadArray (18), Invader Assays (19), and Padlock probe-based assays (20), have been made

Jian-Bing Fan (ed.), *Next-Generation MicroRNA Expression Profiling Technology: Methods and Protocols*, Methods in Molecular Biology, vol. 822, DOI 10.1007/978-1-61779-427-8_3, © Springer Science+Business Media, LLC 2012

available to miRNA researchers. Short size and large dynamic range of miRNAs generally make hybridization-based approaches less desirable. Stem-loop RT-based TaqMan® MicroRNA Assays are widely accepted as the gold-standard method with large dynamic range and high specificity and sensitivity.

1.1. Principle of Stem-Loop RT-qPCR

TaqMan® MicroRNA Assays use novel stem-loop primers for reverse transcription, followed by real-time TaqMan® qPCR in wells or array cards (7, 10). The stem-loop RT primer includes 3′ overhang sequence, a stem, and a loop. The 3′ overhang sequence is short, ranging from 5 to 8 nucleotides. The stem-loop structure, which is specific to the 3′ end of the mature miRNA, extends from the very short, mature miRNA molecule and then adds a universal 3′ priming site for follow-up qPCR. This new primer design overcomes a fundamental problem in miRNA quantification because the short length of mature miRNAs prohibits a conventional qPCR assay design. There are several advantages about stem-loop RT primers. First, by annealing a short RT priming sequence to the 3′ miRNA, it gives RT a better specificity for discriminating similar miRNAs. Second, the stem-loop structure prevents hybridization of its RT primer to miRNA precursors, other long RNAs, as well as genomic DNA. Third, the base stacking of the stem enhances the thermal stability of miRNA and DNA heteroduplex, further improving the RT efficiency for short RT primers. Finally, the stem-loop extends the 3′ end of the miRNA by RT. The resulting longer RT product presents a template amenable to real-time TaqMan® qPCR with high sensitivity and specificity that are largely due to specific PCR primers and the TaqMan® probe.

1.2. Use of Preamplification in MicroRNA Profiling from Limited RNA Samples

Preamplification is a powerful technology for preamplifying a large number of DNA or cDNA targets using a specific set of PCR primers at lower primer concentration prior to TaqMan®. Megaplex™ RT and PreAmp primer pools have been developed to detect and quantify up to 380 miRNA genes per pool. When assay sensitivity is of utmost importance, Megaplex™ PreAmp Primers can significantly enhance the ability to detect low-expressed miRNAs, enabling the generation of a comprehensive expression profile using as little as 1 ng of input total RNA or even single cells (8). Advantages of Megaplex™ RT and preamplification include (1) lower RNA input; (2) detection of low-abundant genes; (3) simplified workflow; and (4) lower RT cost.

1.3. Two Major qPCR Platforms for MicroRNA Expression Profiling

1.3.1. TaqMan® Array Cards

TaqMan® Array Cards (Fig. 1) are microfluidic devices containing 384 reaction chambers for individual TaqMan® real-time qPCR reactions. Relative to standard 96- or 384-well plates, TaqMan® Array Cards offer multiple practical and experimental advantages for profiling. Primary among these is an extremely simple and robust workflow with a minimal number of pipetting steps. The assays are preloaded in the reaction chambers at the time of

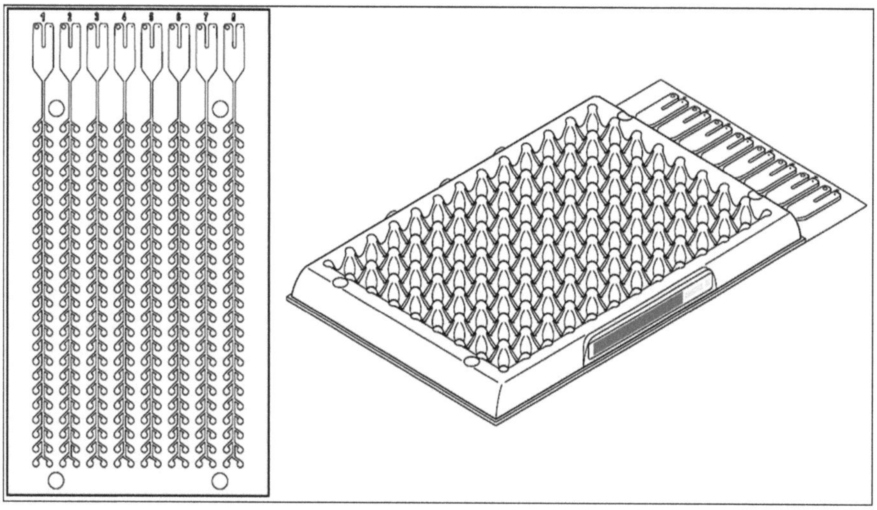

Fig. 1. TaqMan® Array Cards have loading chambers, microfluidic channels, and reaction chambers laid out on a flat surface. Structural stability of the card is provided by a plastic carrier mounted above the wells. The reaction chambers are loaded via centrifugation and sealed by collapsing the microfluidic channels. Each of the eight loading chambers feeds 48 individual reaction chambers. The reaction chamber spacing and overall dimensions of the TaqMan® Array Card match the spacing and dimensions of a 384-well plate. The reaction volume is 1.7 μl.

manufacture, eliminating the need to aliquot and track large numbers of assays. Sample and TaqMan® Master Mix are added through eight loading ports and then distributed to all reaction chambers using centrifugation, minimizing opportunities for experimental errors. Cost savings, relative to 96- or 384-well-based experiments, are significant since reagent volumes are lower, and there is no need to purchase the individual assays for the large number of targets common to profiling experiments (21). The real-time qPCR reactions for TaqMan® Array Cards are performed on either the Applied Biosystems 7900HT Fast Real-Time PCR System or the Applied Biosystems ViiA 7 Real-Time PCR System using compatible thermal blocks and lids.

TaqMan® Array Cards are available with both fixed and custom panels of miRNA targeting assays for real-time qPCR using samples generated with the stem-loop reverse transcription, as well as pre-amplification products. The contents of the fixed panels are exactly matched to the Megaplex™ RT and PreAmp primer pools. They target either human or rodent miRNAs and there are two panels for each, an "A" panel with assays which target well-studied, broadly expressed miRNAs selected for their interest to the research community and a "B" panel targeting miRNAs which have not been as extensively characterized or are narrowly expressed. When using TaqMan® Array Cards for miRNA profiling experiments, these fixed panel cards are usually the preferred format, as they allow characterization of up to 380 different miRNAs per sample on each instrument run (21, 22).

1.3.2. OpenArray® Plates

The OpenArray® plate (Fig. 2) is a thin, stainless steel plate (25×75×0.3 mm) that has been etched with 3,072 through-holes and polymer coated to make the external surface hydrophobic and the inside surface of each hole hydrophilic and PCR compatible.

Assays are preloaded onto the OpenArray® plate and dried at the time of manufacture and shipped to the customer along with a plate file that encodes the cycling protocol and assay information. Plates are then loaded with master mix and sample. The reaction chambers are arranged in 48 blocks of 64 through-holes, and the blocks or subarrays are spaced at a 4.5-mm pitch equal to that of the wells of a 384-well plate. This allows the sample mix to be transferred with a passive microfluidic device from a 384-well plate onto the OpenArray® plate. The grouping of reaction chambers into subarrays also reduces costs in master mix and sample usage since 56 targets can be addressed with the same volume that would be used for a single target in a 384-well plate.

The loaded OpenArray® plate is encased in a glass chamber containing a fluorescently inert, immiscible liquid that prevents evaporation during cycling and sealed with a UV-curable epoxy. The qPCR reactions are performed on the OpenArray® Real-Time PCR Instrument. The OpenArray® is loaded into the instrument and the plate file is loaded into the software, driving the thermal cycling of the block and mapping the assays to their respective through-holes.

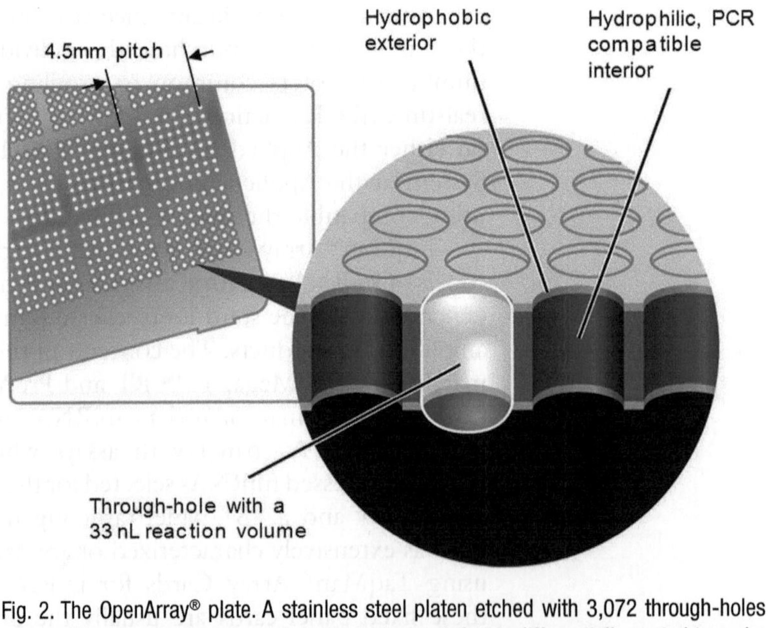

Fig. 2. The OpenArray® plate. A stainless steel platen etched with 3,072 through-holes arranged in 48 sets of 64 through-holes (*subarrays*) that is differentially coated to make the exterior hydrophobic and the interior hydrophilic and PCR compatible. The reaction volume is 33 nl.

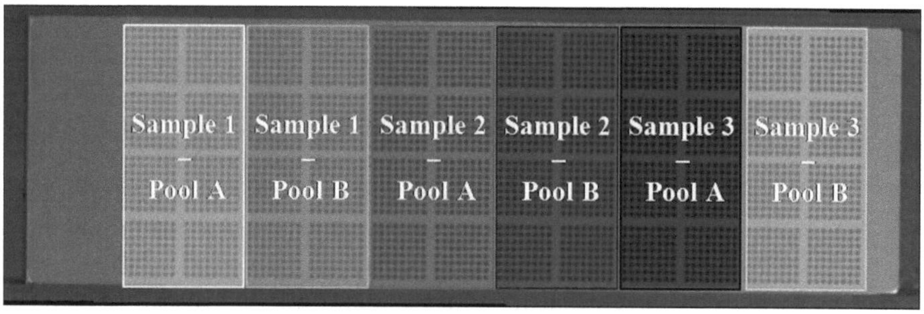

Fig. 3. Layout of Human miRNA Assays on the OpenArray® plate: Assays are arranged separating the assays represented by different pools in different subarrays. Assays for each pool represented in eight subarrays and each pool is represented three times in each OpenArray® plate.

The 48 × 64 format of the OpenArray® plate lends itself to flexibility in experimental design allowing a large number of assay–sample combinations to be analyzed from one sample and 3,072 assay targets to 144 samples and 16 assay targets. As such, the OpenArray® plate is either custom configured or supplied as a fixed content panel. The TaqMan® miRNA OpenArray® panel is the human complete set of assays represented in Megaplex™ pools A and B. A total of 754 human miRNA and four control assays are arranged on the OpenArray® plate such that three samples or replicates can be interrogated on a single plate (Fig. 3). The assays are arranged so that assays from different pools are located in different subarrays with one replicate of each control assay present in every subarray. The scientist prepares samples for pools A and B separately. The instrument can run up to three OpenArrays at one time and three runs can be completed in a single work day such that 27 samples/replicates can be screened for approximately 750 miRNAs in a day.

1.4. Applications of qPCR-Based MicroRNA Expression Profiling

Expression profiling of miRNAs using the stem-loop RT-PCR method can be used for numerous basic and applied applications. One important basic application for miRNA expression profiling is deciphering gene regulatory networks and identifying the key molecular players in cell fate decisions. Also, researchers are gaining a richer understanding of the complexities of disease by including miRNA expression profiling as an additional tool in the genomics arsenal. Finally, miRNA expression profiling is used to identify biomarkers that predict disease progression to classify cancer types and predict response to therapy.

Understanding the connections and complexity of gene regulatory networks has recently become possible by using "omics" methods. Expression profiling of miRNAs is one such method and provides an essential dimension to the understanding of gene regulation. Recently, our understanding of the gene regulatory networks involved in cell fate decisions has progressed significantly by using miRNA expression profiling (23–25). Cell differentiation is very tightly regulated and it is no surprise that the first miRNA described was found as a key regulator of cell lineage determination in *Caenorhabditis elegans* (26, 27). This work has been expanded to other model systems and miRNAs have been found to play an important role in embryonic stem cell differentiation, cell cycle regulation, neurogenesis, muscle differentiation, hematopoeisis, and bone formation (28–33).

Given that miRNAs are thought to regulate a majority of the human protein-coding genes, understanding their role in disease pathogenesis represents a very wide field of study. Specifically, miRNAs have been identified to play key roles in cancer, hematopoeisis, neurological disorders, cardiovascular disease, metabolic disorders, renal disease, and autoimmune disorders. It has been hypothesized that miRNAs can act as oncogenes when upregulated and as tumor suppressors when downregulated (34). The elucidation of the role of individual miRNAs in oncogenesis and metastasis is ongoing and will progress with high-throughput RT-qPCR miRNA expression profiling methods.

Identification of miRNA biomarkers that predict disease progression or response to therapy is another rapidly growing area of research. Recently, researchers have identified expression-based miRNA signatures that predict melanoma post-recurrence survival, recurrence of non-small-cell lung cancer and colon cancer, and patient survival in cervical cancer (35–38). These studies take advantage of sensitivity of miRNA expression profiling, and as work progresses we will see more studies involving the detection of circulating miRNAs in the blood of patients, taking advantage of the relative stability of miRNAs and the broad dynamic range of stem-loop RT-qPCR.

2. Materials

2.1. Total RNA Samples

1. FirstChoice® Human Brain Reference RNA.
2. qPCR reference total RNA.
3. Human liver total RNA.

2.2. Reagents for Megaplex™ RT and Preamplification

1. TaqMan® MicroRNA Reverse Transcription Kit.
2. Megaplex™ Primer Pools, Human Pools Set v3.0.
3. TaqMan® PreAmp Master Mix.
4. Nuclease-free water.
5. Applied Biosystems 7900 Real-Time PCR Systems.

2.3. TaqMan® MicroRNA Array Cards

1. Megaplex™ Primer Pools, Human Pools Set v3.0 (see Notes 1 and 2).
2. TaqMan® Array Human Micro RNA A+B Cards v3.0.
3. TaqMan® 2× Universal Master Mix II.
4. Nuclease-free water.
5. Applied Biosystems ViiA 7 Real-Time PCR Systems (see Note 3).
6. TaqMan® Array Upgrade Kit (see Note 4).
7. Centrifuge compatible with TaqMan® Array Cards (see Note 5).

2.4. TaqMan® MicroRNA OpenArray® Plates

1. TaqMan® OpenArray® Human miRNA Panel (Early Access) (see Note 6).
2. 2× TaqMan® OpenArray® Real-Time PCR Master Mix.
3. TaqMan® OpenArray® Real-Time PCR Accessories Kit.
4. OpenArray® AccuFill™ System Tips.
5. OpenArray® 384-well sample plates.
6. RNase-free water.
7. OpenArray® Real-Time PCR Platform (see Note 7).

3. Methods

The workflow for preparing samples is shown in Fig. 4. All reaction mixes should be prepared on ice.

3.1. Preparing the Reverse Transcription Reactions

1. For each sample, prepare two independent reverse transcription reactions – one each for pools A and B – in separate 1.5-ml tubes as indicated in Table 1. The example multiple reaction mix is sufficient to prepare ten reverse transcription reactions for either pool A or B, with approximately 12.5% additional volume for pipetting overage.
2. Mix the tubes by inversion six times and spin briefly to return all liquid to the bottom of the plate.
3. Mix 4.5 μl reaction mix and 3 μl total RNA (see Note 8).
4. Incubate on ice for 5 min.
5. Cycle 40× (16°C/2 min, 42°C/1 min, 50°C/1 s), 85°C/5 min, and 4°C hold.

3.2. Preparing the Preamplification Reactions

1. Prepare two preamplification reaction mixes, one each for pools A and B in separate 1.5-ml tubes as indicated in Table 2. The example multiple reaction mix is sufficient to prepare ten preamplification reactions for either pool A or B, with approximately 12.5% additional volume for pipetting overage.

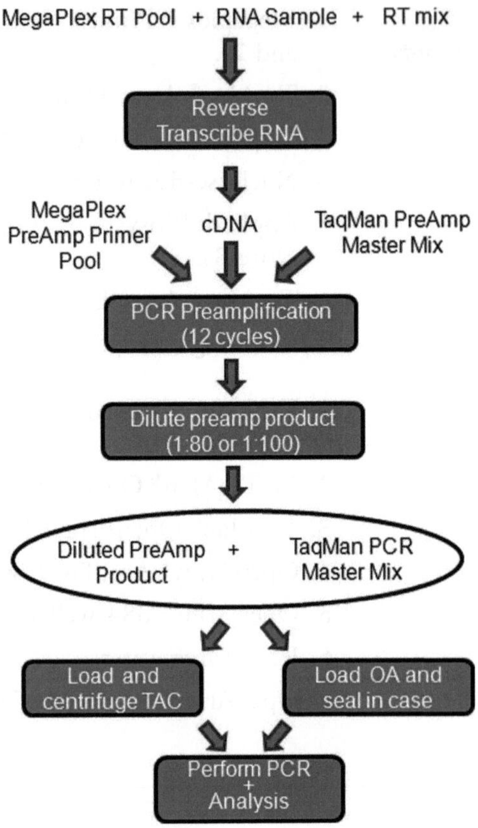

Fig. 4. Overview of the workflow for preparing and loading samples onto either the TaqMan® Array Card or the OpenArray® plate.

Table 1
Preparing reverse transcription reactions

Reagent	Stock	Final	Single reaction volume, μl	Multiple reaction volume, μl
10× RT primer mix, pool A or B	10×	1×	0.75	7.59
25 mM each dNTPs	25 mM	0.5 mM	0.15	1.52
MultiScribe reverse transcriptase	50 U/μl	10	1.50	15.19
10× RT buffer	10×	1×	0.75	7.59
25 mM $MgCl_2$	25 mM	3.0 mM	0.90	9.11
RNase inhibitor	20 U/μl	0.25 U/μl	0.09	0.95
Nuclease-free water			0.35	3.58
Total volume			7.50	45.54

Table 2
Preparing preamplification reactions

Reagent	Stock	Final	Single reaction volume, μl	Multiple reaction volume, μl
2× TaqMan® PreAmp Master Mix	2×	1×	12.5	126.56
10× PreAmp primer mix, pool A or B	10×	1×	2.5	25.31
Nuclease-free water			7.5	75.94
Total volume			25.0	227.81

PreAmp Plate

	1	2	3	4	5	6	7	8	9	10	11	12
A	sample 1 pool A	sample 1 pool B	sample 2 pool A	sample 2 pool B	sample 3 pool A	sample 3 pool B	sample 4 pool A	sample 4 pool B	sample 5 pool A	sample 5 pool B	sample 6 pool A	sample 6 pool B
B	sample 7 pool A	sample 7 pool B	sample 8 pool A	sample 8 pool B	sample 9 pool A	sample 9 pool B	sample 10 pool A	sample 10 pool B	sample 11 pool A	sample 11 pool B	sample 12 pool A	sample 12 pool B
C	sample 13 pool A	sample 13 pool B	sample 14 pool A	sample 14 pool B	sample 15 pool A	sample 15 pool B	sample 16 pool A	sample 16 pool B	sample 17 pool A	sample 17 pool B	sample 18 pool A	sample 18 pool B
D	sample 19 pool A	sample 19 pool B	sample 20 pool A	sample 20 pool B	sample 21 pool A	sample 21 pool B	sample 22 pool A	sample 22 pool B	sample 23 pool A	sample 23 pool B	sample 24 pool A	sample 24 pool B
E	sample 25 pool A	sample 25 pool B	sample 26 pool A	sample 26 pool B	sample 27 pool A	sample 27 pool B	sample 28 pool A	sample 28 pool B	sample 29 pool A	sample 29 pool B	sample 30 pool A	sample 30 pool B
F	sample 31 pool A	sample 31 pool B	sample 32 pool A	sample 32 pool B	sample 33 pool A	sample 33 pool B	sample 34 pool A	sample 34 pool B	sample 35 pool A	sample 35 pool B	sample 36 pool A	sample 36 pool B
G	sample 37 pool A	sample 37 pool B	sample 38 pool A	sample 38 pool B	sample 39 pool A	sample 39 pool B	sample 40 pool A	sample 40 pool B	sample 41 pool A	sample 41 pool B	sample 42 pool A	sample 42 pool B
H	sample 43 pool A	sample 43 pool B	sample 44 pool A	sample 44 pool B	sample 45 pool A	sample 45 pool B	sample 46 pool A	sample 46 pool B	sample 47 pool A	sample 47 pool B	sample 48 pool A	sample 48 pool B

Fig. 5. Suggested preamplification sample plate layout when working with a large set of samples.

2. Mix tubes by inversion six times.
3. Aliquot 22.5 μl pool A or B reaction mix to individual wells in a 96-well plate, as shown in the sample plate layout in Fig. 5.
4. Add 2.5 μl of the respective RT reaction to the preamplification reaction (i.e., pool A RT reaction with pool A preamplification mix).
5. Seal the plate following the addition of sample to each reaction mix.
6. Mix by inversion six times and spin briefly to return all liquid to the bottom of the plate.
7. Incubate on ice for 5 min.
8. Cycle 1× (95°C/10 min, 55°C/2 min, 72°C/2 min).
9. Cycle 12× (95°C/15 s, 60°C/4 min).
10. Cycle 1× 99.9°C/10 min.
11. Complete cycling with a 4°C hold step.
12. Postamplification sample dilution: Replicating the layout as shown in Fig. 5 in Subheading 3.2, step 3, prepare the following dilutions.

- For the OpenArray® plate, dilute 5 μl preamplified sample into 95 μl 0.1× TE, pH 8.0.
- For the TaqMan® array card, dilute 25 μl preamplified sample in 75 μl 0.1× TE, pH 8.0.

3.3. Preparing the OpenArray® Real-Time qPCR Reactions

1. Prepare the reaction master mix in a 1.5-ml tube by mixing 405 μl OpenArray® Real-Time PCR Master Mix and 202.5 μl of nuclease-free water. With approximately 12.5% additional volume for pipetting overage, this is sufficient to load nine samples in singlet or three samples in triplicate (see Note 6).
2. Mix tubes by inversion six times.
3. For each sample/pool combination, pipette 34.5 μl reaction mix into a well of a 96-well plate, as shown in the example plate layout in Fig. 6.
4. Add 11.5 μl diluted preamplified pool A product to one well containing 34.5 μl master mix and add 11.5 μl diluted preamplified pool B product to the adjacent well containing 34.5 μl PCR master mix.
5. Repeat for all experimental samples (according to Fig. 6).
6. Seal the plate.
7. Mix by vortexing and spin briefly to return all liquid to the bottom of the plate.
8. For each sample/pool combination, transfer 5 μl of reaction mixture into eight wells in an OpenArray® 384-well sample plate as shown in Fig. 7 such that the sample originating in well A1 (sample 1 Pool A) in the 96-well plate in Fig. 6 has been transferred into wells A1, A2, B1, B2, C1, C2, D1, and D2 of the 384-well OpenArray® sample plate.
9. Repeat for all remaining wells in the 96-well preamplified sample plate.
10. Cover the plate with a foil seal and seal tightly.

qPCR Mother Build Plate	1	2	3	4	5	6	7	8	9	10	11	12
A	sample 1 pool A	sample 1 pool B	sample 2 pool A	sample 2 pool B	sample 3 pool A	sample 3 pool B	sample 4 pool A	sample 4 pool B	sample 5 pool A	sample 5 pool B	sample 6 pool A	sample 6 pool B
B	sample 7 pool A	sample 7 pool B	sample 8 pool A	sample 8 pool B	sample 9 pool A	sample 9 pool B	sample 10 pool A	sample 10 pool B	sample 11 pool A	sample 11 pool B	sample 12 pool A	sample 12 pool B
C	sample 13 pool A	sample 13 pool B	sample 14 pool A	sample 14 pool B	sample 15 pool A	sample 15 pool B	sample 16 pool A	sample 16 pool B	sample 17 pool A	sample 17 pool B	sample 18 pool A	sample 18 pool B
D	sample 19 pool A	sample 19 pool B	sample 20 pool A	sample 20 pool B	sample 21 pool A	sample 21 pool B	sample 22 pool A	sample 22 pool B	sample 23 pool A	sample 23 pool B	sample 24 pool A	sample 24 pool B
E	sample 25 pool A	sample 25 pool B	sample 26 pool A	sample 26 pool B	sample 27 pool A	sample 27 pool B	sample 28 pool A	sample 28 pool B	sample 29 pool A	sample 29 pool B	sample 30 pool A	sample 30 pool B
F	sample 31 pool A	sample 31 pool B	sample 32 pool A	sample 32 pool B	sample 33 pool A	sample 33 pool B	sample 34 pool A	sample 34 pool B	sample 35 pool A	sample 35 pool B	sample 36 pool A	sample 36 pool B
G	sample 37 pool A	sample 37 pool B	sample 38 pool A	sample 38 pool B	sample 39 pool A	sample 39 pool B	sample 40 pool A	sample 40 pool B	sample 41 pool A	sample 41 pool B	sample 42 pool A	sample 42 pool B
H	sample 43 pool A	sample 43 pool B	sample 44 pool A	sample 44 pool B	sample 45 pool A	sample 45 pool B	sample 46 pool A	sample 46 pool B	sample 47 pool A	sample 47 pool B	sample 48 pool A	sample 48 pool B

Fig. 6. Suggested qPCR sample plate layout when preparing a large set of samples.

OpenArray Load Plate

	1 2	3 4	5 6	7 8	9 10	11 12	13 14	15 16	17 18	19 20	21 22	23 24
A B C D	sample 1 pool A	sample 1 pool B	sample 2 pool A	sample 2 pool B	sample 3 pool A	sample 3 pool B	sample 4 pool A	sample 4 pool B	sample 5 pool A	sample 5 pool B	sample 6 pool A	sample 6 pool B
E F G H	sample 7 pool A	sample 7 pool B	sample 8 pool A	sample 8 pool B	sample 9 pool A	sample 9 pool B	sample 10 pool A	sample 10 pool B	sample 11 pool A	sample 11 pool B	sample 12 pool A	sample 12 pool B
I J K L	sample 13 pool A	sample 13 pool B	sample 14 pool A	sample 14 pool B	sample 15 pool A	sample 15 pool B	sample 16 pool A	sample 16 pool B	sample 17 pool A	sample 17 pool B	sample 18 pool A	sample 18 pool B
M N O P	sample 19 pool A	sample 19 pool B	sample 20 pool A	sample 20 pool B	sample 21 pool A	sample 21 pool B	sample 22 pool A	sample 22 pool B	sample 23 pool A	sample 23 pool B	sample 24 pool A	sample 24 pool B

Fig. 7. Layout of samples required when loading qPCR samples onto an OpenArray® plate.

11. To prepare the sample plate for loading into the OpenArray®, use a razor blade to cut the foil seal into eight equal sections of 48 wells each (e.g., wells A1-D12, wells A13-D24, wells E1-H12, etc.).

3.4. Loading the Real-Time qPCR Reaction Mix onto the OpenArray® Plate

1. OpenArrays are loaded using the AccuFill™ System.
2. Initialize the AccuFill™ System by ensuring that the door is closed.
3. Double click the OpenArray® AccuFill™ icon to launch the software.
4. Open the instrument; ensure that the waste bin is empty and that there are two full tip boxes in the side-by-side recessed rectangular platforms toward the back of the instrument. Close the instrument door.
5. Begin the instrument self-test by clicking *Proceed*. The instrument conducts a series of self-tests and is then ready for loading.
6. Prepare a case for loading an OpenArray® (one OpenArray® per case) by using scissors to open a container of immersion fluid (see Notes 9 and 10).
7. Place the case in a rack and fill approximately ¾ full with immersion fluid.
8. Prepare the AccuFill™ software for loading by clicking *Setup and Load*. The Setup/Load Information window opens.
9. Type the name of the sample plate in the Sample Plate field.
10. Open the door and position the 384-well sample plate on the deck with the foil cover still in place. Press on the plate until you hear it snap into place.
11. Select 1 from the samples per subarray drop-down list (see Note 11).

12. In the Plate Holder Position 1 text field, enter the unique serial number of the OpenArray® plate that is loaded in the first position of the Plate Holder.
13. Place the OpenArray® plate into the Plate Holder with the bar code up and facing to the left.
14. Place up to three additional OpenArray® plates in the other open positions of the Plate Holder.
15. In the AccuFill™ software, select the appropriate section of the sample plate to be loaded on the OpenArray® selected in Subheading 3.4, step 12. Click *Next.*
16. To load the OpenArray® plate, verify that the configuration of the deck in the tip status window of the AccuFill™ software matches the configuration of the tips on the AccuFill™ deck. Manually reconfigure any sections not matching by clicking on the appropriate section of the tip box in the software to toggle between tips present/no tips present.
17. In the AccuFill™ software window, click the box to verify that the tips are configured as pictured.
18. Additionally verify that the waste bin is empty and that the OpenArray® plate chosen to be loaded in Subheading 3.4, step 12, is in the Plate Holder. Click both boxes to confirm.
19. Use forceps to peel off the foil covering from the section of the sample plate that is to be loaded onto the selected OpenArray® plate.
20. In the AccuFill™ software window, click the check box *Remove foil from the highlighted section of the Sample Plate.*
21. Close the instrument door. Click *Load.*
22. When the loading is complete, the software displays a window: Remove OpenArray® plate. Click *OK.*
23. Open the instrument and carefully remove the OpenArray® plate from the plate holder.
24. Insert the OpenArray® into a case previously prefilled with immersion fluid (see Note 12).
25. Slide the OpenArray® completely into the case, ensuring that the bottom of the OpenArray® plate is touching the bottom of the case. If necessary, remove excess immersion fluid until the level of immersion fluid is level with the top of the OpenArray plate.
26. Repeat Subheading 3.4, steps 6 through 25, to fill additional OpenArray® plates as needed.
27. Seal the OpenArray® plate in the case (see Note 13).
28. Apply the frame to the exterior of the case (see Note 14).

3.5. Cycling the Real-Time qPCR Reactions in the OpenArray®

1. Open the cycler door and raise the lid.
2. Place the OpenArray® plates onto the cycler block (see Note 15).
3. Open the OpenArray® Real-Time qPCR Analysis software on the computer attached to the NT cycler. Click the cycle button in the upper right-hand corner of the OpenArray® software window. The "Input Plate Serial Numbers" Window should appear.
4. Click the Locate File button to the right of the position 1 OpenArray®. Locate the plate file for the first OpenArray® on the CD which accompanied the plates (see Note 16).
5. Repeat this process for the second and third OpenArray® plates.
6. Close the cycler lid and door.
7. Click the cycle button in the Input Plate Serial Numbers Window (see Note 17).
8. At the end of the run, save run data as *ncx* file. An example of PCR amplification curve is shown in Fig. 8.

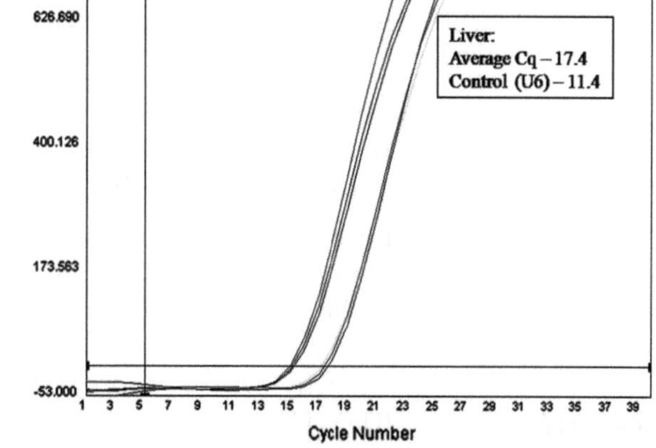

Fig. 8. Tissue-specific amplification has miR-484 in human brain and liver in the OpenArray® plate. The *inlet* shows the average Cq for the target and a respective control target.

3.6. Loading the Preamplified Real-time PCR Reactions into the TaqMan® Low-Density Array

1. Prepare the reaction mix for one TaqMan® Low-Density Array in a 1.5-ml tube by mixing 450 μl 2× TaqMan® Universal Master Mix, 441 μl nuclease-free water, and 9 μl diluted Preamp Pool A sample.
2. Dispense 100 μl qPCR reaction mixture into each TaqMan® Array Card loading port using a hand pipette.
3. Once all eight loading ports have been filled, place the TaqMan® Array Card into the swing bucket centrifuge racks (see Note 5).
4. Repeat for pool B and all other samples to be run.
5. Centrifuge at 331 ×*g* for 1 min. Repeat the centrifugation for a total of two consecutive centrifugations.
6. Prepare the TaqMan® Array Card Sealer by placing it on a firm level surface with the start position close to the user and the carriage in the start position.
7. Place the TaqMan® Array Card into the sealer with the reaction chambers facing down and the loading ports positioned to the side furthest from the carriage start position.
8. Push the sealer carriage across the TaqMan® Array Card using a slow steady motion.
9. Using scissors, trim the loading ports off the TaqMan® Array Card.

3.7. Cycling in the TaqMan® Low-Density Array

1. Open the SDS software on the computer attached to the 7900HT instrument. Open a new file and use the pull-down menus in the New Document Dialog Box to set the plate format to "384 Well TaqMan® Low Density Array" and the analysis method to "ddCq (RQ)." Leave the template field as "Blank template." Also enter the bar code number located on the side of the TaqMan® Array Card.
2. Optional: Using the Import Dialog Box, import the TaqMan® Array setup file from the CD supplied with the TaqMan® Array Cards.
3. Save the file (file type *.sds).
4. Select the (Real-Time) tab and press the (Open/Close) button to rotate the plate tray out of the instrument.
5. Load the TaqMan® Array Card onto the plate tray with reaction chamber A1 on the top left and the bar code toward the front of the instrument. Press "Start Run."

3.8. Data Normalization and Clustering Analysis

3.8.1. Determination of Quantification Cycle

(a) Launch SDS software v2.2 or later version and open SDS RUN files.

(b) Analysis settings: Automatic baseline and manual quantification cycle (Cq) threshold to 0.2.

(c) Analyze and review the amplification plots and any flags.

(d) Export data as tab-delimited text files.

3.8.2. Applying Quantification Cycle Cutoff

(a) It is important to apply right Cq cutoff to data. It can be estimated based on standard deviation of Cq or recommended values in a protocol.

(b) Different preamplification protocol and qPCR platforms (TaqMan® Array Cards vs. OpenArray® plates) may require different Cq cutoffs.

3.8.3. Synthetic Spiking miRNA Controls

(a) Synthetic spiking controls can be used for positive assay controls, RT and plate-to-plate normalization, and standard curve.

(b) Plant miRNA miR159a is highly recommended as a spiking control in RT reactions at 100 pM (1–1,000) for human or rodent samples.

(c) For serum or plasmid samples, the synthetic miRNA must be added *after* mixing your samples with 2× denaturing solution to prevent RNA degradation.

3.8.4. Selection of Endogenous Controls

(a) Test a set of at least three endogenous control genes and select the best candidates using geNorm (http://medgen.ugent.be/~jvdesomp/genorm/) or other similar software.

(b) For genome-wide expression profiling, global normalization is also recommended (39).

3.8.5. Cq Normalization Using a Set of Endogenous Controls

(a) ΔCq (miRNA Cq − averaged endogenous control Cq) (Fig. 9).

(b) ΔΔCq ($\Delta Cq^{(\text{sample1})} - \Delta Cq^{(\text{sample 2})}$) relative to a reference sample.

(c) Fold change ($2^{\Delta\Delta Cq}$).

3.8.6. Clustering Analysis Using Normalized ΔΔCq

(a) Commonly used method – agglomerative hierarchical clustering.

(b) Tools include Dr. Eisen's "Cluster" (40) and Real-Time StatMiner™ from Integromics Inc. according to the manual.

3.8.7. OpenArray® Data Analysis and Export: Determination of Quantification Cycle

(a) Launch OpenArray® Real-Time qPCR Analysis Software and open *ncx* run file.

(b) To perform most routine analyses, leave the settings at the default values: auto thresholding, default algorithm, divide by baseline, cycle 5 baseline start, and a minimum signal of 300.

(c) Analyze and review the amplification plots and quantification cycle values obtained. If necessary, adjustments to the baselining parameters may be made to individual assays.

(d) Export data as comma-delimited text files by Select *File > Export Cycling Data*. Select the *Summary Results* checkbox in the dialog box.

(e) Downstream analysis of OpenArray quantification data can be undertaken as in Subheading 3.6, steps 4–6.

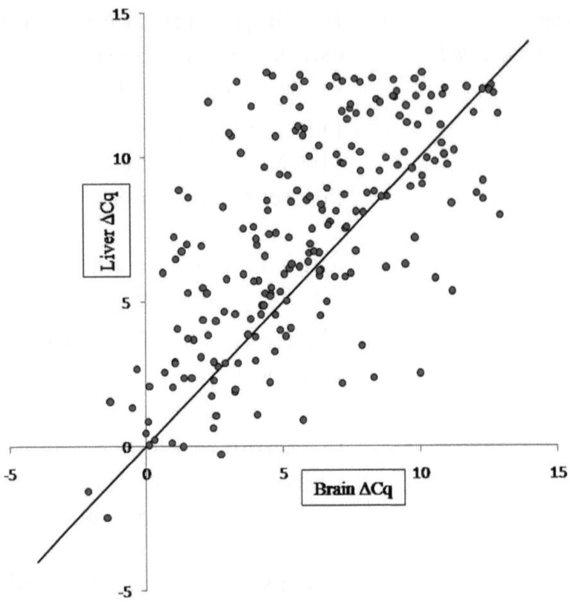

Fig. 9. ΔCq scatter plot for human brain and liver samples in the OpenArray® plate. ΔCq was calculated subtracting the average Cq for each target from the geometric mean of the three control assays: RNU44, RNU48, and U6. Data for Cq greater than 26 were filtered out.

4. Notes

1. Megaplex™ Primer Pool Sets contain prepooled oligos for reverse transcription and preamplification of miRNA targets. Different sets are available for human and rodent. The sets contain oligo pools for both reverse transcription and preamplification. They include separate tubes materials for pools A and B panels. Each of the pools is also available separately. Samples being prepared for the OpenArray® platform utilize all of the components in the set. However, samples being prepared for the TaqMan® Array Card platform may not require preamplification or both panels A and B, depending on experimental design.
2. Both TaqMan® Array Cards and Megaplex™ Primer Pools have been continuously updated as new miRNAs have been discovered, and as miRNA sequence information has been updated in publically available miRNA databases. The version 3 of the Megaplex™ primer pool sets has been updated to capture the content in Sanger miRBase v14 (41). The canonical sequences of individual miRNAs are occasionally changed in the Sanger database as characterization of the miRNAs improves. For instance, the length of some miRNAs has been changed to reflect discoveries about the most commonly expressed variants (42).

It is important that samples are processed using matching versions of the Oligo Pools, Cards, and OpenArray® plates.

3. Due to their unique shape, TaqMan® Array Cards are not compatible with standard 384-well thermocycling blocks. At this time, there are three instruments capable of running these cards: the ABI PRISM® 7900HT Sequence Detection System, its successor the Applied Biosystems 7900HT Fast Real-Time PCR System, and the Applied Biosystems ViiA 7 Real-Time PCR System.
4. The AB7900 and ViiA 7 instruments are compatible with multiple plate and card formats. The TaqMan® Array Upgrade Kits provide the thermal blocks and heated covers which match the shape of the cards. These kits also include the required centrifuge swing buckets and rotor adapters, as well as the TaqMan® Array Card Sealer.
5. In the TaqMan® Array Cards, reaction mix is distributed from loading ports to individual wells via centrifugation. Unlike 384-well plates, which are commonly spun flat, TaqMan® Array Cards are spun standing on end. Swing buckets and rotor adapters for the cards are supplied as part of the 7900HT TaqMan® Array Upgrade Kit. They are compatible with the following families of centrifuge models: Sorvall Legend T+, Legend XT, and ST40; Fisher accuSpin Model 3; Thermo SL40; Heraeus Multifuge 3S+, Multifuge X3, and Megafuge 40.
6. Pools A and B are represented in separate subarrays, so samples for pools A and B should be prepared separately. There are a total of eight subarrays for each pool; therefore, enough sample volume for eight subarrays should be prepared allowing for pipetting overage.
7. OpenArray® plates can only be cycled and imaged in the OpenArray® Real-Time PCR Instrument.
8. The recommended starting material for the TaqMan® Array Card is 30 ng/reaction while the 100 ng/reaction is recommended for samples to be run in the OpenArray® system.
9. Immersion fluid must not be exposed to air for more than 60 min since the fluid shipped under vacuum to prevent gas bubbles from forming inside the case. Immersion fluid exposed to air for more than 60 min should be discarded.
10. All accessories required to encase, seal, and run the OpenArray® plate are included with the OpenArray® accessory kit.
11. To perform AccuFill™ loading using more advanced features, such as sample tracking integration, refer to OpenArray® AccuFill™ User's Guide.
12. To handle the OpenArray® and facilitate its insertion into the case, hold the OpenArray® with the thumb and index finger of your left hand, placing your fingers on each side near the end

with the bar code. Keep the side with the bar code facing you. Switch hands and now hold the OpenArray® plate by the sides using your index and thumb of your right hand. Hold the case at the top with the thumb and index finger of your left hand so that the black-framed side is facing you. Slide the OpenArray® plate into the case so that it sits in the rails along both inside edges of the case.

13. To seal the OpenArray® in a case, place a drop of UV glue on one edge of the case opening; fill until it reaches the top of the OpenArray® plate. Place a similar amount of glue on the other side of the case opening and add glue on each side of the case until the glue runs together in the middle. Fill the case to the top and make sure that both the left and right sides of the case are completely covered with glue. With a laboratory wipe, wipe any excess glue from the surface of the glass. To cure the glue, place the OpenArray® case into the Case Sealing Station. Place up to two cases into the device and close the door. Turn the switch to *ON*. Allow the glue to cure for 3 min and then turn the switch to *OFF*. Open the door and turn the cases over. Cure the glue for an additional 3 min. Remove the OpenArray® cases. If there is glue on the side of the case, carefully remove it with a razor blade. Clean the case with a laboratory wipe that has been thoroughly sprayed with ethanol.

14. To apply the frame, position the case on a laboratory wipe so that the black-framed side faces up. Use scissors to cut two 3-mm long rectangles from the frame adhesive (see Note 10). With forceps, remove the backing from one side of the tape of one rectangle, exposing a sticky side. Stick the tape in the middle of a long side of the frame. Remove the backing from the other rectangle of adhesive and place on the other long side of the case. Using forceps, peel the backing off both pieces of adhesive. Align the frame with two corners of the case and bring the other corners of the frame down to gently touch the case surface. Flatten the frame against the case surface for 10 s.

15. Position 1 is to the back of the thermal block and position 3 to the front. OpenArray® plates should be positioned with the bar code facing up and to the right with the case pushed against the right edge of the block. OpenArray® 1 should be pushed all the way to the back and 3 all the way to front.

16. Each OpenArray® plate is accompanied by a plate file that contains all the cycling and assay information required to run the plate in the NT Cycler system. Plate files can be read directly from the CD or transferred to a convenient location of the control computer.

17. If the camera has not cooled down below 1°C, then the cycling will not begin immediately. A warning appears. Press OK and the run starts as soon as the camera has cooled to 1°C.

Acknowledgments

The authors would like to thank members of TaqMan MicroRNA Assays R&D and OpenArray R&D teams at Life Technologies for their support.

References

1. Bartel, D. P. (2004) MicroRNAs: Genomics, biogenesis, mechanism, and function. *Cell* **116**, 281–97.
2. Lee, R. C., Feinbaum, R. L., and Ambros, V. (1993) The C. elegans heterochronic gene lin-4 encodes small RNAs with antisense complementarity to lin-14. *Cell* **75**, 843–54.
3. Lu, J., Getz, G., Miska, E. A., Alvarez-Saavedra, E., Lamb, J., Peck, D., et al. (2005) MicroRNA expression profiles classify human cancers. *Nature* **435**, 834–8.
4. Mitchell, P. S., Parkin, R. K., Kroh, E. M., Fritz, B. R., Wyman, S. K., Pogosova-Agadjanyan, E. L., et al. (2008) Circulating microRNAs as stable blood-based markers for cancer detection. *Proc Natl Acad Sci USA* **105**, 10513–8.
5. Wang, K., Zhang, S., Marzolf, B., Troisch, P., Brightman, A., Hu, Z., et al. (2009) Circulating microRNAs, potential biomarkers for drug-induced liver injury. *Proc Natl Acad Sci USA* **106**, 4402–7.
6. Heidbrink, C., Hausler, S. F., Buttmann, M., Ossadnik, M., Strik, H. M., Keller, A., et al. (2010) Reduced cortisol levels in cerebrospinal fluid and differential distribution of 11beta-hydroxysteroid dehydrogenases in multiple sclerosis: implications for lesion pathogenesis. *Brain Behav Immun* **24**, 975–84.
7. Chen, C., Ridzon, D. A., Broomer, A. J., Zhou, Z., Lee, D. H., Nguyen, J. T., et al. (2005) Real-time quantification of microRNAs by stem-loop RT-PCR. *Nucleic Acids Res* **33**, e179.
8. Mestdagh, P., Feys, T., Bernard, N., Guenther, S., Chen, C., Speleman, F., et al. (2008) High-throughput stem-loop RT-qPCR miRNA expression profiling using minute amounts of input RNA. *Nucleic Acids Res* **36**, e143.
9. Schmittgen, T. D., Lee, E. J., Jiang, J., Sarkar, A., Yang, L., Elton, T. S., et al. (2008) Real-time PCR quantification of precursor and mature microRNA. *Methods* **44**, 31–8.
10. Liu, D. Z., Tian, Y., Ander, B. P., Xu, H., Stamova, B. S., Zhan, X., et al. (2010) Brain and blood microRNA expression profiling of ischemic stroke, intracerebral hemorrhage, and kainate seizures. *J Cereb Blood Flow Metab* **30**, 92–101.
11. Calin, G. A., Liu, C. G., Sevignani, C., Ferracin, M., Felli, N., Dumitru, C. D., et al. (2004) MicroRNA profiling reveals distinct signatures in B cell chronic lymphocytic leukemias. *Proc Natl Acad Sci USA* **101**, 11755–60.
12. Krichevsky, A. M., King, K. S., Donahue, C. P., Khrapko, K., and Kosik, K. S. (2003) A microRNA array reveals extensive regulation of microRNAs during brain development. *RNA* **9**, 1274–81.
13. Nelson, P. T., Baldwin, D. A., Scearce, L. M., Oberholtzer, J. C., Tobias, J. W., and Mourelatos, Z. (2004) Microarray-based, high-throughput gene expression profiling of microRNAs. *Nat Methods* **1**, 155–61.
14. Sarkar, D., Parkin, R., Wyman, S., Bendoraite, A., Sather, C., Delrow, J., et al. (2008) Quality Assessment and Data Analysis for microRNA Expression Arrays. *Nucleic Acids Res* **37**, e17.
15. Raymond, C. K., Roberts, B. S., Garrett-Engele, P., Lim, L. P., and Johnson, J. M. (2005) Simple, quantitative primer-extension PCR assay for direct monitoring of microRNAs and short-interfering RNAs. *RNA* **11**, 1737–44.
16. Sharbati-Tehrani, S., Kutz-Lohroff, B., Bergbauer, R., Scholven, J., and Einspanier, R. (2008) miR-Q: a novel quantitative RT-PCR approach for the expression profiling of small RNA molecules such as miRNAs in a complex sample. *BMC Mol Biol* **9**, 34.
17. Shi, R., and Chiang, V. L. (2005) Facile means for quantifying microRNA expression by real-time. PCR *Biotechniques* **39**, 519–25.
18. Chen, J., Lozach, J., Garcia, E. W., Barnes, B., Luo, S., Mikoulitch, I., et al. (2008) Highly sensitive and specific microRNA expression profiling using BeadArray technology. *Nucleic Acids Res* **36**, e87.
19. Allawi, H. T., Dahlberg, J. E., Olson, S., Lund, E., Olson, M., Ma, W.-P., et al. (2004) Quantitation of microRNAs using a modified invader assay. *RNA* **10**, 1153–61.

20. Jonstrup, S. P., Koch, J., and Kjems, J. (2006) A microRNA detection system based on padlock probes and rolling circle amplification. *RNA* **12**, 1747–52.
21. Goulter, A. B., Harmer, D. W., and Clark, K. L. (2006) Evaluation of low density array technology for quantitative parallel measurement of multiple genes in human tissue. *BMC Genomics* 7, 34.
22. Cervigne, N. K., Reis, P. P., Machado, J., Sadikovic, B., Bradley, G., Galloni, N. N., et al. (2009) Identification of a microRNA signature associated with progression of leukoplakia to oral carcinoma. *Hum Mol Genet* **18**, 4818–29.
23. Herranz, H., and Cohen, S. M. (2010) MicroRNAs and gene regulatory networks: managing the impact of noise in biological systems. *Genes Dev* **24**, 1339–44.
24. Ivey, K. N., and Srivastava, D. (2010) MicroRNAs as regulators of differentiation and cell fate decisions. *Cell Stem Cell* 7, 36–41.
25. Martinez, N. J., and Gregory, R. I. (2010) MicroRNA gene regulatory pathways in the establishment and maintenance of ESC identity. *Cell Stem Cell* 7, 31–5.
26. Lee, R. C., Feinbaum, R. L., and Ambros, V. (1993) The C. elegans heterochronic gene lin-4 encodes small RNAs with antisense complementarity to lin-14. *Cell* **75**, 843–54.
27. Wightman, B., Ha, I., and Ruvkun, G. (1993) Posttranscriptional regulation of the heterochronic gene lin-14 by lin-4 mediates temporal pattern formation in C. elegans. *Cell* **75**, 855–62.
28. Chivukula, R. R., and Mendell, J. T. (2008) Circular reasoning: microRNAs and cell-cycle control. *Trends Biochem Sci* **33**, 474–81.
29. Fineberg, S. K., Kosik, K. S., and Davidson, B. L. (2009) MicroRNAs potentiate neural development. *Neuron* **64**, 303–9.
30. Gangaraju, V. K., and Lin, H. (2009) MicroRNAs: key regulators of stem cells. *Nat Rev Mol Cell Biol* **10**, 116–25.
31. Garzon, R., and Croce, C. M. (2008) MicroRNAs in normal and malignant hematopoiesis. *Curr Opin Hematol* **15**, 352–8.
32. Hassan, M. Q., Gordon, J. A., Beloti, M. M., Croce, C. M., Wijnen, A. J., Stein, J. L., et al. (2010) A network connecting Runx2, SATB2, and the miR-23a~27a~24-2 cluster regulates the osteoblast differentiation program. *Proc Natl Acad Sci USA* **107**, 19879–84.
33. Williams, A. H., Liu, N., van Rooij, E., and Olson, E. N. (2009) MicroRNA control of muscle development and disease. *Curr Opin Cell Biol* **21**, 461–9.
34. Garzon, R., Marcucci, G., and Croce, C. M. (2010) Targeting microRNAs in cancer: rationale, strategies and challenges. *Nat Rev Drug Discov* **9**, 775–89.
35. Gallardo, E., Navarro, A., Vinolas, N., Marrades, R. M., Diaz, T., Gel, B., et al. (2009) miR-34a as a prognostic marker of relapse in surgically resected non-small-cell lung cancer. *Carcinogenesis* **30**, 1903–9.
36. Hu, X., Schwarz, J. K., Lewis, J. S., Jr., Huettner, P. C., Rader, J. S., Deasy, J. O., et al. (2010) A microRNA expression signature for cervical cancer prognosis. *Cancer Res* **70**, 1441–8.
37. Schepeler, T., Reinert, J. T., Ostenfeld, M. S., Christensen, L. L., Silahtaroglu, A. N., Dyrskjot, L., et al. (2008) Diagnostic and prognostic microRNAs in stage II colon cancer. *Cancer Res* **68**, 6416–24.
38. Segura, M. F., Belitskaya-Levy, I., Rose, A. E., Zakrzewski, J., Gaziel, A., Hanniford, D., et al. (2010) Melanoma MicroRNA signature predicts post-recurrence survival. *Clin Cancer Res* **16**, 1577–86.
39. Mestdagh, P., Van Vlierberghe, P., De Weer, A., Muth, D., Westermann, F., Speleman, F., et al. (2009) A novel and universal method for microRNA RT-qPCR data normalization. *Genome Biol* **10**, R64.
40. Eisen, M. B., Spellman, P. T., Brown, P. O., and Botstein, D. (1998) Cluster analysis and display of genome-wide expression patterns. *Proc Natl Acad Sci USA* **95**, 14863–8.
41. Griffiths-Jones, S., Saini, H. K., van Dongen, S., and Enright, A. J. (2008) miRBase: tools for microRNA genomics. *Nucleic Acids Res* **36**, D154–8.
42. Wu, H., Neilson, J. R., Kumar, P., Manocha, M., Shankar, P., Sharp, P. A., et al. (2007) miRNA profiling of naive, effector and memory CD8 T cells. *PLoS One* **2**, e1020.

Chapter 4

Poly(T) Adaptor RT-PCR

Rui Shi, Ying-Husan Sun, Xing-Hai Zhang, and Vincent L. Chiang

Abstract

Reverse transcription PCR (RT-PCR) is one of the most important techniques for analyzing RNA abundance. MicroRNAs (miRNAs) are a group of 20- to 24-nucleotide regulatory small RNAs which play an important role in plants and animals. However, the small size of miRNAs makes them difficult to be detected and quantified by conventional RT-PCR techniques. Here, we describe a poly(T) adaptor RT-PCR method specifically designed for quantifying miRNAs. In this method, total RNAs, including miRNAs, are extended by a poly(A) tailing reaction using poly(A) polymerase and ATP. The miRNA with a poly(A) tail is converted into cDNA through reverse transcription primed by a poly(T) adaptor, and then PCR-amplified using a miRNA-specific forward primer and a universal poly(T) adaptor reverse primer. The RT-PCR amplification can be monitored by real-time detection or by end-point detection for quantifying the miRNA transcript level. The PCR amplicons can be sequenced for validating the expression of the specific miRNA gene.

Key words: MicroRNA, Poly(A) tailing, Poly(A) polymerase, Poly(T) adaptor, Reverse transcription, RT-PCR, Real time, End point, Detection, Quantitation

1. Introduction

MicroRNAs (miRNAs) are single-stranded small RNAs of 20–24 nucleotides (nt) long. They are processed from their hairpin-structured precursors encoded by endogenous genes. MiRNAs are believed to be incorporated into the RNA-induced silencing complex (RISC) and guide the RISC to identify and bind to messenger RNAs (mRNAs) with sequences complementary to the miRNA. The RISC then proceeds to cleave the mRNA or block the translation process (1, 2). MiRNA-mediated gene regulation has been extensively studied for various developmental, metabolic,

Jian-Bing Fan (ed.), *Next-Generation MicroRNA Expression Profiling Technology: Methods and Protocols*, Methods in Molecular Biology, vol. 822, DOI 10.1007/978-1-61779-427-8_4, © Springer Science+Business Media, LLC 2012

and cellular processes in plants and animals (1, 2). Information of spatio-temporal expression patterns of miRNAs is important for understanding their functions. However, the small size of miRNAs presents a great challenge for their detection and quantitation by conventional RNA techniques. For example, Northern blot analysis is often insensitive and lacks specificity in analyzing miRNA transcripts (3). Although Reverse transcription PCR (RT-PCR) is a method of choice for RNA study (4), it cannot be readily applied because miRNAs are too short to accommodate both forward and reverse primers without the two primers overlapping. To overcome this problem, we developed a poly(T) adaptor RT-PCR strategy (5). In the method, miRNA sequences within total RNAs are extended by adding a poly(A) tail to their 3′ terminus. Subsequently, the poly(A) tailed miRNAs are converted to cDNAs in reverse transcription reaction primed by a poly(T) adaptor. Amplification of miRNA-derived cDNA by PCR can be then accomplished using a miRNA-specific forward primer along with a universal reverse primer complementary to the poly(T) adaptor (Fig. 1).

This poly(T) adaptor RT-PCR is more sensitive and specific, and less laborious than Northern blotting analysis. In this method, cDNA can be synthesized from as little as 100 ng of total RNA, and cDNA from only 100 pg total RNA is sufficient for one PCR for a specific miRNA. This method can also discriminate between miRNAs with only 1 nt difference in sequence, allowing the quantification of individual miRNA family members sharing high sequence homology (5). Northern hybridization, on the other hand, needs at least 5 μg of total RNA per gel lane for an abundant miRNA to be detected. Furthermore, to increase sensitivity and specificity, Northern hybridization also requires the use of hazardous or expensive chemicals, such as radioisotope (P^{32}) (3) or Lock Nucleic Acid (6), for probe preparation.

Other RT-PCR methods such as stem-loop RT-PCR (7) or primer extension RT-PCR (8) can also be used for miRNA analysis. Compared to these methods, poly(T) adaptor RT-PCR is less expensive because the reagents required are readily available, and the primers needed are based on regular nucleotides instead of the expensive modified nucleotides. In addition, poly(T) adaptor RT-PCR can also work as 3′ rapid amplification of cDNA ends (3′RACE) to obtain miRNA 3′ partial sequence. Thus, this method is particularly useful for validating the expression of specific miRNAs (9) and other small RNAs such as artificial miRNAs (Figs. 1 and 2) (10).

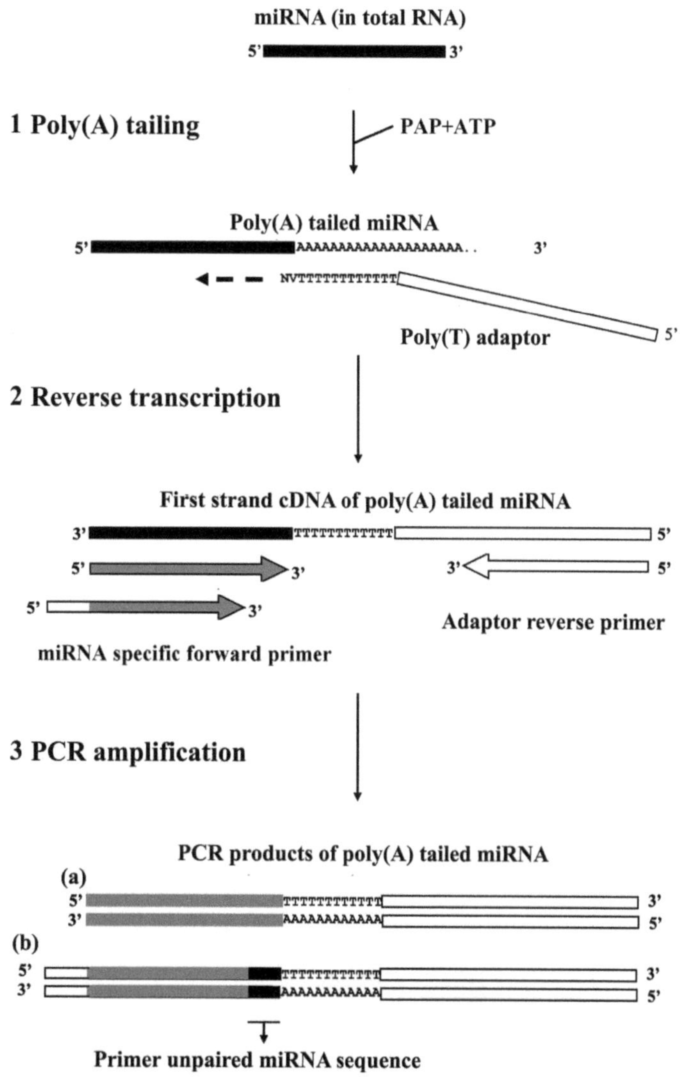

Fig. 1. Scheme for poly(T) RT-PCR for miRNAs. *1*: Poly(A) tailing of miRNA. *2*: Reverse transcription of poly(A) tailed miRNA primed by poly(T) adaptor. *3*: First-strand cDNAs of poly(A) tailed miRNAs are amplified in PCR by using miRNA-specific forward primer and reverse primer complementary to poly(T) adaptor. Depending on miRNA primer designs, two types of miRNA PCR products could be generated, one is illustrated in (**a**), where all sequences come from primer and adaptor. Another type of PCR products contains primer unpaired sequence, which should reflect a portion of the "original" sequence of detected miRNA.

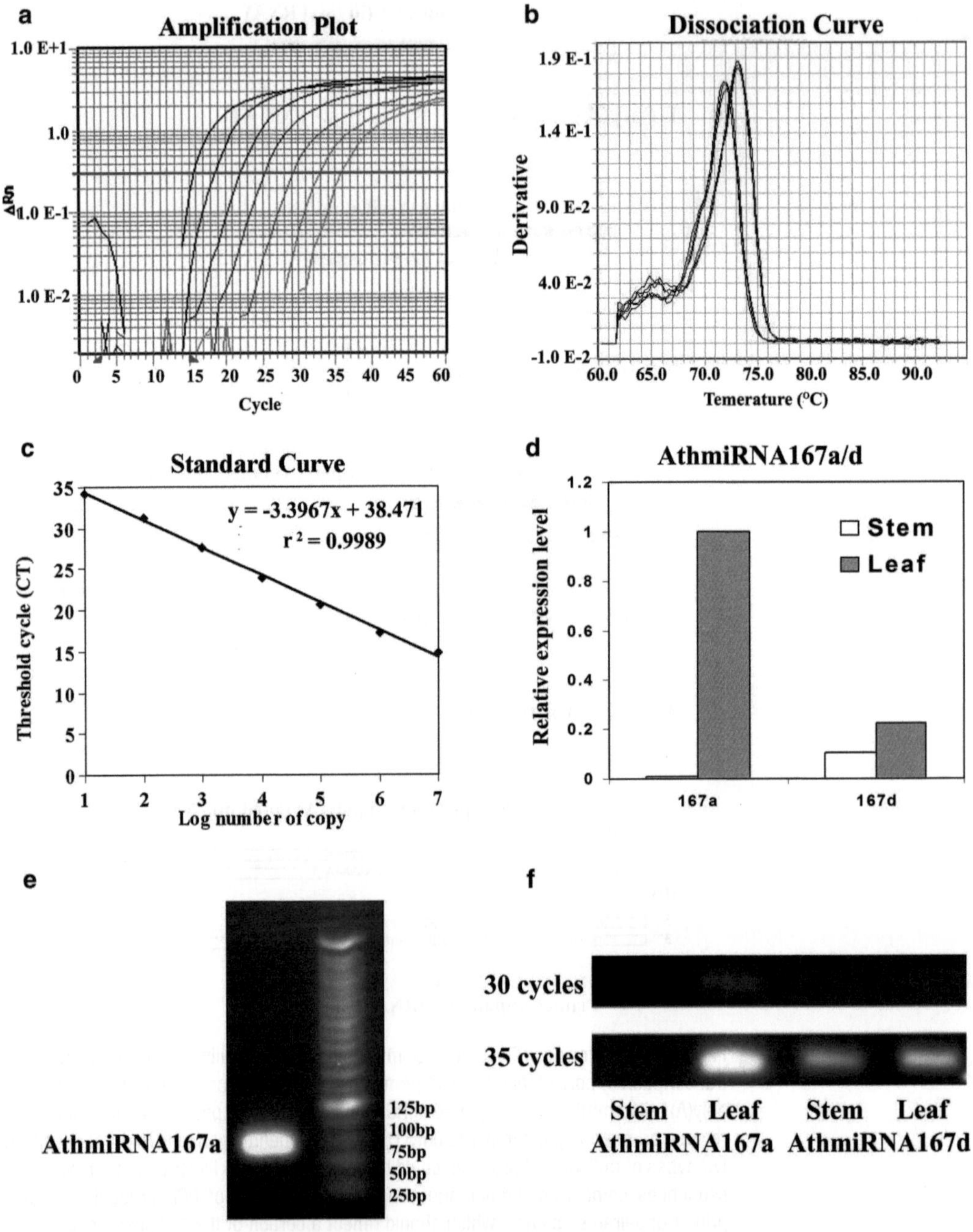

g

```
Primer          5'    ttggtgaagctgccagcat
AthmiRNA167a    5'         UGAAGCUGCCAGCAUGAUCUA
PCR product     5'...ttggtgaagctgccagcatGATCTAaaa...
AthmiRNA167d    5'         UGAAGCUGCCAGCAUGAUCUGG
PCR product     5'...ttggtgaagctgccagcatGATCTGGaaa...
```

2. Materials

2.1. Plant Material

1. Collect plant tissues and store in liquid nitrogen before RNA extraction.

2.2. RNA Extraction

1. TRIzol® Reagent or Plant RNA Reagent (Invitrogen) (see Note 1). Other reagents and solutions required for TRIzol® or Plant RNA Reagent protocol: chloroform, 5 M sodium chloride, isopropanol, 100 and 75% ethanol, RNase-free water (see Note 2).
2. RNase-free DNase I (Promega) (see Note 3).
3. Reagents and solutions required for purification of RNA after enzyme treatment include acid chloroform: phenol saturated solution (Ambion), chloroform, 3 M sodium acetate, 100 and 75% ethanol, RNase-free water (see Note 2).
4. 1.2% MOPS gel for RNA analysis: the gel is prepared by adding 1.2 g agarose to 72 ml water, 10 ml 10× MOPS buffer (0.4 M MOPS, pH 7.0, 0.1 M sodium acetate, 10 mM EDTA), and heating until agarose is dissolved. When the gel solution cools down to ~60°C, add 18 ml formaldehyde (37%; molecular biology grade), mix well, and pour the gel.
5. 2× RNA sample buffer: 1 ml buffer contains 110 μl 10× MOPS, 538 μl formamide, 250 μl formaldehyde, 100 μl loading dye (50% glycerol, 1 mM EDTA, pH 8.0, 0.25% bromophenol blue, 0.25% xylene cyanol FF) (11), 2 μl 10 μg/μl ethidium bromide (EB), mix well, and store at −80°C.

2.3. RNA PolyA Tailing and cDNA Synthesis

1. Poly(A) tailing kit (Ambion).
2. TaqMan® Reverse Transcription Reagents (Applied Biosystems).

Fig. 2. Detection and quantitation of miRNA by poly(T) adaptor RT-PCR. (**a**) Real-time PCR amplification plot of AthmiRNA167a using diluted PCR products equal to 10, 10^2, 10^3, 10^4, 10^5, 10^6, and 10^7 copies of AthmiR167a (from *right* to *left*), showing more miRNA requires less PCR cycles to reach a threshold and a low C_T value. (**b**) Dissociation curve analysis showing the peak for amplicons for AthmiRNA167a (*left*) and AthmiRNA167d (*right*), which indicates the T_ms are 71.8 and 73.2°C, respectively. (**c**) Standard curve showing a plot of log copy number of input miRNA cDNA via C_T value revealed by real-time PCR. The equation ($y = -3.3967x + 38.471$) derived from this plot can be used for calculating the absolute copy number of miRNA template in PCR. (**d**) Quantifying two miRNA species AthmiRNA167a and AthmiRNA167d from stem and leaf tissues of Arabidopsis by real-time qPCR. (**e**) Detection of AthmiRNA167 amplicon from poly(T) adaptor RT-PCR (45 PCR cycles) by 2% agarose-TAE gel electrophoresis. With a forward primer for AthmiRNA167a (Table 1) and a poly(T) adaptor primer, the PCR amplicon is 64 bp long. (**f**) Quantitation of two miRNA species AthmiRNA167a and AthmiRNA167d from stem and leaf tissues of Arabidopsis by end-point PCR at 30 cycles and 35 cycles. PCR products were separated by 2% agarose gel electrophoresis and shown a similar expression pattern as that revealed by real-time PCR in (**c**). (**g**) Sequence analyses of amplicon clone for poly(T) RT-PCR using a forward primer for all members of the miR167 family (Table 1). The sequences of primer unpaired region (*underlined*) on the amplicons (*lowercase*) match the real sequence (*highlighted*) in AthmiRNA167a and AthmiRNA167d (*uppercase*), the two abundant members in the AthmiR167 family.

Table 1
Sequences of primers and adaptor for the analysis of Arabidopsis AthmiRNA167s in poly(T) RT-PCR

Oligonucleotides ID	Sequence (5′–3′)
Poly(T) adaptor	GCGAGCACAGAATTAATACGACTCACTATAGG(T)12VN[a]
Poly(T) adaptor reverse primer	GCGAGCACAGAATTAATACGAC
Forward primer for AthmiRNA167a	TGAAGCTGCCAGCATGATCTA
Forward primer for AthmiRNA167d	TGAAGCTGCCAGCATGATCTGG
Forward primer for AthmiR167 family	TTGGTGAAGCTGCCAGCAT
Forward primer for 5.8S rRNA	ACGTCTGCCTGGGTGTCACAA

[a]*V*=G, C, A; *N*=G, A, C, T

3. Poly(T) adaptor: this is a 3′RACE adaptor adapted from RLM-RACE kit (Ambion). It contains 2 nt (VN) as selective and anchor sequence at its 3′ (V=G, A, C, and N=G, A, T, C). This adaptor can be ordered commercially from companies such as Eurofins MWG Operon (see Note 4).

2.4. Polymerase Chain Reaction

1. SYBR® Green PCR Master Mix (Applied Biosystems) or FastStart Universal SYBR Green Master (Roche Diagnostics).
2. Taq DNA polymerase (Qiagen or Roche).
3. Primers for miRNA and poly(T) adaptors (Table 1, see Note 4).
4. 2% Agarose gel prepared with 0.5× Tris–borate–EDTA (TBE) buffer or 1× Tris–acetate–EDTA (TAE) buffer (11) and respective buffer for electrophoresis (11).

2.5. Cloning of Amplicons (PCR Products)

1. QIAquick® PCR Purification Kit (Qiagen).
2. TOPO TA Cloning® Kit for sequencing (Invitrogen).
3. QIAprep® Spin Miniprep Kit (Qiagen).

2.6. Equipment

1. A desktop refrigerated centrifuge such as the Eppendorf 5417C/R (Eppendorf).
2. A spectrophotometer such as the NanoDrop ND-100 spectrophotometer (NanoDrop Technologies) for RNA quantification.
3. An UV transilluminator such as the TM-36 benchtop UV transilluminater (UVP) for visualizing EB-stained RNA or DNA after gel electrophoresis.
4. A thermal cycler, such as the MJ Research PTC-200 (PCR) Thermal cycler PCR system (Bio-Rad), for DNase treatment, poly(A) tailing, reverse transcription, and end-point PCR.

5. A real-time thermal cycler such as the Applied Biosystems 7900HT sequence detection system.
6. Horizontal electrophoresis apparatus set.

3. Methods

3.1. RNA Purification

1. Total RNAs of plant materials such as leaf or stem of tobacco and Arabidopsis are extracted using TRIzol® Reagent following the product manual. Total RNAs of other plant samples such as tissues of poplar and pine are extracted using Plant RNA Reagent (Invitrogen) according to the kit manual (see Note 1). Usually, RNAs are extracted in small scale from 50 to 100 mg samples either freshly collected or stored in liquid N_2 (see Note 1).
2. Remove possible DNA contamination in the isolated total RNAs by RNase-free DNase treatment following the product manual (see Note 3).
3. RNase-free DNase-treated RNAs are extracted with equal volume of phenol–chloroform: after thorough mix and centrifugation at 12,000×*g* for 5 min at 4°C, the supernatant is transferred to a new tube. Then add equal volume of chloroform and extract again. Transfer the supernatant into a new tube, add 1/10 volume of 3 M sodium acetate and 2–2.5 times volumes of ethanol, mix thoroughly, and keep the tube in −80°C for at least 1 h for RNA precipitation. The precipitated RNA is collected by centrifugation at a maximum speed (≥16,000×*g*) for 30 min at 4°C. The RNA pellet is washed with 75% precooled ethanol once, briefly air dried, and dissolved in suitable volume of RNase-free water.

3.2. Evaluation of RNA Quality and Quantity

1. RNA quality is determined by 1.2% MOPS gel electrophoresis. Prerun MOPS gel in tank filled with 1× MOPS buffer. Mix RNA sample with an equal volume of 2× RNA sample buffer and heat at 65°C for 5–10 min to denature RNA, cool on ice, and then load to the gel and run electrophoresis at 5–6 V/cm.
2. Visualize EB-stained RNA on an UV transilluminator.
3. If RNA quality is good, showing clearly visible rRNA bands, then RNA quantity will be determined by using UV spectrometer (such as NanoDrop). Adjust RNA solutions with RNase-free water to appropriate concentrations, such as 100 ng/μl.

3.3. MiRNA Poly(A) Tailing and Reverse Transcription (see Note 5)

1. Add 100–200 ng of total RNA and 25 pmol of poly(T) adaptor into a mixture including 1 μl 10× RT buffer, 2.5 μl 25 mM $MgCl_2$, 2 μl dNTPs, 0.2 μl RNase inhibitor, 0.625 μl MultiScribe reverse transcriptase from the TaqMan® reverse

transcription kit (Applied Biosystems), 0.5 μl *Escherichia coli* polyA polymerase (E-PAP), and 0.25 μl 10 mM ATP from the PolyA tailing kit (Ambion). The final reaction volume is 10 μl (see Note 6).

2. Incubate the reaction mixture at 37°C for 1 h.
3. Dilute cDNA reaction 20 times with water and store at –20°C (see Note 7).

3.4. Adaptor, Primers, and Melting Temperature (T_m) Evaluation

1. The reverse primer for PCR is the poly(T) adaptor primer adapted from 3′RACE outer primer in RLM-RACE kit (Ambion). This primer is complementary to 3′ portion of the poly(T) adaptor (Fig. 1; Table 1) (see Note 8).
2. The forward primer is designed based on the miRNA sequence to be studied. In general, miRNA-specific forward primers should encompass the whole mature miRNA sequence (Fig. 1) (5). To validate specific miRNA expression, forward primer can also be designed to include only the 5′ part of miRNA and leave 3–6 nt in the 3′ portion unpaired. Arbitrary sequence can be added to 5′ of such primer to retain primer's T_m. (Figs. 1 and 2; Table 1).
3. Primers for reference RNA, such as 5.8S rRNA, are designed and amplified following the similar strategy for miRNA, i.e., using only one 5.8S rRNA-specific forward primer and a universal poly(T) adaptor reverse primer (Table 1) (see Note 9).
4. T_m of miRNA forward primer to miRNA can be calculated using online programs, such as the Web application – oligoanalyzer of Integrative DNA Technologies (http://www.idtdna.com/analyzer/applications/oligoanalyzer). The input sequence is the forward primer sequence that encompasses the miRNA (see Note 10). Such T_m can also be experimentally determined by carrying out dissociation curve analysis on a real-time PCR machine: mix 100 pmol of primer and its corresponding antisense oligonucleotides with 12.5 μl 2× SYBR green mix in a total volume of 25 μl. Run ABI7900HT with a program of 5 min at 95°C followed by a dissociation curve step with the low temperature set to 45°C. After the program is completed, the T_m can be determined by the SDS software (Applied Biosystems) (5).

3.5. Quantitation of miRNA by Real-Time PCR (see Note 11)

3.5.1. Real-Time PCR

1. 1 μl Diluted template cDNA (usually equivalent to 100 pg total RNA) (see Note 12) is mixed with 12.5 μl 2× SYBR Green PCR master mix and 5 pmol of each forward and reverse primers in a final volume of 25 μl. Triplicate reactions should be carried out for assaying miRNA or reference RNA.
2. PCR is carried out according to the standard program on ABI Prism 7000HT Sequence Detection System: 10 min at

95°C to activate Taq DNA polymerase, 45 cycles of 15 s at 95°C, and 1 min at 60°C, and then a thermal denaturing step to generate the dissociation curves to verify amplification specificity. This program can quantify transcript levels of miRNAs differing by two or more nucleotide overall or miRNAs differing by 1 nt at their 3′ terminus.

3. A more stringent PCR condition can be used for quantifying transcript levels of miRNAs with 1 nt difference. The PCR is carried out for 10 min at 95°C, followed by 45 cycles of 15 s at 95°C, 15 s at a temperature 5°C below the primer's estimated T_m to miRNA (see Subheading 3.4, step 4), and 20 s at 72°C, and a thermal denaturing step as described above.
4. To further verify the results of miRNA expression, particularly for validating the newly identified putative miRNAs (9) or artificial miRNAs (10), the forward (miRNA specific) primer is designed to only encompass part of the miRNA sequence, leaving up to 6 nt of miRNA 3′ out of the forward primer. To adjust for the primer' lower annealing ability to miRNA, a lower annealing temperature is used for PCR. A suggested PCR condition: 10 min at 95°C, followed by 45 cycles of 15 s at 95°C, 30 s at 53–57°C, and 1 min at 60°C, and finally a dissociation curve step.

3.5.2. Detection of miRNA

1. Detection of miRNA in real-time PCR is mainly based on the amplified signal, which is presented as a low threshold cycle (C_T) value (Fig. 2a), i.e., fluorescence of SYBR green on double-stranded DNA products reaching a certain (arbitrary) detection threshold for all samples. If no or less amount of miRNA is detected, there is virtually no SYBR green signal or the signal is detected at very later cycles (such as after 35th cycle). In contrast, successful detection of miRNA usually results in the detection of SYBR green signals in early cycles (i.e., low C_T) and the more copies of miRNA template in PCR, the lower C_T value (Fig. 2a).
2. It is strongly recommended to perform a dissociation curve analysis of amplicons at the end of the real-time PCR amplification step. A dissociation curve that shows a single peak between 70 and 80°C indicates a specific amplification (Fig. 2b).

3.5.3. Absolute Quantitation

1. For absolute quantitation of the miRNA copy number by real-time PCR, a standard curve based on a dilution series of assayed miRNA template, such as the purified amplicons or plasmid DNA of cloned amplicons for the assayed miRNA, is used to establish the linear relationship of log copy numbers of the assayed miRNA at different dilutions and the corresponding C_T (5). For example, in the first PCR, add 1×10^7 copies of purified AthmiRNA167a amplicons (~0.35 pg), then add

further ten times diluted template for next PCR until ten copies of AthmiRNA167a amplicons in a PCR. After real-time PCR, set an arbitrary threshold level to extract C_T numbers for all these PCRs.

2. Input all extracted C_T and respective log copy numbers of template into two columns in an Excel spreadsheet, then draw an *XY* (scatter) plot on the work sheet with the log input amount as *X* values and C_T as the *Y* values. Use the "Open trendline" option to plot a line through the data point, and select boxes in the option for display equation and R^2 value on chart (Fig. 2c). The equation will be used for calculating miRNA copy number in PCR, and correlation coefficient (r) is used for evaluating the quality of the standard curve. If r is near 1, then the standard curve is in good quality.

3. The copy number of the assayed miRNA within input RNA can be calculated using equation of the trendline as mentioned above (Fig. 2c). For example, in a real-time PCR using a template of 100 pg total RNA, and the resulted C_T value is 30 using the same threshold as that for standard curve, then input this C_T value 30 as *Y* for equation derived from the plot (Fig. 2c) for calculating the value of *X*, the log of template copy number. Based on such calculation, *X* is ~2.4939. Therefore, the template number in PCR should be $10^{\sim 2.4939}$, which means that there are about 311.82 copies of this miRNA per 100 pg total RNA of the examined sample. Assume there are 10 pg of total RNA in one cell, then, there would be about 31 copies of the assayed miRNA in a single cell.

3.5.4. Relative Quantitation

1. Quantitation of miRNA can also be presented as a relative abundance to a reference gene such as 5.8S rRNA. A simple way is to assume that PCR amplification efficiency for miRNA or reference RNA is close to the ideal efficiency of 2, which means the amount of PCR amplicons doubles in each PCR cycle. Therefore, the quantity of miRNA relative to a reference gene (5.8S rRNA here) can be calculated using the formula $2^{-\Delta C_T}$, where $\Delta C_T = (C_T$ miRNA − C_T reference RNA) (12).

2. Comparison of miRNA expression between different samples can be made based on a comparative C_T method ($\Delta\Delta C_T$) (13). The relative miRNA expression can be quantified according to the formula of $2^{-\Delta\Delta C_T}$, where $\Delta\Delta C_T = (C_T$ miRNA − C_T reference RNA) − (C_T calibrator − C_T reference RNA) (13). The miRNA sample with the lowest C_T value, and thus the highest expression level, is selected as the calibrator, of which expression level is designated 100% for normalization in each comparison. For instance, in Fig. 2d, expression of AthmiRNA167a in leaf is the most abundant, which is used as calibrator and illustrated as 100%. The expression of AthmiRNA167a in stem and expression

of AthmiRNA167d in stem and leaf are compared and presented as relative abundance to AthmiRNA167a (Fig. 2d).

3.6. Detection of miRNA by End-Point PCR (see Note 13)

3.6.1. PCR Using Conventional Taq DNA Polymerase

1. Add cDNA amounts equivalent to 0.1–1 ng total RNA (see Note 11) to reaction mixture containing 2 μl 10× PCR buffer, 1 μl 10 mM dNTPs (2.5 mM for each dNTP), 1 U Taq (Qiagen), and 5 pmol each of forward and reverse primers in a final volume of 20 μl.
2. Carry out the PCR using a "hot start" program. Prepare all PCRs on ice. Preheat the block of thermal cycler to 85°C and then put in PCR tubes and start the PCR. PCR condition for end-point RT-PCR is similar to real-time PCR except that preheating is shorter and without a dissociation curve step. For instance, the standard program can be 95°C predenaturation for 2 min, followed by 45 cycles of 15 s at 95°C and 1 min at 60°C.

3.6.2. End-Point Analysis of miRNA Amplicons

1. Amplicons are examined by gel electrophoresis. We use a 2% agarose-TAE gel and EB staining. Detection of products at the expected size is a main proof for the expression of the detected miRNA. The amplicons for poly(A) tailed miRNAs are usually 60–70 bp long if the primers and adaptor described here are used (Table 1; Fig. 2e).
2. To quantify miRNA transcripts, PCR amplicons are separated by 2% agarose gel electrophoresis. After the gel is stained with EB, the relative amount of amplicons of different samples under the same RT-PCR conditions after the same PCR cycles can be evaluated according to fluorescence strength of bands under UV light. Brighter EB-stained band means higher miRNA expression as compared to samples with weaker bands (Fig. 2f).
3. For semiquantitative measurement, perform an end-point PCR using a dilution series of purified amplicons or plasmid DNA of amplicon clone of the assayed miRNA. The abundance of amplified products is measured by using any commercially available gel documentation system to estimate approximate copy number of miRNA in PCR.

3.7. Cloning and Sequencing of Amplicons (see Note 14)

1. Amplicons are purified using QIAquick® PCR Purification Kit (Qiagen) following manufacturer's manual.
2. Eluted DNAs are cloned into TA-vector using TOPO TA Cloning® Kit (Invitrogen) as described in the product manual.
3. Plasmid DNAs are purified with QIAprep® Spin Miniprep Kit and sequenced using the common primers derived from the cloning vectors (such as T7 promoter or M13 primers).
4. The sequence of primer unpaired region is compared to the expected miRNA sequence for verification of the assayed miRNA as illustrated in Fig. 2g.

4. Notes

1. Total RNAs containing small RNAs can be purified using many commercially available precipitation-based reagents or kits, such as TRIzol® Reagent and plant RNA reagent (Invitrogen). Animal RNAs can also be isolated using TRIzol® Reagent. For analyzing miRNA expression, we do not recommend column-based RNA purification kits, unless such kits are specifically designed for small RNAs. Most of the column-based kits are designed for extracting longer RNA transcripts. In addition, we suggest using total RNAs as start material for miRNA analysis in poly(T) adaptor RT-PCR. Fractioned small-sized RNAs seem not to be suitable for this poly(T) adaptor RT-PCR protocol, since in addition to the expected miRNA amplicons, high molecular weight (HMW) unspecific amplicons may also be generated. Production of HMW products may competitively consume PCR reagents, and therefore affect the accuracy of quantifying assayed miRNA (ref. 14, our unpublished data).
2. Care should be taken to avoid RNase contamination. Water or solution that does not contain primary amines should be treated with 0.1% diethylpyrocarbonate (DEPC) at room temperature overnight, followed by autoclave at 120°C for 20 min to remove the residual DEPC.
3. DNase treatment of RNA is optional if only miRNAs are to be analyzed. All RNA transcripts, including miRNAs and reference RNA such as 5.8S rRNA, are amplified in PCR using only one gene-specific forward primer and a universal poly(T) adaptor reverse primer. In theory, genomic DNA cannot be amplified because double-stranded DNA cannot be poly (A) tailed by PAP and ATP.
4. Adaptor and primers are oligonucleotides synthesized with unmodified nucleotides.
5. The protocol is modified from our previous protocol (5) and described in a recent publication (10). It combines components for poly(A) tailing and reverse transcription into one tube as one step "spontaneous" reaction. Therefore, it is less labor intensive because the step for purification of poly(A) tailed RNAs is eliminated. This protocol is suitable for analyzing limited number of miRNAs from a large number of samples. Although less cDNA is generated by this protocol compared to our early version (5), there is enough cDNA generated to assay the abundance of most miRNAs.
6. The reaction can be scaled down to 5 μl for 100 ng total RNAs for cost saving.

7. The cDNA from miRNAs has been reported to be easily degraded (15). However, we found that cDNA diluted in water is quite stable at −20°C. Although we have not assayed miRNA abundance using undiluted cDNA template, the qRT-PCR results are comparable for cDNA samples that have been stored in −20°C for over 2 years and for freshly prepared cDNA (our unpublished data).
8. Although we adapt the poly(T) adaptor and corresponding adaptor primer from RLM-RACE kit (Ambion), other similar types of poly(T) adaptor and corresponding adaptor reverse primer can also be used. However, to produce specific size of miRNA amplicons, it is recommended that the poly(T) adaptor should have at least one anchor base, such as *V* (= G, C, A), at its 3′ terminus.
9. Amplification of reference RNA using two gene-specific primers is acceptable. However, complete removal of genomic DNA contamination by DNase treatment of the RNA samples is required.
10. Primer's T_m to miRNA reflects primer's annealing ability to miRNA template in PCR, and it is not the same as primer's own T_m.
11. Quantitative real-time PCR is highly recommended for accurate quantitation of miRNA transcripts.
12. It is recommended to use less amount of template for poly(T) adaptor RT-PCR. Based on our experience, cDNA derived from 100 pg total RNA is enough for analyzing abundant miRNAs. For low abundant miRNAs, the amounts of cDNA can be increased to, however, no more than 5 ng RNA per PCR. Too much template may result in nonspecific HMW products, and therefore compromise the performance of quantitation. This is especially true for end-point PCR detection using normal Taq polymerase.
13. End-point PCR is not as accurate as real-time PCR for evaluating miRNA expression. However, it can still provide semiquantitation data, and is more applicable for labs without the real-time PCR machine (14).
14. Sequencing of poly(A) tailed miRNA amplicons is not always necessary, especially if the forward primer complementary to the whole sequence of the miRNA is used for the assay. However, if forward primer is designed to use part of the miRNA sequence and leave 3–6 nt miRNA at 3′ terminus unpaired, then the sequencing analysis of miRNA amplicons would be useful for validating miRNA expression, especially for artificial miRNA expression. This process is similar to 3′ RACE for sequencing of mRNA in that the 3′ sequence of miRNA is revealed by sequencing.

Acknowledgments

This work was supported by research grants from the US Department of Energy Division of Energy Biosciences DE-FG02-03ER15442 and the National Science Foundation, Plant Genome Research Program DBI-0922391 to VLC, and from the North Carolina State University Forest Biotechnology Industrial Research Consortium (FORBIRC) to VLC and RS. X-HZ wishes to thank the Chiang Laboratory for hosting his sabbatical leave.

References

1. Bartel, D. P. (2004) MicroRNAs: genomics, biogenesis, mechanism, and function. *Cell* **116**, 281–97.
2. Ambros, V. (2004) The functions of animal microRNAs. *Nature* **431**, 350–5.
3. Lu, S., Sun, Y.H., Shi, R., Clark, C., Li, L. and Chiang, V. L. (2005) Novel and mechanical stress-responsive MicroRNAs in *Populus trichocarpa* that are absent from Arabidopsis. *Plant Cell* **17**, 2186–203.
4. Wong, M. L. and Medrano, J. F. (2005) Real-time PCR for mRNA quantitation. *Biotechniques* **39**, 75–85.
5. Shi, R. and Chiang, V. L. (2005) Facile means for quantifying microRNA expression by real-time PCR. *Biotechniques* **39**, 519–25.
6. Válóczi, A., Hornyik, C., Varga, N., Burgyán, J., Kauppinen, S. and Havelda, Z. (2004) Sensitive and specific detection of microRNAs by northern blot analysis using LNA-modified oligonucleotide probes. *Nucleic Acids Res* **32**, e175.
7. Chen, C., Ridzon, D. A., Broomer, A. J., Zhou, Z., Lee, D. H., Nguyen, J. T., et al. (2005) Real-time quantification of microRNAs by stem-loop RT-PCR. *Nucleic Acids Res* **33**, e179.
8. Raymond, C. K., Roberts, B. S., Garrett-Engele, P., Lim, L. P. and Johnson, J. M. (2005) Simple, quantitative primer-extension PCR assay for direct monitoring of microRNAs and short-interfering RNAs. *RNA* **11**, 1737–44.
9. Oh, T. J., Wartell, R. M., Cairney, J. and Pullman, G. S. (2008) Evidence for stage-specific modulation of specific microRNAs (miRNAs) and miRNA processing components in zygotic embryo and female gametophyte of loblolly pine (Pinus taeda). *New Phytol* **79**, 67–80.
10. Shi, R., Yang, C., Lu, S., Sederoff, R. and Chiang, V. L. (2010) Specific down-regulation of *PAL* genes by artificial microRNAs in *Populus trichocarp*. *Planta* **232**, 1281–8.
11. Sambrook, J. and Russell, D. W. (2001) Molecular cloning: a laboratory manual, 3rd edn. Cold Spring Harbor, New York.
12. Schmittgen, T. D., Jiang, J., Liu, Q. and Yang, L. (2004) A high-throughput method to monitor the expression of microRNA precursors. *Nucleic Acids Res* **32**, e43.
13. Schmittgen, T. D. and Livak, K. J. (2008) Analyzing real-time PCR data by the comparative $C_{(T)}$ method. *Nat Protoc* **3**, 1101–8.
14. Fu, H. J., Zhu, J., Yang, M., Zhang, Z. Y., Tie, Y., Jiang, H. et al. (2006) A novel method to monitor the expression of microRNAs. *Mol Biotechnol* **32**, 197–204.
15. Bravo, V., Rosero, S., Picordi, C. and Pastori, R. L. (2007) Instability of miRNA and cDNAs derivatives in RNA preparations. *Biochem Biophys Res Commun* **353**, 1052–5.

Chapter 5

MicroRNA In Situ Hybridization

Boye Schnack Nielsen

Abstract

In situ hybridization (ISH) is a technology that allows detection of specific nucleic acid sequences in tissue samples at the cellular level. For detection of individual microRNAs (miRNAs) and mRNAs, the ISH technology determines the cellular origin of expression and provides information on expression levels in different tissue compartments and cell populations. This histological expression analysis is of crucial importance for elucidating roles particularly of miRNAs in molecular and biological processes. mRNA expression analyses can partly be replaced by immunohistochemical detection of the protein encoded by the mRNA. Combined with the short sequences of the miRNAs (18–22 bp), this leaves miRNA ISH as an indispensable yet challenging technology in terms of detection and specificity analysis. In this chapter, a simple miRNA ISH protocol using chromogenic detection is presented. I touch upon critical steps in the ISH protocol, different applications on ISH technology platforms, advantageous use of locked nucleic acids (LNA™) in miRNA detection probes, qualification of clinical paraffin samples, and specificity analyses and quantification of the ISH signal.

Key words: Digoxigenin, Formalin fixed and paraffin embedded (FFPE), In situ hybridization, Locked nucleic acid, MicroRNA

1. Introduction

The microRNA (miRNA) in situ hybridization (ISH) technology enables the determination of miRNA expression at the cellular level, and therefore addresses the question: What cell population in a tissue expresses the miRNA? A precise answer to that question is of high importance since it directs the biological interpretation for further functional studies in disease models and in molecular or cellular assays.

Although mRNA ISH is a well-established technique, implementation of mRNA or miRNA ISH in a molecular or histology laboratory is challenging, and the reason for frequent failure in

Jian-Bing Fan (ed.), *Next-Generation MicroRNA Expression Profiling Technology: Methods and Protocols*, Methods in Molecular Biology, vol. 822, DOI 10.1007/978-1-61779-427-8_5, © Springer Science+Business Media, LLC 2012

setting up RNA ISH technology cannot be attributed to a single process or reagent, but rather a combination of technical prerequisites. First of all, miRNA ISH is an RNA-dependent technology indicating that the laboratory should support an RNase-free environment, and the laboratory technician or scientist executing the experiments should be aware of working RNase free by using gloves during the handling steps. Second, the ISH protocol contains many hands-on steps involving a variety of reagents and buffers. Each step has a risk of failure, and tissue sections can be fragile and therefore a person skilled in histology techniques, such as immunohistochemistry, may be helpful. Third, the lack of good and robust positive and negative controls during setup has prevented the establishment of a reproducible protocol. Finally, after the basics are in place, the ISH protocol includes a number of critical steps, some of which need optimization related to the type of tissue and the detection probe used.

Several detailed miRNA ISH protocols have been reported that cover paraffin sections (1–3), cryostat sections (4–7), whole-mount specimens (8, 9), and cultured cells (6, 10, 11). Replacing DNAs (or RNAs) with locked nucleic acids (LNAs, see Note 1) in certain nucleotide positions increases the binding affinity of the antisense oligo, and the use of LNA:DNA chimeric probes has significantly improved specific miRNA detection (8, 12). Using LNA™ probes, the miRNA ISH protocol on paraffin sections can be cut down to less than 20 essential steps and can be completed within a working day (see protocol below), Fig. 1. Two steps that were found not to be essential for the performance of the protocol but rather would increase the level of complexity include (a) acetylation with acetic anhydride after postfixation or the proteolytic digestion step (3, 5, 6), which is considered to reduce unspecific binding of the probe to amino groups and (b) prehybridization in hybridization buffer as reported in other protocols (3, 5, 6).

The two most critical steps in the miRNA ISH protocol are the target demasking step, often known as the *predigestion* or the *proteinase-K step*, and the hybridization step. The proteolytic digestion step is based on the use of relatively unspecific cleavage of tissue components by proteases, like proteinase-K or pepsin. Proteolytic digestion of the tissue provides access to the miRNAs in the tissue matrix, and the optimal level of proteinase-K treatment is dependent on the extent of fixation. The other step that needs particular attention is the probe hybridization step. Specific probe hybridization needs a well-defined and stable medium (hybridization buffer), a stable hybridization temperature, and an optimal probe concentration. The hybridization buffer must have an optimal level of salts, oligonucleotide stabilizers (such inert RNA or DNA), viscosity-reducing agents (such as dextran and Denhardt's solution), and denaturing agents (like formamide). Successful miRNA ISH using LNA™ probes has been reported with various concentrations (25–70%) of formamide (3, 10, 13).

Process*	Step	Equipment	Time /accumulated	Temperature
Deparaffination	1	Coplin Jars	40 min. /	Room Temperature
Proteinase-K	2	Hybridizer	10 min. /	37°C
Dehydration	4	Coplin Jars	20 min. /	Room Temperature
In Situ hybridization	5	Hybridizer	60 min. / 3 hours	55°C
Stringent washes	7	Water bath	30 min. /	55°C
Blocking	9	IHC staining racks	15 min. /	Room Temperature
Anti-DIG/AP	10		60 min. /	
AP reaction	12	IHC staining racks in Oven	120 min. /	30°C
Counter stain	15	IHC staining racks	10 min. / 6½ hours	Room Temperature
Dehydration	17	Coplin Jars	10 min. /	
Mounting	18		5 min. / 7 hours	

Fig. 1. Outline of the microRNA *in situ* hybridization protocol. Sequence of required steps in the miRNA *in situ* hybridization protocol on FFPE tissue section. *Asterisk*: Some of the washing and handling steps are not included.

For the use of LNA™ probes and to avoid the use of potentially harmful formamide, a formamide-free hybridization buffer that allows detection of miRNAs at high signal-to-noise ratio has been developed (2). To obtain a stable medium during hybridization, where the buffer does not dry out, the probe solution applied to the tissue is protected with a cover glass and then sealed with rubber cement (see protocol below). With this approach, there is no need to establish humidifying conditions like using humidifying chambers. Sufficiently stable hybridization conditions are also obtained in flow-through chambers using a Tecan Genepaint system (14). A constant and precise hybridization temperature is essential for optimal and specific hybridization. Optimizing the T_{hyb} for a given probe, therefore, adds some requirements to the laboratory equipment, and hybridization in an ordinary hybridization oven may not provide sufficiently reproducible data.

The protocol below is prepared for the detection of miR-126 (as the example in Fig. 3c) in clinical routinely processed formalin-fixed and paraffin-embedded tissue samples. The miR-126 probe can be replaced by other double-digoxigenin (DIG)-labeled LNA™ detection probes (Exiqon), which may involve optimization of the probe concentration and T_{hyb}. Other probe labels alternative to DIG are possible (see Note 2), but are not supported by the current protocol.

2. Materials

2.1. Tissue Samples

Human formalin-fixed and paraffin-embedded (FFPE) tissue specimens from pathology departments or equivalent mouse tissue samples can be used. Tissue fixation is an important parameter to consider for ISH analyses and the performance of individual blocks can vary accordingly (see Note 3).

2.2. Commercial Reagents

1. Double-DIG-labeled LNA™ miR-126 probe: ISH Optimization Kit 5 (Exiqon); this Kit also contains a Double-DIG-labeled LNA™ 22-mer negative control probe (with a random scrambled sequence with no complementary sequences among human and mouse transcripts), a probe specific for U6 snRNA as an assay optimization and control probe, hybridization buffer, and proteinase-K.
2. Xylene and ethanol for deparaffination and dehydration.
3. PBS, sterile.
4. Tween-20.
5. 20× SSC buffer, ultrapure.
6. Dig Wash and Blocking reagent (Roche).
7. Sheep anti-DIG-AP (Roche).
8. Sheep serum.
9. NBT/BCIP ready-to-use tablets (Roche).
10. Levamisole.
11. Rubber cement, such as Fixogum (MP Biomedicals).
12. Nuclear Fast Red (Vector Laboratories).
13. Eukitt.
14. Hybridizer: Dako Hybridizer, Vysis' ThermoBrite, or Invitrogen's Spotlight Hybridizer. These have a temperature and time adjustable and programmable heating plate. Other ISH platforms can be used, including simple hybridization ovens (see Note 4) and automated systems (see Note 5).
15. Humidifying chambers: Shandon Sequenza® slide racks (Thermo Scientific) or equivalent for immunohistochemical detection.
16. RNase-depleted Milli-Q water (Millipore).
17. RNase ZAP (Ambion).
18. SuperFrost®Plus (Thermo Fisher Scientific).

2.3. Buffers and Reagent Solutions

1. *Proteinase-K buffer*: 1 L of Proteinase-K buffer: To 900 ml milli-Q water, add 5 ml of 1 M Tris–HCl (pH 7.4), 2 ml 0.5 M EDTA, and 0.2 ml 5 M NaCl. Adjust volume to 1,000 ml. Autoclave.

2. *Proteinase-K reagent*: Proteinase-K reagent: For 10 ml proteinase-K reagent with a concentration 15 μg/ml (range 5–20 μg/ml): To 10 ml proteinase-K buffer, add 7.5 μl proteinase-K stock of 20 mg/ml.
3. *SSC buffers*: 1 L of 5× SSC: To 750 ml Milli-Q water, add 250 ml 2× SSC. One liter of 1× SSC: To 950 ml Milli-Q water, add 50 ml 20× SSC. One liter of 0.2× SSC: To 990 ml Milli-Q water, add 10 ml 20× SSC. All SSC buffers should be autoclaved.
4. *PBS-T*: 1 L of PBS-T: To 1 L of PBS (pH 7.4), add 1 ml 0.1% Tween-20.
5. *KTBT*: 50 mM Tris–HCl, 150 mM NaCl, 10 mM KCl. One liter of KTBT buffer: To 900 ml Milli-Q water, add 7.9 g Tris–HCl, 8.7 g NaCl, and 0.75 g KCl. Adjust volume to 1,000 ml. Do not adjust pH. Autoclave.
6. *Levamisol stock*: 100 mM. Prepare 100 mM stock by adding 10 ml milli-Q water to 250 mg Levamisole.

3. Methods

3.1. Glassware

All glassware, including coplin jars, glass-staining racks, and stacks of cover glass and bottles for buffers, should be heat treated in an oven at 180°C for 8 h. The items can be covered by aluminum foil before being placed in the oven in order to prevent contamination when removing the items afterward.

3.2. Tissue Sectioning

Prevent RNase contamination by wearing gloves and using RNase-depleted water. Clean the sectioning workstation (bench top, microtome, blade holder, brushes, tweezers, cooling plate, water bath, etc.) with RNase ZAP. Place the FFPE blocks on a cooling plate at approximately −15°C. Prepare a water bath with room-temperature RNase-free water and a warm water bath with RNase-free water to 40–50°C. Trim blocks and discard the first couple of sections, and then collect 6-μm sections (not thinner) in the room-temperature water bath for manual unfolding. Transfer sections to the heated water bath to briefly allow tissue stretching. Mount sections immediately on SuperFrost®Plus slides obtained from a noncontaminated package. Air dry the paraffin sections for 1–2 h at room temperature (RT). Paraffin sections can be stored as such at 4°C for at least 2 weeks. Melt paraffin sections in an oven at 60°C for 45 min the day prior to the ISH experiment and store overnight at 4°C.

3.3. microRNA In Situ Hybridization on Human FFPE Samples

1. Deparaffinize slides in xylene and ethanol solutions in coplin jars ending up in PBS. In parallel, prepare a water bath and SSC buffers to be heated to 55°C (or the hybridization

temperature). Deparaffination: Place slides in Xylene for 15 min (through 2–3 coplin jars) and then hydrate through ethanol solutions 99% (three coplin jars), 96% (two coplin jars), and 70% (two coplin jars) to PBS (two coplin jars). Each solution should include one 5-min incubation.

2. Apply 300 μl/slide proteinase-K reagent at 15 μg/ml (see Note 6) directly on the slide over the tissue and incubate for 10 min at 37°C in the Dako hybridizer or equivalent.
3. Discard the proteinase-K reagent and wash twice with PBS.
4. Dehydrate slides through ethanol solutions 70% (two coplin jars), 96% (two coplin jars), and 99% (two coplin jars). Keep the dehydration short by one immersion and a 1-min incubation in the two coplin jars at each step. Air dry the slides on clean paper towels for approximately 15 min. In parallel, denature LNA™ probe and dilute the probe in Exiqon ISH buffer. Example for 2 ml hybridization mix containing 50 nM double-DIG-labeled miR-126 LNA™ probe: From 25 μM probe stock, transfer 4 μl into the bottom of a 2-ml nonstick RNase-free tube and place the tube at 90°C for 4 min. Spin down shortly using a tabletop centrifuge, and immediately add 2 ml ISH buffer into the tube. LNA™ probes for other miRNAs may require optimization of the concentration (see Note 7).
5. Then, apply 25 μl hybridization mix containing the double-DIG-labeled LNA™ probe (now at 50 nM) on each tissue section. Gently cover with heat-treated cover glass and seal with Fixogum. Place the slides in the hybridizer and start a program hybridizing for 1 h at 55°C (see Note 8). The hybridization step should not exceed 2 h.
6. Remove Fixogum using a tweezers and carefully detach cover glass and place the slides into 5× SSC at RT. Place slides in a staining rack.
7. The stringent washing steps are performed by placing the staining racks with slides in coplin jars containing 5× SSC (1 × 5 min), 1× SSC (2 × 5 min), and 0.2× SSC (2 × 5 min) at the hybridization temperature (55°C) and 0.2× SSC (1 × 5 min) at RT.
8. Transfer slides to PBS-T and mount into Shandon Sequenza® slide racks. Avoid air bubbles during mounting.
9. Incubate with 200 μl blocking solution for 15 min at RT. Blocking solution: Prepare blocking solution according to the manufacturer's recommendations. For 1,600 μl blocking reagent, add 32 μl sheep serum (final concentration 2%).
10. Apply sheep anti-DIG-AP at 1:800 diluted in blocking solution containing 2% sheep serum and incubate for 60 min at RT (preferably two times for 30 min). Anti-DIG reagent: Example: for 1,600 μl anti-DIG reagent with a dilution of 1:800: To 1,600 μl blocking solution, add 2 μl sheep-anti-DIG AP conjugated.

11. Wash each slide with 300 μl PBS-T two times for 3 min.
12. Incubate freshly prepared NBT/BCIP substrate reagent containing 0.2 mM Levamisole and incubate 300 μl for 2 h at 30°C in the incubation racks. Protect from light during development. For 10 ml AP substrate: In 10 ml Milli-Q water, add one NBT–BCIP tablet according to the manufacturer's recommendations. Add 20 μl Levamisol stock to a final concentration of 0.2 mM. Keep protected from light.
13. Incubate slides with 300 μl KTBT buffer two times for 5 min.
14. Wash with water two times for 1 min.
15. Counter stain with Nuclear Fast Red for 1 min (optional step). Shake bottle with Nuclear Fast Red gently before use. Draw the expected volume and optionally lead through filter paper to remove undissolved precipitates. Immediately incubate 200–300 μl per slide.
16. Remove slides from the Shandon racks, place in staining racks, and rinse in tap water for 10 min.
17. Dehydrate with ethanol solutions through ethanol solutions 70% (two coplin jars), 96% (two coplin jars), and 99% (two coplin jars). Keep the dehydration short by one immersion and a 1-min incubation in the two coplin jars at each step. An optional xylene step may be included before mounting.
18. Mount slides directly with Eukitt. Avoid air drying sections at this step.
19. Allow overnight settlement of the precipitate and analyze results by light microscopy the subsequent day.
20. During light microscopy evaluation, the miR-126 signal is prevalent in endothelial cells. Certain diseased or activated tissues may divert from this expression pattern. The evaluation of miRNA ISH results requires considerations on the specificity of the signal. Specificity control analysis is an inherent part of molecular histology in general. For miRNA ISH analyses, there are several ways to resolve such questions (see Note 9).

3.4. Advanced miRNA In Situ Hybridization Applications

Having settled a miRNA ISH protocol with good signal-to-noise ratio, a number of questions evolve, for example: Can the miRNA-positive cells be better characterized in terms of specific cell markers? Is it possible to detect the expression of potential mRNA targets and compare directly with the miRNA ISH signal at the cellular level? Can the subcellular localization of the miRNA be better described? Is it possible to make quantitative estimates of the ISH signal for diagnostic or prognostic studies? Also, although the abundant expression of miR-143 and miR-145 in smooth muscle cells in normal tissues is evident (Fig. 4), it may be important to know how their expression is affected in diseased tissues.

In terms of characterizing the miRNA-expressing cell population by combining ISH with immunohistochemistry, different approaches have been taken, including double immunofluorescence and combined chromogenic ISH and immunohistochemical fluorescence detection (3, 15–18). Such studies would preferably be done on cryostat sections, where autofluorescence often is a minor problem. Advanced fluorescence microscopy (3) or laser scanning microscopy combined with emission fingerprinting (19) may facilitate fluorescence studies in FFPE sections. Identification of specific miRNA targets by superimposing expression of the miRNA and its putative target mRNA/protein is probably possible only through ISH analyses (3, 4, 20–24).

The ISH technology is merely a qualitative assay rather than a quantitative assay. Particularly, in the light of variation in tissue samples caused by tissue processing and fixation (see Note 3), it may seem obscure to even think of performing quantitative ISH especially in archival samples. However, it is appealing to obtain semiquantitative estimates from molecular histology analyses similar to the well-known semiquantitative scoring in immunohistochemistry. To justify semiquantitative estimates in molecular histology, a certain level of assay reproducibility is required. One simple parameter that, for example, may influence the score is the section thickness. Tissue sections are cut manually on microtomes and even the micrometer is set, the final thickness is affected by parameters, such as the type of paraffin, the temperature of the block during sectioning, time span, and temperature in the recovery water bath. Therefore, when preparing a semiquantitative study in ISH or immunohistochemistry, all manual steps must be controlled to the best possible standard. Introducing an automated slide processor, such as the Tecan Genepaint system, strongly reduces the variability otherwise caused by manual handling during the ISH protocol. Image analysis employed to obtain quantitative estimates of relative expression levels has been a long-existing technology in molecular histology, which however never has reached a level confidence partly due to the lack of strong normalization parameters. Indeed, relative estimates are to be obtained particularly when automated ISH is combined with systematic random acquisition of image fields using a Visiopharm system (14). A robust ISH platform may enable miRNA ISH on tissue microarrays (25) and thereby provide comprehensive localization and expression data in clinical FFPE samples.

The fast-growing knowledge on miRNA expression from array and qPCR profiling studies needs to be supplemented by solid, reliable, and sensitive ISH platforms and protocols. In this report, an overview of the challenges in the technology and of the current technology platforms is presented. The technology may need to reach a higher sensitivity level in order to allow robust detection of low-expressed miRNAs, and thereby address subtle questions linked to the specific molecular functions of individual miRNAs.

4. Notes

1. *LNA ISH probes*: ISH is performed with sequence-specific antisense probes. Because of the small size of miRNAs compared to mRNAs, miRNA ISH is limited to the use of small-size probes. The use of LNA:DNA chimeric probes has significantly improved specific miRNA detection. In locked nucleic acid, the ribose ring is locked into a C3′-endo conformation by a 2′-O, 4′-C methylene bridge (26, 27). Replacing DNAs (or RNAs) with LNAs in certain nucleotide positions increases the binding affinity of the antisense oligo as measured by UV melting analyses (12). The binding efficiency or the ***hybridization affinity*** of oligonucleotides is, thus, measured in melting temperature T_m, which is defined as the temperature, where 50% of the oligo is bound to its complementary sequence. In practice, the T_m is dependent of the solution's physicochemical properties, like pH, salt content, and denaturing agents. The T_m of LNA™ probes can be measured against a synthetic complementary DNA or RNA oligo or estimated using T_m prediction algorithms, and gives rise to RNA-T_m or DNA-T_m values. Replacing a single DNA with an LNA increases the T_m 2–8 (~5)°C depending on the nucleic acid being an A, T, C, or G (27). This means that for a 22-mer DNA probe the T_m increases 20–40°C by replacing 5–8 of the DNAs with LNAs. The specific and high affinity binding facilitates fast and specific binding of the probe within a complex tissue compartment, and the superior performance of LNA™ probes compared to DNA and RNA probes in ISH analyses has been demonstrated in whole-mount specimens (8). Typically, a 22-mer oligonucleotide with 50% GC content has RNA-T_m of 50°C, whereas the same oligo with a content of 20–40% LNA would have a T_m in the range of 80–90°C. In addition to the high affinity binding, a significant advantage of using the LNA-modified oligos is the increased mismatch discrimination (12, 28, 29). Two to three mismatches reduce the T_m with up to 20°C (8), which may enable discrimination of highly similar sequences, such as the miRNA family members in the Let7 or miR-200 families.
2. For most of the miRNA ISH studies reported, LNA™-modified antisense probes have been employed, whereas the use of traditional RNA antisense probes generated by ***in vitro*** transcription have been used in only a few miRNA ISH studies (6, 30–33). In such case, nonbound probe can be digested using RNase before the stringent wash steps (6) to increase the signal-to-noise ratio.
3. ***Probe labeling***: Oligonucleotides can be labeled with a variety of molecules appropriate for their use. For ISH, the choice of

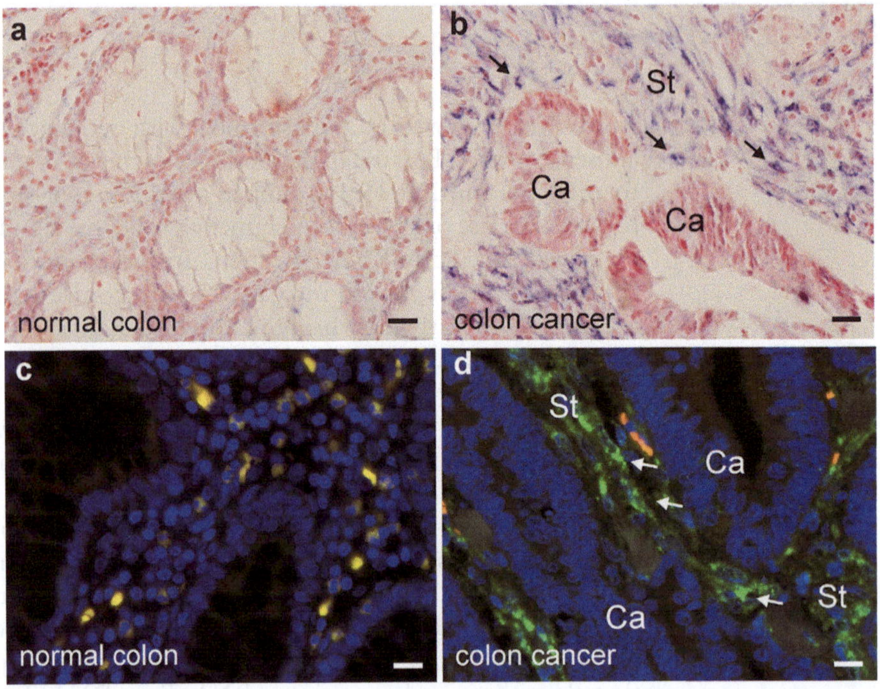

Fig. 2. Chromogenic and fluorescence detection of miR-21 by *in situ* hybridization. *In situ* hybridization for miR-21 in colon cancer samples containing areas of normal colon. *In situ* hybridization was performed following the steps in Fig. 1, but submitted to a Tecan *in situ* hybridization system (14). The miR-21 probe was detected with alkaline phosphatase-conjugated sheep-anti-DIG followed by NBT–BCIP substrate [*blue stain* in (**a**) and (**b**)] or peroxidase-conjugated sheep-anti-DIG followed by TSA-FITC substrate [*green stain* in (**c**) and (**d**)]. High miR-21 expression (examples indicated by *arrows*) is seen in the colon cancer area in stromal cells (St) while cancer cells (Ca) are negative, and virtually no miR-21 signal is seen in the normal colon. The fluorescence images were exposed in both the *red* and *green* filters to show the presence of autofluorescence (*yellow–orange stain*). The FFPE samples were purchased from Proteogenex. Bars: (**a**–**b**) 20 μm; (**c**–**d**) 10 μm.

label provides different detection approaches and subsequent signal enhancement. Since direct ISH detection using fluorescence-labeled probes may not be sufficiently sensitive for the detection of most miRNAs, labeling of LNA™ oligos with hapten molecules is recommended and allows for subsequent antibody-based detection, amplification, and visualization. Examples include DIG, carboxyfluorescein (6-FAM), bromodeoxyuridine (BrdU), and biotin. The subsequent visualization can be either using chromogene or fluorescence (Fig. 2). A preferred labeling of the oligos has been the use of DIG, a steroid synthesized in a few members of the Digitalis plant family. Different from DIG, using biotin as hapten can provide potential cross-reaction with endogenous biotin if binding to the endogenous biotin is not prevented using appropriate biotin-blocking reagents. 6-FAM can work both as a direct fluorophore and as a hapten and is like DIG nonexisting in animals. It should be noted that not all antibodies against fluorescein may cross-react with similar affinity to 6-FAM. To increase sensitivity in ISH, it is beneficial to add as many haptens on the detecting probe as possible. A critical balance would be that the haptens do not prevent probe mobility or cause unspecific adherence to the tissue matrix. It has been reported that double-hapten-labeled LNA™ probes produce significantly higher signal than single-DIG-labeled probes (3, 34). In the excellent study by Sempere et al. (3), several probe labeling strategies were tested.

4. *Tissue qualification for ISH*: MiRNA ISH is applicable to whole-mount animal tissues (8, 9, 13, 22, 35–39), plant material (33, 40, 41), and cells (1, 10, 17, 42). Different from clinical tissue samples, fixation and processing of these samples can be specifically optimized for the miRNA ISH analysis. In contrast, ISH on patient samples or other clinical samples most often need to adjust to the quality of the samples in tissue banks. MiRNA ISH can be performed in routinely processed and archived clinical samples (1–3), which to a wide extent are formalin-fixed tissue specimens embedded in paraffin also known as FFPE samples. Fortunately, miRNAs can be demasked and detected in the standard FFPE samples if the tissue specimens have been fixed properly. For a standard-sized tissue specimen measuring 4 × 10 × 10 mm, a window of 1–3 days in standard formalin would be adequate. Processing of the tissue specimens often follows standard formalin fixation and paraffin embedding – that is, overnight fixation in neutral-buffered formalin followed by overnight embedding into paraffin, a step that involves heating the tissue specimens for a couple of hours in 60°C liquid paraffin.

   Despite the hospitals' standard procedures, the clinical tissue material is a heterogeneous collection of individually processed tissue specimens. Factors that may influence the final ISH result in clinical FFPE samples include (a) surgery method and time, (b) time and storage temperature after resection until processing into formalin, (c) formalin fixation time and temperature, and (d) storage of paraffin samples. Taking these steps and factors into consideration, it should not be surprising that samples may perform differently in ISH studies. Therefore, it is recommended always to test several samples when optimizing conditions for miRNA ISH. It may be advantageous to screen a series of blocks for best retained RNA. In order to do this, different RNA targets can be considered: U6 snRNA, T oligo, β-actin, miR-223, and miR-126.

   U6 small nuclear RNA (snRNA) is a noncoding RNA transcript used in pre-mRNA splicing and as such U6 is expressed in all cells. ISH for U6 reveals an intense ISH signal in nuclei (Fig. 3a), and diluting the probe concentration to the level of sensitivity may facilitate qualification of individual FFPE samples (2). An oligo consisting of T's only, detects the poly-A tail of all RNA transcripts. Similarly, as for the U6

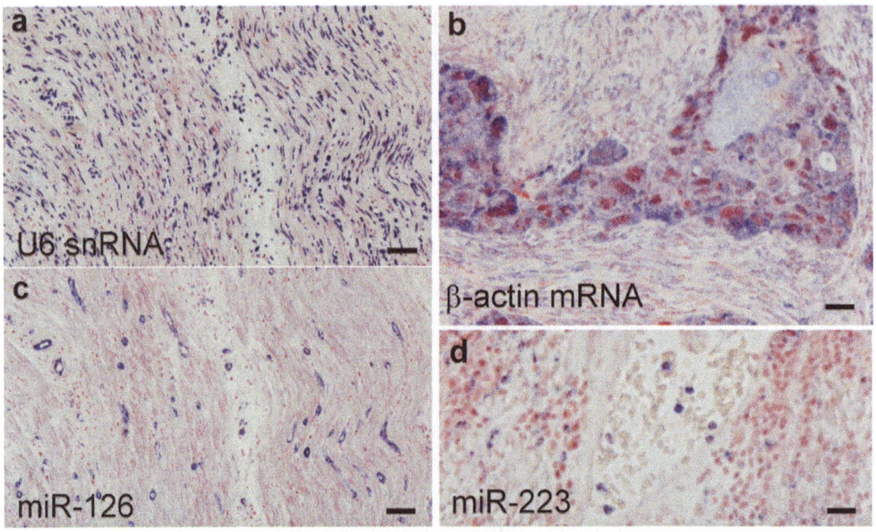

Fig. 3. *In situ* hybridization probes for qualification of FFPE tissue samples. (**a**) Nuclear localization of U6 snRNA in virtually all cells in the area of muscularis externa from a human colon. (**b**) *In situ* hybridization signal for human β-actin mRNA in a human esophagus cancer using an LNA probe (designed by Niels Tolstrup, Exiqon). (**c**) miR-126 expression in vessels located in the area of muscularis externa from a human colon [adjacent to section in (**a**)]. (**d**) miR-223 expression in granulocytes located in a vessel and in the surrounding inflamed tissue. The FFPE samples were purchased from Proteogenex. Bars: (**a**) and (**c**) 60 μm; (**b**) 40 μm, and (**d**) 20 μm.

probe, dilution of such probe to the level of sensitivity may allow qualification of individual FFPE samples. More sensitive may be a specific mRNA probe to a household protein, such as β-actin. β-actin is a non-muscle actin that takes part in the formation of filaments comprising a major component of the cytoskeleton, and due to its general and widespread expression β-actin expression is often used for normalization in Northern and Western blotting. The number of specific binding sites to β-actin mRNA would be far below that of the U6 and T-oligo probes. Applying a 22 bp LNA™ probe specific to β-actin in a cancer sample reveals a dynamic expression pattern present in both cancer cells and cells in the stromal compartment (Fig. 3b).

One miRNA, which is interesting in terms of tissue qualification, is miR-223. miR-223 is expressed in granulocytes (43, 44) in all humans and the sequence is conserved in many species. Circulating granulocytes are loaded with miR-223 and are easily detected using the 1-day ISH protocol presented here; see Fig. 3d. Granulocytes are recruited to inflamed tissues and therefore abundant in, e.g., cancers. Assuming that the content of miR-223 is similar in different individuals, the presence of miR-223 in paraffin samples may be a parameter to qualify (or disqualify) FFPE samples for further studies. miR-126 is another attractive miRNA candidate (Fig. 3c), which, in contrast to miR-223, is detectable in most normal tissues; however, since miR-126 expression in endothelium is sensitive to environmental factors (45), it is more likely to be differentially expressed among individuals. Thus, having access to a larger series of heterogeneous group of FFPE samples, it may be an advantage to select samples based on their overall performance with one or more of these well-known RNA transcripts.

5. *Hybridization oven*: In case you need to use a hybridization oven during the hybridization step, then seal cover slides with Fixogum as specified in step 5. The slides can be placed as such in the hybridization oven without humidifying conditions. In order to establish a more stable hybridization temperature, it is suggested to place a metallic plate, e.g., the inserts from a multiblock heater, in the oven. Place the slides on the plate and hybridize for 1–2 h.
6. *ISH platforms*: The actual hybridization step in the ISH protocol does not require any specific hybridization equipment. All from simple hybridization ovens to advanced automated hybridizers can be used, most critical being the precision of the hybridization temperature as mentioned earlier. A simple platform is based on a regular programmable heating plate system, such as the hybridizers from Dako, Invitrogen, and Vysis, which allow high precision temperature settings. For such a manual setup, most other steps can be performed using ordinary water bath-based heat-controlled stringent washes and immunohistochemistry incubation racks.

   Among automated systems are Tecan's Genepaint system, Ventana's Discovery system, and Intavis' InsituPro system. Tecan's Genepaint system is a high-throughput ISH platform based on vertical flow-through chambers. It has been used to systematically study mRNA and miRNA expression in mouse tissues (46, 47), and miRNA in a large prognostic study (14). Ventana's Discovery platform uses horizontal incubations and has been used in multiple applications, including immunohistochemistry, ISH, and FISH (48–50). Intavis' application has major advantages in whole-mount ISH (9, 51), but also provides applications for tissue sections.
7. *Enzymatic predigestion*: The degree of proteolytic digestion needed is partly related to the degree of fixation. In general terms, weak fixation needs low proteolytic treatment and strong fixation needs high proteolytic treatment. However, proteolytic digestion leads to partial loss of miRNAs at least in frozen sections (7) and excess digestion leads to tissue damage. In addition, tissue specimens may not be sufficiently fixed or become overfixed indicating some degree of fixation restrictions for ISH analyses. Routinely fixed and paraffin-embedded tissue

samples may not all be optimal for ISH (see Note 3). For most human archival samples, the optimal proteinase-K concentration is in the range of 10–20 μg/ml when incubated for 10 min at 37°C.

8. *Probe concentration*: The probe concentration should be kept in the range of 20–80 nM depending on the sensitivity of the assay. Too low concentration prevents detection, and too high concentration may allow cross hybridization and detection of highly similar and abundant sequences.

9. *Hybridization temperature*: The optimal T_{hyb} is linked to the T_m of the probe. The hybridization temperature, T_{hyb}, should be kept in the range of 50–60°C using the Exiqon ISH buffer or as a rule of thumb approximately 30°C below RNA-T_m. If the T_{hyb} is set too low, it will facilitate binding to highly similar sequences present in other miRNAs, cross hybridization to family members or mRNAs. At too high T_{hyb}, the amount of specifically bound probe is reduced. Therefore, the optimal range of T_{hyb} for a given LNA™ probe is always essential to establish.

10. *Specificity analyses in ISH*: An inevitable and challenging aspect of miRNA ISH is a firm and conclusive specificity analysis. In ISH reports, we would like to address questions like: What is the evidence that the probe recognizes the target sequence? Or, can it be proved that the ISH signal observed really represents the miRNA probed for? In mRNA ISH, the use of RNA probes obtained from *in vitro* transcription specific for different parts of the mRNA has been used to verify the origin of the ISH signal. In parallel, sense probes are applied to neighboring sections and used as negative controls. Due to the restricted size of the mature miRNA sequence, microRNA ISH requires other approaches to verify the origin of the ISH signal. Positive-control probes that can be used include alternative probe designs with varying LNA patterns and lengths (8, 14). Some miRNAs are expressed in clusters of two or more hairpins, such as the miR-143/145 cluster and the miR-17–92 cluster. Probes specific for the individual miRNAs in the same transcript would work as positive controls linked to the fact that they recognize different sequences on the same transcript. For miR-143 and miR-145, this is indeed observed. Both miRNAs are expressed in smooth muscle cells, including muscularis mucosa (Fig. 4) and vascular smooth muscle cells, which are a cellular component of arteries. The use of a scrambled probe (labeled identical to the miRNA probe) is an essential negative control in any ISH study. Together with a no-probe control (incubating hybridization buffer only), the ISH assay setup identifies any problematic unspecific probe binding and endogenous activity, particularly of alkaline phosphatase (or peroxidase).

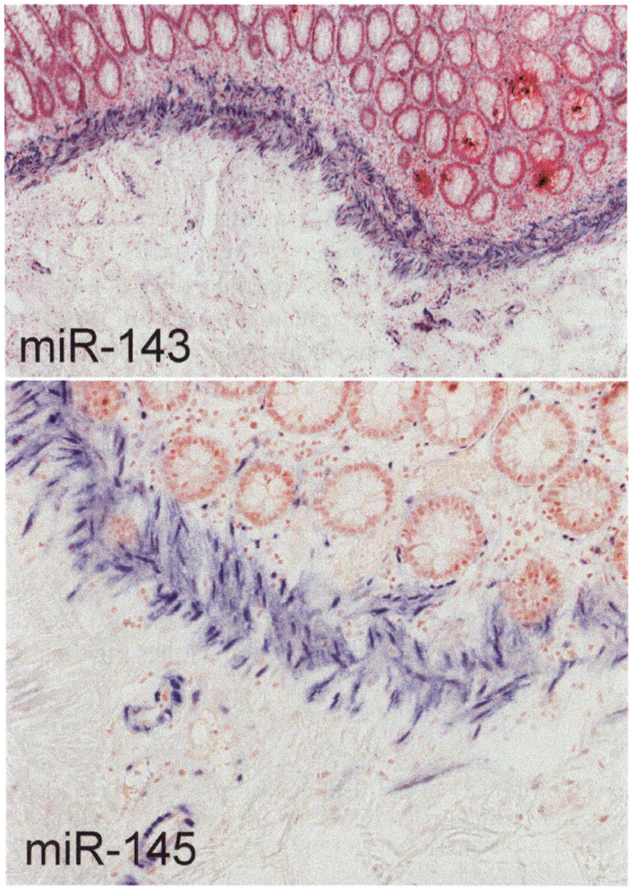

Fig. 4. *In situ* hybridization for the two miRNAs in the miR-143/145 cluster. *In situ* hybridization using probes for miR-143 (**a**) and miR-145 (**b**) in normal human colon is showing the expression in smooth muscle cells in the lamina muscularis and the vascular smooth muscle cells in arteries located in submucosa. An identical *in situ* hybridization staining pattern is obtained with the two probes recognizing the two miRNA neighbors. Expression of the miR-143/145 cluster in smooth muscle is well-described by others (54, 55). The FFPE samples were purchased from Proteogenex. Bars: (**a**) 80 μm; (**b**) 40 μm.

Unspecific staining may also be caused by the detection system employed and is tested by omitting the detecting antibody, where any stain obtained is related to endogenous activity. For example, encountering endogenous alkaline phosphatase activity, it may be useful to change to alternative detection approaches, such as peroxidase staining or fluorescence detection. Among other specificity control probes are mismatch probes (5, 8, 14, 30). Introducing three mismatches at non-LNA positions in the probe sequence was found to be sufficient to prevent binding of the miR-21 LNA™ probe (14). Also competition experiments are interesting in specificity analysis. Prehybridizing nonlabeled oligos having an identical LNA pattern as a labeled probe would prevent subsequent binding of a labeled probe. Finally, mix-incubating the nonlabeled oligo with the labeled probe would prevent binding of the labeled probe dependent on the concentration balance between the oligo and the probe (14, 52). It is also noteworthy that the availability of highly sensitive miRNA qPCR assays (53) enables miRNA detection in laser capture microdissected tissue areas – even obtained from paraffin sections. In our study of miR-21 in colorectal cancers, where miR-21 is predominantly expressed in the stromal compartment, laser capture microdissection data were in agreement with our observations in the ISH analyses; thus, we measured a much higher miR-21 expression level in microdissected stromal tissue than in microdissected cancer cells (14).

Acknowledgments

Special thanks to ISH team at Exiqon: Stine Jørgensen, Mette Carlsen Mohr, Tina Bisgaard Sørensen, Marie-Louise Lunn, Adam Baker, and Søren Møller for their dedication and support.

References

1. Nuovo, G.J. (2008) In situ detection of precursor and mature microRNAs in paraffin embedded, formalin fixed tissues and cell preparations. *Methods* **44**, 39–46.
2. Jorgensen, S., Baker A., Moller S., and Nielsen B.S. (2010) Robust one-day in situ hybridization protocol for detection of microRNAs in paraffin samples using LNA probes. *Methods* **52**, 375–81.
3. Sempere, L.F., Preis M., Yezefski T., Ouyang H., Suriawinata A.A., Silahtaroglu A., et al. (2010) Fluorescence-based codetection with protein markers reveals distinct cellular compartments for altered MicroRNA expression in solid tumors. *Clin Cancer Res* **16**, 4246–55.
4. Obernosterer, G., Martinez J., and Alenius M. (2007) Locked nucleic acid-based in situ detection of microRNAs in mouse tissue sections. *Nat Protoc* **2**, 1508–14.
5. Silahtaroglu, A.N., Nolting D., Dyrskjot L., Berezikov E., Moller M., Tommerup N., et al. (2007) Detection of microRNAs in frozen tissue sections by fluorescence in situ hybridization using locked nucleic acid probes and tyramide signal amplification. *Nat Protoc* **2**, 2520–8.
6. Thompson, R.C., Deo M., and Turner D.L. (2007) Analysis of microRNA expression by in situ hybridization with RNA oligonucleotide probes. *Methods* **43**, 153–61.
7. Pena, J.T., Sohn-Lee C., Rouhanifard S.H., Ludwig J., Hafner M., Mihailovic A., et al. (2009) miRNA in situ hybridization in formaldehyde and EDC-fixed tissues. *Nat Methods* **6**, 139–41.
8. Kloosterman, W.P., Wienholds E., de Bruijn, E., Kauppinen S., and Plasterk R.H. (2006) In situ detection of miRNAs in animal embryos using LNA-modified oligonucleotide probes. *Nat Methods* **3**, 27–9.
9. Pearson, B.J., Eisenhoffer G.T., Gurley K.A., Rink J.C., Miller D.E., and Sanchez A.A. (2009) Formaldehyde-based whole-mount in situ hybridization method for planarians. *Dev Dyn* **238**, 443–50.
10. Lu, J., Tsourkas A. (2009) Imaging individual microRNAs in single mammalian cells in situ. *Nucleic Acids Res* **37**, e100.
11. Nuovo, G., Lee E.J., Lawler S., Godlewski J., and Schmittgen T. (2009) In situ detection of mature microRNAs by labeled extension on ultramer templates. *Biotechniques* **46**, 115–26.
12. You, Y., Moreira B.G., Behlke M.A., and Owczarzy R. (2006) Design of LNA probes that improve mismatch discrimination. *Nucleic Acids Res* **34**, e60.
13. Christodoulou, F., Raible F., Tomer R., Simakov O., Trachana K., Klaus S., et al. (2010)

Ancient animal microRNAs and the evolution of tissue identity. *Nature* **463**, 1084–8.

14. Nielsen, B.S., Jorgensen S., Fog J.U., Sokilde R., Christensen I.J., Hansen U., et al. (2011) High levels of microRNA-21 in the stroma of colorectal cancers predict short disease-free survival in stage II colon cancer patients. *Clin Exp Metastasis* **28**, 27–38.
15. de Planell-Saguer, M., Rodicio M.C., and Mourelatos Z. (2010) Rapid in situ codetection of noncoding RNAs and proteins in cells and formalin-fixed paraffin-embedded tissue sections without protease treatment. *Nat Protoc* **5**, 1061–73.
16. Maiorano, N.A., Mallamaci A. (2009) Promotion of embryonic cortico-cerebral neuronogenesis by miR-124. *Neural Dev* **4**, 40.
17. Larsson, E., Fredlund F.P., Heldin J., Barkefors I., Bondjers C., Genove G., et al. (2009) Discovery of microvascular miRNAs using public gene expression data: miR-145 is expressed in pericytes and is a regulator of Fli1. *Genome Med* **1**, 108.
18. Nelson, P.T., Dimayuga J., and Wilfred B.R. (2010) MicroRNA in Situ Hybridization in the Human Entorhinal and Transentorhinal Cortex. *Front Hum Neurosci* **4**, 7.
19. Nielsen, B.S., Rank F., Illemann M., Lund L.R., and Dano K. (2007) Stromal cells associated with early invasive foci in human mammary ductal carcinoma in situ coexpress urokinase and urokinase receptor. *Int J Cancer* **120**, 2086–95.
20. Yamamichi, N., Shimomura R., Inada K., Sakurai K., Haraguchi T., Ozaki Y., et al. (2009) Locked nucleic acid in situ hybridization analysis of miR-21 expression during colorectal cancer development. *Clin Cancer Res* **15**, 4009–16.
21. Wei, T., Orfanidis K., Xu N., Janson P., Stahle M., Pivarcsi A., et al. (2010) The expression of microRNA-203 during human skin morphogenesis. *Exp Dermatol* **19**, 854–6.
22. Hoesel, B., Bhujabal Z., Przemeck G.K., Kurz-Drexler A., Weisenhorn D.M., Angelis M.H., et al. (2010) Combination of in silico and insitu hybridisation approaches to identify potential Dll1 associated miRNAs during mouse embryogenesis. *Gene Expr Patterns* **10**, 265–73.
23. Nuovo, G.J. (2010) In situ detection of microRNAs in paraffin embedded, formalin fixed tissues and the co-localization of their putative targets. *Methods* **52**, 307–15.
24. Mansfield, J.R. (2010) Cellular context in epigenetics: quantitative multicolor imaging and automated per-cell analysis of miRNAs and their putative targets. *Methods* **52**, 271–80.
25. Dillhoff, M., Liu J., Frankel W., Croce C., and Bloomston M. (2008) MicroRNA-21 is overexpressed in pancreatic cancer and a potential predictor of survival. *J Gastrointest Surg* **12**, 2171–6.
26. Nielsen, C.B., Singh S.K., Wengel J., and Jacobsen J.P. (1999) The solution structure of a locked nucleic acid (LNA) hybridized to DNA. *J Biomol Struct Dyn* **17**, 175–91.
27. Petersen, M., Bondensgaard K., Wengel J., and Jacobsen J.P. (2002) Locked nucleic acid (LNA) recognition of RNA: NMR solution structures of LNA:RNA hybrids. *J Am Chem Soc* **124**, 5974–82.
28. Neely, L.A., Patel S., Garver J., Gallo M., Hackett M., McLaughlin S., et al. (2006) A single-molecule method for the quantitation of microRNA gene expression. *Nat Methods* **3**, 41–6.
29. Denys, B., El H.H., Nollet F., Verhasselt B., and Philippe J. (2010) A Real-Time Polymerase Chain Reaction Assay for Rapid, Sensitive, and Specific Quantification of the JAK2V617F Mutation Using a Locked Nucleic Acid-Modified Oligonucleotide. *J Mol Diagn* **12**, 512–9.
30. Deo, M., Yu J.Y., Chung K.H., Tippens M., and Turner D.L. (2006) Detection of mammalian microRNA expression by in situ hybridization with RNA oligonucleotides. *Dev Dyn* **235**, 2538–48.
31. Yamasaki, K., Nakasa T., Miyaki S., Ishikawa M., Deie M., Adachi N., et al. (2009) Expression of MicroRNA-146a in osteoarthritis cartilage. *Arthritis Rheum* **60**, 1035–41.
32. Niimoto, T., Nakasa T., Ishikawa M., Okuhara A., Izumi B., Deie M., et al. (2010) MicroRNA-146a expresses in interleukin-17 producing T cells in rheumatoid arthritis patients. *BMC Musculoskelet Disord* **11**, 209.
33. Kidner, C., Timmermans M. (2006) In situ hybridization as a tool to study the role of microRNAs in plant development. *Methods Mol Biol* **342**, 159–79.
34. Darnell, D.K., Stanislaw S., Kaur S., and Antin P.B. (2010) Whole mount in situ hybridization detection of mRNAs using short LNA containing DNA oligonucleotide probes. *RNA* **16**, 632–7.
35. Ason, B., Darnell D.K., Wittbrodt B., Berezikov E., Kloosterman W.P., Wittbrodt J., et al. (2006) Differences in vertebrate microRNA expression. *Proc Natl Acad Sci USA* **103**, 14385–9.
36. Darnell, D.K., Kaur S., Stanislaw S., Konieczka J.H., Yatskievych T.A., and Antin P.B. (2006) MicroRNA expression during chick embryo development. *Dev Dyn* **235**, 3156–65.

37. Karali, M., Peluso I., Marigo V., and Banfi S. (2007) Identification and characterization of microRNAs expressed in the mouse eye. *Invest Ophthalmol Vis Sci* **48**, 509–15.
38. Sweetman, D., Goljanek K., Rathjen T., Oustanina S., Braun T., Dalmay T., and Munsterberg A. (2008) Specific requirements of MRFs for the expression of muscle specific microRNAs, miR-1, miR-206 and miR-133. *Dev Biol* **321**, 491–9.
39. Sweetman, D., Rathjen T., Jefferson M., Wheeler G., Smith T.G., Wheeler G.N., et al. (2006) FGF-4 signaling is involved in mir-206 expression in developing somites of chicken embryos. *Dev Dyn* **235**, 2185–91.
40. Wheeler, G., Valoczi A., Havelda Z., and Dalmay T. (2007) In situ detection of animal and plant microRNAs. *DNA Cell Biol* **26**, 251–5.
41. Havelda, Z. (2010) In situ detection of miRNAs using LNA probes. *Methods Mol Biol* **592**, 127–36.
42. Yuasa, K., Hagiwara Y., Ando M., Nakamura A., Takeda S., and Hijikata T. (2008) MicroRNA-206 is highly expressed in newly formed muscle fibers: implications regarding potential for muscle regeneration and maturation in muscular dystrophy. *Cell Struct Funct* **33**, 163–9.
43. Fazi, F., Rosa A., Fatica A., Gelmetti V., De Marchis M.L., Nervi C., et al. (2005) A minicircuitry comprised of microRNA-223 and transcription factors NFI-A and C/EBPalpha regulates human granulopoiesis. *Cell* **123**, 819–31.
44. Johnnidis, J.B., Harris M.H., Wheeler R.T., Stehling-Sun S., Lam M.H., Kirak O., et al. (2008) Regulation of progenitor cell proliferation and granulocyte function by microRNA-223. *Nature* **451**, 1125–9.
45. Wang, S., Aurora A.B., Johnson B.A., Qi X., McAnally J., Hill J.A., et al. (2008) The endothelial-specific microRNA miR-126 governs vascular integrity and angiogenesis. *Dev Cell* **15**, 261–71.
46. Yaylaoglu, M.B., Titmus A., Visel A., varez-Bolado G., Thaller C., and Eichele G. (2005) Comprehensive expression atlas of fibroblast growth factors and their receptors generated by a novel robotic in situ hybridization platform. *Dev Dyn* **234**, 371–86.
47. Diez-Roux, G., Banfi S., Sultan M., Geffers L., Anand S., Rozado D., et al. (2011) A high-resolution anatomical atlas of the transcriptome in the mouse embryo. *PLoS Biol* **9**, e1000582.
48. Byers, R.J., Di V.D., O'connell F., Tholouli E., Levenson R.M., Gossage K., et al. (2007) Semiautomated multiplexed quantum dot-based in situ hybridization and spectral deconvolution. *J Mol Diagn* **9**, 20–9.
49. Nitta, H., Hauss-Wegrzyniak B., Lehrkamp M., Murillo A.E., Gaire F., Farrell M., et al. (2008) Development of automated brightfield double in situ hybridization (BDISH) application for HER2 gene and chromosome 17 centromere (CEN 17) for breast carcinomas and an assay performance comparison to manual dual color HER2 fluorescence in situ hybridization (FISH). *Diagn Pathol* **3**, 41.
50. Nitta, H., Zhang W., Kelly B.D., Miller M., Pestic-Dragovich L., Bieniarz C., et al. (2010) Automated brightfield break-apart in situ hybridization (ba-ISH) application: ALK and MALT1 genes as models. *Methods* **52**, 352–8.
51. Hultman, K.A., Bahary N., Zon L.I., and Johnson S.L. (2007) Gene Duplication of the zebrafish kit ligand and partitioning of melanocyte development functions to kit ligand a. *PLoS Genet* **3**, e17.
52. Song, R., Ro S., and Yan W. (2010) In situ hybridization detection of microRNAs. *Methods Mol Biol* **629**, 287–94.
53. Andreasen, D., Fog J.U., Biggs W., Salomon J., Dahslveen I.K., Baker A., and Mouritzen P. (2010) Improved microRNA quantification in total RNA from clinical samples. *Methods* **50**, S6–9.
54. Boettger, T., Beetz N., Kostin S., Schneider J., Kruger M., Hein L., et al. (2009) Acquisition of the contractile phenotype by murine arterial smooth muscle cells depends on the Mir143/145 gene cluster. *J Clin Invest* **119**, 2634–47.
55. Cordes, K.R., Sheehy N.T., White M.P., Berry E.C., Morton S.U., Muth A.N., et al. (2009) miR-145 and miR-143 regulate smooth muscle cell fate and plasticity. *Nature* **460**, 705–10.

Chapter 6

Agilent MicroRNA Microarray Profiling System

Petula N. D'Andrade and Stephanie Fulmer-Smentek

Abstract

MicroRNA (miRNA) profiling is of great interest because of the significant roles these short noncoding RNA molecules play in cellular regulation. Signature profiles, usually involving several miRNAs, have also been associated with dysfunctional cellular regulation such as in cancer. Profiling miRNAs can be done using the Agilent Technologies miRNA profiling system, which is a sensitive and accurate miRNA microarray assay. The assay is based on a highly efficient labeling method linked to a novel probe design strategy. The labeling method uses a simple, single-vial approach where 100 ng of nonfractionated total RNA is directly labeled by ligation of a Cy3 labeled pCp molecule to the 3′ end of the RNA. The labeled cytosine interacts with the guanidine at the 5′ end of the probe which adds stability to the hybridization complex. In addition, the probes have been designed to provide both sequence and size discrimination, generally resulting in highly specific detection of closely related mature miRNAs. The labeling and probe design strategies allow for a precise and accurate measurement that spans a linear dynamic range of greater than four orders of magnitude from at least 0.2 amol to 2 fmol of miRNA and a detection limit of less than 0.1 amol. The assay works over a wide range of sample types including FFPE samples. Agilent's microarray technology is a flexible design platform allowing quick array design iterations and incorporation of the latest miRBase content.

Key words: Agilent, microRNA, miRNA, Microarray, Mature microRNA, miRNA profiling, Probe design and RNA labeling

1. Introduction

MicroRNAs regulate messenger RNAs (mRNAs) in a highly complex manner where an individual microRNA can regulate many mRNAs and a single mRNA can be regulated by many microRNAs. This regulation is achieved by either degrading the mRNAs or by repressing translation. The fate of the mRNA is determined by the homology of the microRNA to the 3′ untranslated region of the targeted mRNA. When there is extensive pairing of the microRNA to its targeted mRNA, a cut of a single

Jian-Bing Fan (ed.), *Next-Generation MicroRNA Expression Profiling Technology: Methods and Protocols*, Methods in Molecular Biology, vol. 822, DOI 10.1007/978-1-61779-427-8_6, © Springer Science+Business Media, LLC 2012

phosphodiester bond is made, triggering the destruction of the mRNA. Partial pairing between the targeted mRNA and the microRNA does not allow the phosphodiester bond to be cut but instead blocks translation of the protein (1). It is estimated that approximately 60% of human genes are putative targets of one or more microRNA (2).

The short, 22 nt length of mature microRNAs along with a high degree of homology among different microRNAs make this class of RNAs challenging to measure in a multiplex assay. The presence of precursor microRNAs, which are longer molecules from which the mature microRNA are processed, also poses a challenge for measurement of the mature microRNA. However, with a highly efficient direct labeling method and a novel probe design strategy, the Agilent microRNA microarray assay can measure hundreds of mature microRNAs simultaneously (3). The labeling reaction requires a small quantity of total RNA and minimizes sample biases, in part by minimizing sample manipulation steps, such as fractionation and amplification. Labeling via ligation of a Cy3-labeled pCp molecule to the 3′ end of the RNA molecule after heating in the presence of DMSO results in negligible sequence bias. The probe design allows for both sequence and size selection of mature microRNAs as much of the sequence variation among related microRNAs occurs near the 3′ end. The probes are Tm balanced by shortening the probe at the 5′ end of the mature microRNA. The probes also include a hairpin sequence at the 5′ end of the probe, which acts as a size discriminator (Fig. 1). Overall, the assay is simple to perform and results in highly sensitive accurate measurements which span greater than 10^4 linear dynamic range.

Agilent's microarray technology is flexible enough to allow rapid array iterations to accommodate the increasing number of microRNAs being discovered. The assay also utilizes labeling and hybridization spike-ins as process controls. The use of the spike-ins assists in determining the quality of the data and is especially beneficial for samples with few expressed microRNAs. Comparison to QRT-PCR has demonstrated that the two methodologies are highly concordant (4, 5).

2. Materials

2.1. Spike-In Solution

1. MicroRNA Spike-In Kit (Agilent); store at −80°C for up to 12 months.

   Kit components:

   (a) Labeling Spike-In.

   (b) Hyb Spike-In.

   (c) Dilution Buffer.

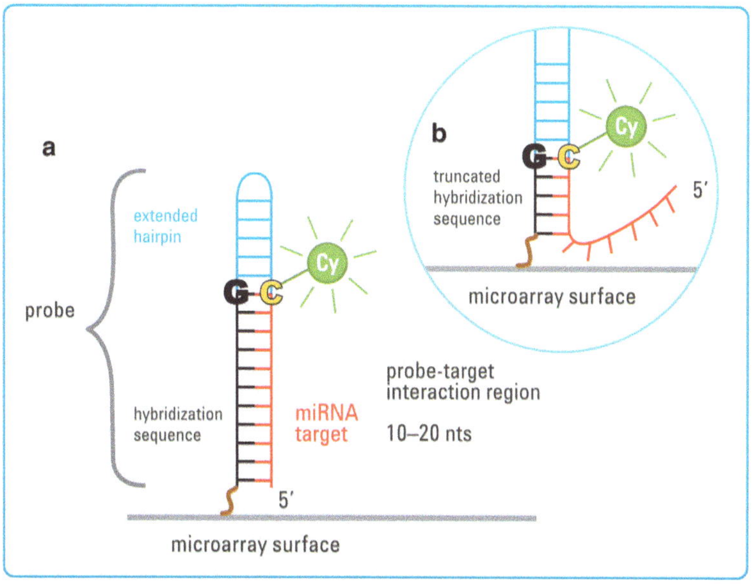

Fig. 1. Agilent probe design strategy. An unmodified microarray probe (*black*) is a synthesized sequence that hybridizes to the target microRNA (*red*). Probes are anchored to the slide surface by a stilt (*brown*). (**a**) Inclusion of a G residue (*black*) to the 5′ end of the hybridization sequence complements the 3′ end C residue (*yellow*) introduced in labeling. This additional G-C pair in the probe–target interaction region stabilizes targeted miRNAs relative to homologous RNAs. In addition, all probes contain a 5′ hairpin (*blue*), abutting the probe–target region, to stabilize binding and increase size specificity. (**b**) T_m reduction of a probe to T_m match all probes on the microarray. For probes requiring T_m reduction, truncation is achieved through sequential elimination of bases to the appropriate length from the 3′ end of the hybridization sequence such that the probe–target interaction region is 9–20 nts long.

2. This kit should be supplemented with DNase- and RNase-free water that has not been treated with DEPC. Prepare the working solution by performing a dilution series just prior to each use (see Note 1).

2.2. Sample Labeling and Hybridization

1. miRNA Complete Labeling and Hyb Kit (Agilent).
   (a) Labeling Kit components should be stored at −20°C until the expiration date.
   - Nuclease-free water.
   - Calf Intestinal Phosphatase.
   - 10× CIP Buffer.
   - DMSO (Dimethly sulfoxide); harmful by inhalation, ingestion, and skin absorption.
   - T4 RNA Ligase.
   - 10× T4 RNA Ligase Buffer.
   - Cyanine-3 pCp Dye; light sensitive: store in the dark.

(b) Hyb Kit components should be stored at room temperature until the expiration date.
   - 2× Hi-RPM Buffer; considered hazardous by OSHA Harzard Communication Standard.
   - 10× Blocking Agent; once re-suspended should be stored at −20°C for up to 2 months.
2. MicroBioSpin 6 Columns (Bio-Rad); store at 4°C. This component is optional.
3. Hybridization Chamber, stainless (Agilent); store at room temperature.
4. Hybridization Chamber gasket slides 8 microarrays/slide, 5 slides/box (Agilent); store in original packaging at room temperature for up to 3 months.
5. Hybridization Oven: 20 rpm and 55°C capability (Agilent).
6. Hybridization Oven Rotator for Agilent Microarray Hybridization Chambers (Agilent).

2.3. Microarray Wash

1. Gene Expression Wash Buffer Kit (Agilent); store at room temperature for up to 12 months.
   (a) Gene Expression Wash Buffer 1.
   (b) Gene Expression Wash Buffer 2.
2. Slide-staining dish, with slide rack (×3) (Thermo Shandon).
   (a) First slide-staining dish is the "Disassembly dish."
   (b) Second slide-staining dish is the "Wash 1 dish."
   (c) Third slide-staining dish is the "Wash 2 dish."
3. 1.5 L Glass dish in which the "Wash 2 dish" should fit.
4. Nitrogen purge or desiccator box (see Note 2).
5. Dye filter (Sigma). This component is optional.

2.4. Scanning and Feature Extraction

1. Ozone-Barrier Slide Cover Kit (Agilent).
2. Microarray Scanner (Agilent) with Scan Control software, version A. 8.0 or later.
3. Feature Extraction software 10.7.3 or later (Agilent).

2.5. Characterization of Total RNA (Optional)

1. 2100 Bioanalyzer (Agilent).
   (a) RNA 6000 Nano Kit (Agilent) or
   (b) RNA 6000 Pico Kit (Agilent).
   (c) Small RNA Kit (Agilent).

3. Method

3.1. Spike-In Solution Preparation

3.1.1. First Dilution of Serial Dilution

1. Just prior to use, thaw the provided spike-in stock on ice.
2. Prepare tubes for the first dilution by adding 198 μL of Dilution Buffer (provided in the kit) to each.
3. Add 2 μL of the spike-in stock to the prepared tube and mix well (see Note 1).

3.1.2. Second and Third Dilutions of Serial Dilution

1. Prepare tubes for the second and third dilution by adding 198 μL of nuclease-free water (not provided in the kit) to each.
2. For the second dilution, add 2 μL of the first dilution to the prepared tube and mix well.
3. For the third dilution, add 2 μL of the second dilution to the prepared tube and mix well. This dilution is to be used in the following step and should be discarded after use (see Note 1).

3.2. Sample Labeling

The sample labeling is dependent on the efficiency of the ligation reaction; therefore, it is essential that the ligation reaction be performed as described (see Note 3). Sample labeling is a simple, one tube process with an optional purification step.

3.2.1. Dephosphorylation

1. The quality and accurate quantification of total RNA is important for generation of good results (see Note 4). The total RNA sample should be diluted to 50 ng/μL in 1× TE pH 7.5 or DNase/RNase-free water. Add 2 μL of the diluted total RNA to a microcentrifuge tube and maintain on ice until the Calf Intestinal Alkaline Phosphatase Master Mix is added.
2. Prepare the Calf Intestinal Alkaline Phosphatase (CIP) Master Mix immediately prior to use, in the order indicated in Table 1 and maintain on ice.
3. Add 2 μL of the CIP Master Mix to each tube containing the total RNA for a total reaction volume of 4 μL and mix gently by pipetting.

Table 1
CIP master mix

Components	Volume (μL) per reaction	Volume (μL) per 9 reactions
10× Calf Intestinal Phosphatase Buffer	0.4	3.6
Labeling Spike-In (third dilution)	1.1	9.9
Calf Intestinal Phosphatase	0.5	4.5
Total volume	2.0	18.0

4. Dephosphorylate the sample by incubating the reaction at 37°C in a circulating water bath or heat block for 30 min.

3.2.2. Denaturation

1. Add 2.8 μL of 100% DMSO to each sample and incubate at 100°C in a circulating water bath or heat block for 5–10 min. The length of this incubation period is critical for the labeling efficiency.
2. Immediately transfer to an ice-water bath (see Note 5).

3.2.3. Ligation

1. Warm the 10× T4 RNA Ligase Buffer at 37°C and vortex until all precipitate has dissolved. Ensure that the 10× T4 RNA Ligase Buffer has cooled to room temperature before adding it to the Ligation Master Mix; failure to do so will impact the T4 RNA Ligase activity and thus the labeling efficiency.
2. Immediately prior to use, prepare the Ligation Master Mix by gently mixing the components listed in Table 2 and maintain on ice. Within 15 min of mixing all the components together, add 4.5 μL to each sample tube, gently mix by pipetting and briefly centrifuge. The total reaction volume should now be 11.3 μL.
3. Incubate at 16°C in a light eliminating circulating water bath or cool block for 2 h.

3.2.4. Purify the Labeled RNA (Optional, see Note 6)

Micro Bio-Spin 6 Column Preparation

1. Invert the column sharply several times to resuspend the settled gel and to remove any air bubbles. Once the gel has been re-suspended, snap off the tip at the base of the column and place the column into a 2 mL microcentrifuge tube supplied with the columns from Bio-Rad.
2. Remove the green cap from the column which should initiate the gravitational dripping of the buffer. If the buffer does not start to drip into the 2 mL microcentrifuge tube, press the green cap back onto the column and remove it again. Allow the buffer to drain for approximately 2 min. Discard the drained buffer from the 2 mL microcentrifuge tube and then place the column back into the tube.

Table 2
Ligation master mix for T4 RNA Ligase

Components	Volume (μL) per reaction	Volume (μL) per 9 reactions
10× T4 RNA Ligase Buffer	1.0	9.0
Cyanine3-pCp	3.0	27.0
T4 RNA Ligase	0.5	4.5
Total volume	4.5	40.5

3. Spin the microcentrifuge tube containing the column for 2 min at 1,000 × *g* in a centrifuge (see Note 6). Transfer the column from the 2 mL microcentrifuge tube to a clean 1.5 mL microcentrifuge tube and discard the 2 mL microcentrifuge tube.

Sample Purification (see Note 6)

1. Add 38.7 μL of RNase-Free Water or 1× TE, pH 7.5, to the labeled sample from the Ligation step (Subheading 3.2 step 3) for a total volume of 50 μL. Without disturbing the gel bed, pipette the 50 μL of sample onto the gel bed of the Micro Bio-Spin 6 prepared column.
2. Elute the purified sample by spinning the microcentrifuge tubes containing the columns for 4 min at 1,000 × *g* in a centrifuge (see Note 7). Discard the columns and keep the miRNA sample-containing flow-through on ice.
3. Check that the final flow-through is slightly pink, does not contain any gel material, and has a volume of approximately 50 μL.

3.2.5. Dry the Sample

1. After the 16°C labeling reaction or optional sample purification, completely dry the samples using a vacuum concentrator at 45–55°C or on the medium heat setting. Depending on the vacuum concentrator model, this step may take up to 1 h after column purification or up to 3 h without column purification.
2. Ensure that the samples are completely dry by flicking each tube to check that the pellets do not move or spread (see Note 8).

3.3. Sample Hybridization

3.3.1. Prepare the 10× Blocking Agent

1. Add 125 μL of nuclease-free water to the vial containing lyophilized 10× GE Blocking Agent supplied with the Agilent miRNA Complete Labeling and Hyb Kit. Gently vortex to resuspend and dissolve the pellet. If the pellet does not go into solution completely after vortexing, heat the solution for 4–5 min at 37°C. After heating, drive down any materials adhering to the tube walls or cap by briefly centrifuging and return the 10× Blocking Agent to room temperature. After use, store at −20°C.

3.3.2. Prepare Sample for Hybridization

1. Sample preparation (Table 3).
   (a) Resuspend the dried sample in 17 μL of nuclease-free water.
   (b) Add 1.0 μL of the Hyb Spike-In solution (third dilution) to each sample.
   (c) Add 4.5 μL of the reconstituted 10× GE Blocking Agent to each sample.
   (d) Add 22.5 μL of 2× Hi-RPM Hybridization Buffer to each sample for a total of 45 μL and carefully mix by gently vortexing.
2. Ensure that the SureHyb chamber, gasket slides, and microarray slides are easily available to prepare the hybridization assembly.

Table 3
Hybridization mix for miRNA microarrays

Components	Volume (μL) per reaction
Labeled miRNA sample	17
Hyb Spike-In	1
10× Blocking Agent	4.5
2× Hi-RPM Hybridization Buffer	22.5
Total volume	45

Incubate the samples at 100°C for 5 min. Immediately transfer to an ice-water bath for 5 min (see Note 5).

3. Quickly spin in a centrifuge to collect any condensation at the bottom of the tube and immediately proceed to "Prepare the hybridization assembly." Do not leave the samples in the ice-water bath for more than 15 min as longer incubations may adversely affect the hybridization results.

3.3.3. Prepare the Hybridization Assembly

1. Place a clean gasket slide with the label facing up into the Agilent SureHyb chamber base, and ensure that the gasket slide is flush with the chamber base and not ajar.
2. Slowly dispense the entire hybridization sample into the gasket well in a "drag and dispense" manner. The final sample volume will vary between 40 and 45 μL due to heating and cooling the sample (see Notes 9 and 10).
3. Apply the microarray by slowly placing the microarray slide "active side" down onto the SureHyb gasket slide, so that the "Agilent"-labeled barcode is facing down and the numeric barcode is facing up and ensure that the sandwich-pair is properly aligned (see Note 11).
4. Place the SureHyb chamber cover onto the sandwiched slides and position the clamp assembly onto both pieces. Hand-tighten the clamp and rotate the assembled chamber to ensure the bubbles move freely. If necessary, tap the assembly to free stationary bubbles. Then place the assembled slide chamber in the hybridization oven rotisserie at 55°C. Set the hybridization rotator to 20 rpm (see Note 12). Hybridize at 55°C for 20 h (see Note 13).

3.4. Microarray Wash

3.4.1. Add Triton X-102 to Gene Expression Wash Buffers

Perform this step to unopened containers of both Gene Expression Wash Buffer 1 and 2 before use.

1. Open the cardboard box containing the cubitainer of wash buffer and carefully remove the outer and inner caps from the cubitainer. Then add 2 mL of the provided 10% Triton X-102

into the wash buffer solution. Replace the original inner and outer caps and mix the buffer thoroughly by inverting the container 5–6 times. Carefully remove the caps and install the spigot provided. Prominently label the wash buffer box to indicate that Triton X-102 has been added and indicate the date of addition.

3.4.2. Pre-warm Gene Expression Wash Buffer 2

The Gene Expression Wash Buffer must be at 37°C for optimal performance.

Warm the *Gene Expression Wash Buffer 2* to 37°C as follows:

1. Dispense sufficient Gene Expression Wash Buffer 2 directly into sterile 1,000 mL bottles. Place the 1,000 mL bottles in a 37°C water bath at least 12 h prior to washing arrays. Alternatively, remove the plastic cubitainer from the box and place it in a 37°C water bath at least 12 h before washing the arrays.
2. Prepare the "Wash 2 dish" by placing it into a 37°C oven or incubator set to 37°C (see Note 14).
3. Also pre-warm water at 37°C for use in the 1.5 L glass dish which is used as secondary containment for the "Wash 2 dish."

3.4.3. Prepare the Equipment

1. It is critical that the dishes be clean when washing the microarray slides. To ensure that the wash dishes are clean, run copious amounts of Milli-Q water through the dish. Be sure to also wash the slide racks and the stir bars with copious amounts of Milli-Q water. Then empty the water and repeat several times to remove any traces of contaminating materials (see Note 15).

3.4.4. Wash the Microarray Slides

1. The dish in which the microarray-gasket slides will be disassembled ("Disassembly dish") should be completely filled with Gene Expression Wash Buffer 1 at room temperature. The slide rack should be placed into a second staining dish filled with Gene Expression Wash Buffer 1 ("Wash 1 dish") at room temperature. A magnetic stir bar should be added to the "Wash 1 dish" and the dish placed on a magnetic stir plate.
2. On a second magnetic stir plate with a heating element, place the pre-warmed 1.5 L glass dish. Place the "Wash 2 dish" into the 1.5 L glass dish. Fill the 1.5 L glass dish, but not "Wash 2 dish," with 37°C water. Fill the "Wash 2 dish" approximately three-fourths full with Gene Expression Wash Buffer 2 (warmed to 37°C). Turn on the magnetic stir plate heating element to maintain the temperature of Gene Expression Wash Buffer 2 in "Wash 2 dish" at 37°C.
3. Remove one hybridization chamber from the incubator and record the time. Record if bubbles formed during hybridization and if all bubbles are rotating freely. Make note of any significant loss of hybridization volume.

4. Disassemble the hybridization chamber assembly by turning the thumbscrew counterclockwise, sliding the clamp off, and removing the chamber cover. With gloved fingers, grip the array-gasket sandwich from the ends and remove it from the chamber base. Keep the microarray slide numeric barcode facing up as you quickly transfer the sandwich to the "Disassembly dish." Submerge the array-gasket sandwich into the "Disassembly dish" containing Gene Expression Wash Buffer 1 and separate the two slides using forceps.
   (a) Insert one of the blunt ends of the forceps between the slides and gently turn the forceps to separate the slides.
   (b) Allow the gasket slide to drop while removing the microarray slide. Then place the slide into the slide rack in "Wash 1 dish" containing Gene Expression Wash Buffer 1 at room temperature. Minimize exposure of the slide to air and touch only the barcode portion or edges of the slide.
5. Repeat steps 3 and 4 for up to five additional slides in the group. For uniform washing, process a maximum of six microarray slides in a group.
6. When all slides in the group are placed in the slide rack in the "Wash 1 dish," stir using setting 4 (moderate speed) for 5 min.
7. Transfer the slide rack to "Wash 2 dish" containing Gene Expression Wash Buffer 2 at 37°C. Stir using setting 4 (moderate speed) for 5 min. Maintain Wash 2 at 37°C for the duration of the wash step; failure to do so may result in a change in stringency of the wash, which can reduce the consistency of experimental results.
8. Slowly remove the slide rack, minimizing droplets on the slides. It should take 5–10 s to remove the slide rack (see Note 16).
9. Discard the used Gene Expression Wash Buffer 1 and Gene Expression Wash Buffer 2 (see Note 17).
10. Repeat steps 1 through 11 for the next group of six slides using fresh Gene Expression Wash Buffer 1 and Gene Expression Wash Buffer 2 pre-warmed to 37°C.
11. Scan slides immediately to minimize the impact of environmental oxidants on signal intensities. If necessary, store slides in storage slide box in a nitrogen purged box, in the dark.

3.5. Scanning and Feature Extraction

This section describes how to generate microarray images and how to convert the image into signal data by using the Agilent scanner and Feature Extraction software, respectively, from miRNA microarrays (see Note 18).

3.5.1. Scan the Slides

1. Assemble the slides into a slide holder. Place the slides into the slide holder such that the numeric barcode side is visible when the ozone barrier cover has been applied and the slide holder is

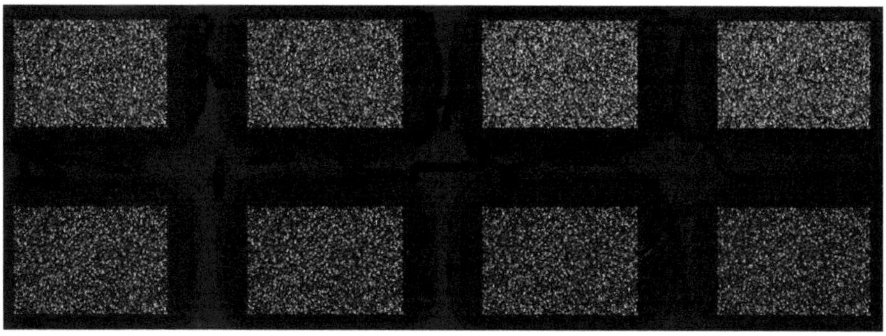

Fig. 2. Image of a microRNA microarray slide with eight miRNA microarrays hybridized with different samples.

closed (*not* the "Agilent"-labeled barcode side). Place assembled slide holders into a scanner carousel.

2. Select the appropriate profile from the drop down list, e.g., *AgilentHD_miRNA* in the Scan control software (see Note 19). Ensure that the setting for the automatic file naming is selected.
3. Select the appropriate folder for the scanned images to be stored.
4. Verify that the Scanner status in the main window says *Scanner Ready* before initialing the scanning of the slides.
5. Click *Scan Slot m-n* on the Scan Control main window where the letter *m* represents the Start slot where the first slide is located and the letter *n* represents the End slot where the last slide is located.
6. After scanning, the scanned image will be found in the folder selected (Fig. 2).

3.5.2. Extract Data Using Agilent Feature Extraction Software

Feature Extraction (FE) is the process by which pixel data is converted to signal data from microarray scan data, allowing researchers to measure miRNA expression in their experiments.

1. Open the Agilent Feature Extraction (FE) program, version 10.7.3 or later (see Note 19).
2. Add the images (.tif) to be extracted to the FE Project.
   (a) Click *Add New Extraction Set(s)* icon on the toolbar or right-click the *Project Explorer* and select *Add Extraction…*
      - Browse to the location of the tiff files, select the tiff file(s) and click *Open*. The FE program automatically assigns a default grid template and protocol for each extraction set when the *default miRNA protocol* has been specified in the FE Grid Template properties. To access the FE Grid Template properties, double-click on the grid template in the Grid Template Browser.
3. Set FE Project Properties.
   (a) Select the *Project Properties* tab.
   (b) In the *General* section, enter the user name in the *Operator* text box.

4. Check the Extraction Set Configuration.
   (a) Select the Extraction Set configuration tab.
   (b) Verify that the correct grid template is assigned to each extraction set in the Grid Name columns. If a grid template is not available to select from the pull down menu, you must add it to the Grid Template Browser.
      - To add a grid template, right-click inside the Grid Template Browser, select *Add*. Browse for the design file (.xml) and click *Open* to load grid template into the FE database. To update to the latest grid templates via Online Update, right-click *Grid Template Browser* and select *Online Update*. You can also download the latest grid templates from Agilent Web site at http://www.agilent.com/chem/downloaddesignfiles. After downloading, you must add the grid templates to the Grid Template Browser. After a new grid template is added to the Grid Template Browser, remember to specify the default protocol for the new grid template if you want the Feature Extraction program to automatically assign a FE protocol to an extraction set.
   (c) Verify that the correct protocol is assigned to each extraction set in the *Protocol Name* column. For Agilent miRNA microarrays, select *miRNA_107_Sept09* or later version.

If a protocol is not available to select from the pull down menu, you must import it to the FE Protocol Browser.

To import,

(a) Right-click *FE Protocol Browser*, select *Import*.
(b) Browse for the FE protocol (.xml) and click *Open* to load the protocol into the FE database.

Visit the Agilent Web site at http://www.agilent.com/chem/feprotocols to download the latest protocols.

5. Verify that the icons for the image files in the FE Project Window no longer have a red X through them. If there is still a red X, reselect the extraction protocol for that image file.
6. Save the FE Project (.fep) by selecting *File > Save As* and browse to the desired location.
7. Select *Project > Start Extracting* to initiate the data extraction process.
8. After the extraction is completed successfully, view the QC report for each extraction set by double-clicking the QC Report link in the *Summary Report* tab (Fig. 3). Review the QC metric table to evaluate the gridding and the data quality. A QC Chart is also generated which displays the values of each metric for all extractions in a given FE Project (Fig. 4).
9. Feature Extracted array data is now available for downstream analysis (see Note 20).

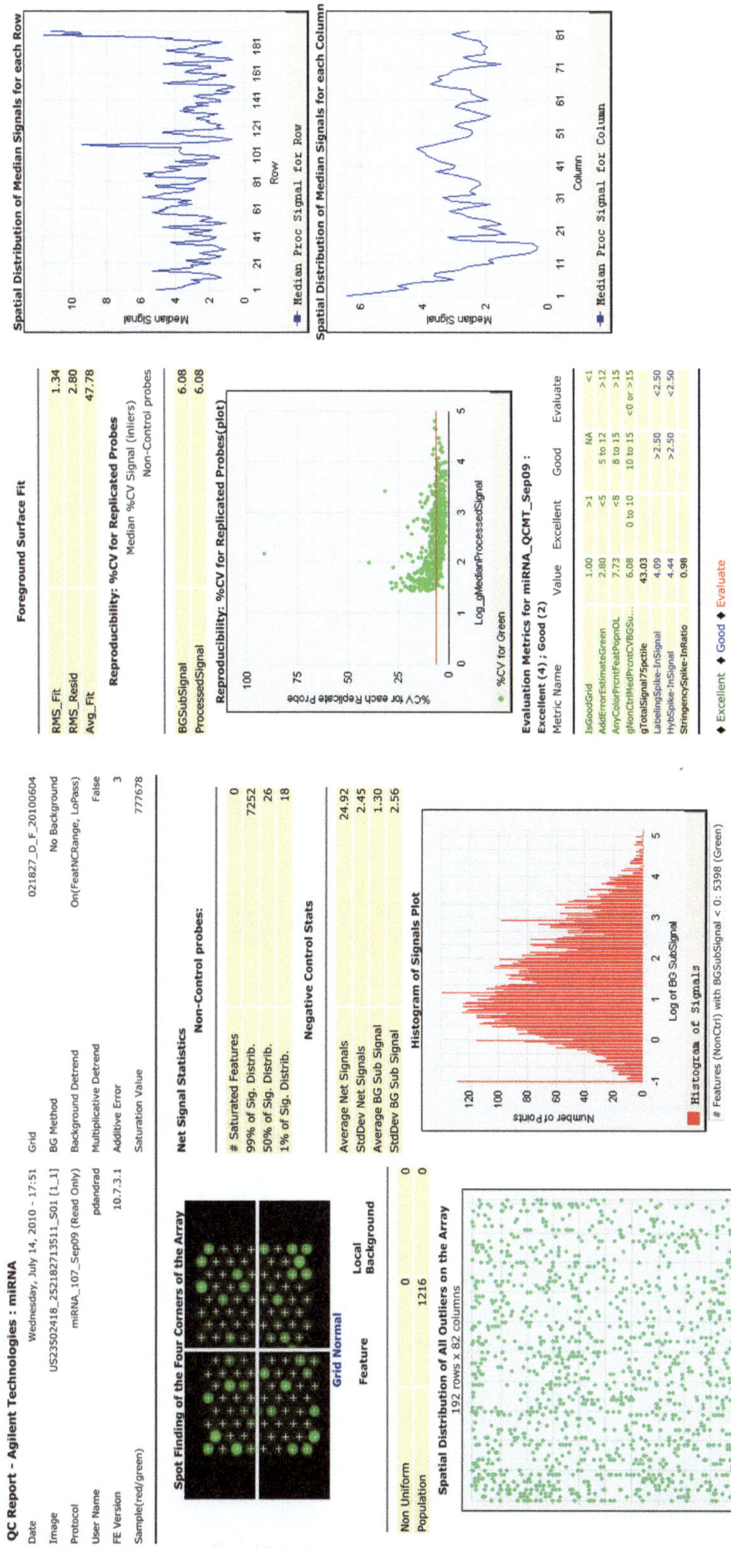

Fig. 3. Example of a QC report for 8 × 15 K miRNA microarray, generated by Feature Extraction 10.7.3.

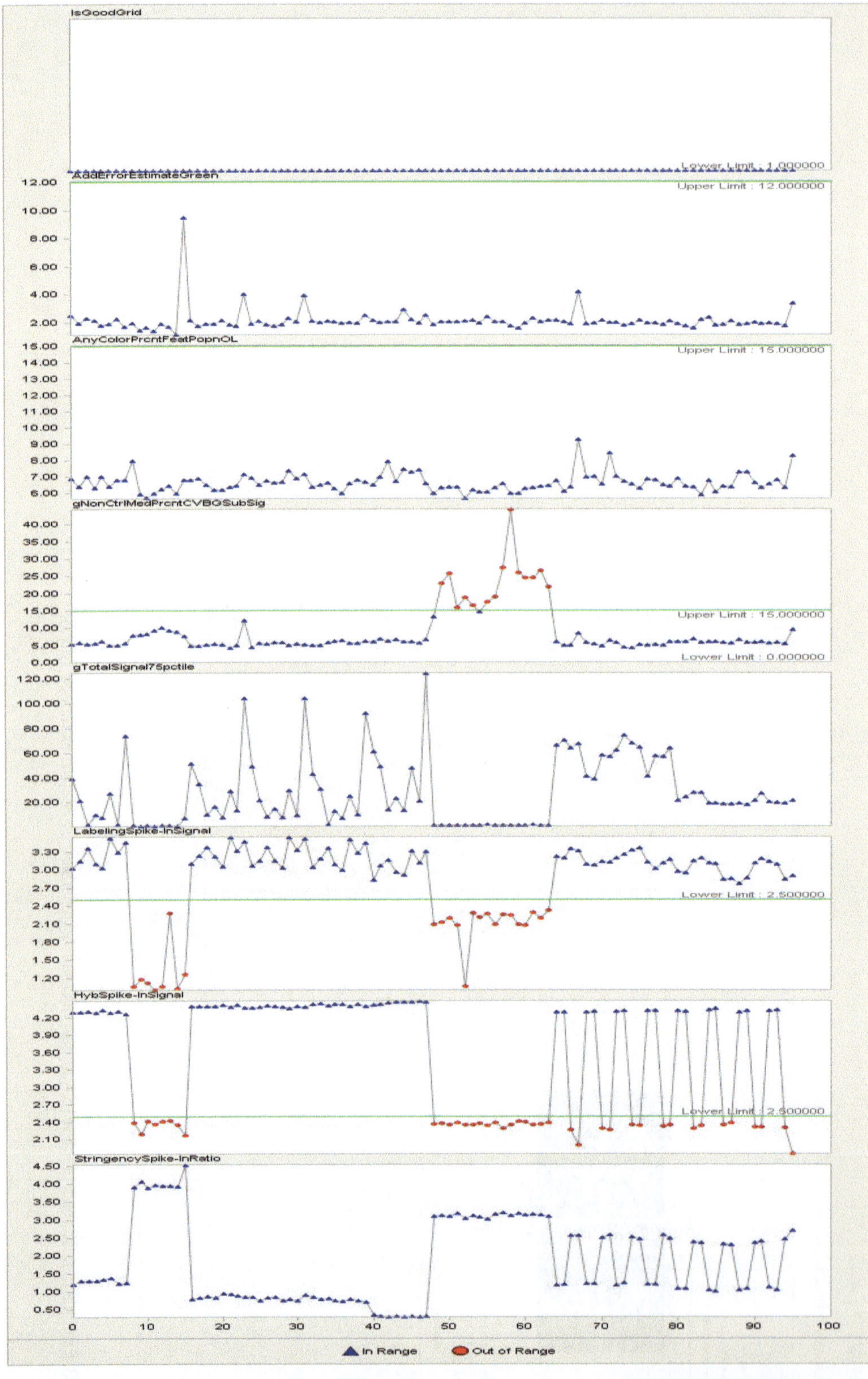

Fig. 4. Example of a QC Chart for multiple miRNA microarray slides demonstrating both passing arrays (*blue*) and arrays needing further evaluation (*red*).

4. Notes

1. The first dilution can be stored at −80°C for reuse with up to two freeze–thaw cycles. To avoid more than two freeze–thaw cycles, the original and first dilution Spike-In solution should be aliquoted and stored in volumes appropriate for the studies being performed. The second and third spike-in dilutions should be discarded after use because they are not sufficiently concentrated to be effectively stored for reuse.
2. A nitrogen purge or desiccator box is used to store the microarray slides prior to hybridization and posthybridization before scanning if necessary to preserve the microarray and maintain signal intensities.
3. Processing the samples and washing the microarrays for Agilent's miRNA platform must be done in environments where ozone levels are 50 ppb or less.
4. Recommended RNA extraction methods. These extraction methods have worked successfully with Agilent's miRNA microarray system:
   (a) Absolutely RNA miRNA Kit – Agilent
   (b) miRNeasy Mini Kit – Qiagen
   (c) mirVana RNA Isolation Kit – Applied Biosystem
   (d) TRIzol Reagent (100 mL) – Invitrogen

   When using these kits, use the total RNA isolation protocol. Do not use the size fractionation or small RNA enrichment protocol because the assay has been optimized for the use of 100 ng total RNA. Use the same total RNA extraction method to obtain consistent results for comparative experiments since different total RNA extraction methods may result in slightly different miRNA profiles. Check the absorption spectrum from 220 nm to about 300 nm to detect possible contaminants. In the absorption spectra, there should only be one peak with an absorption maximum at 260 nm. You may also see a shoulder of an additional peak at <220 nm that overlaps with and inflates the absorption at 260 nm. Calculate the ratio of absorbance at 260/230 nm to estimate the level of contamination. This ratio should be greater than 1.8. If the ratio is less than 1.8, then you need to purify the sample further by doing additional extractions with chloroform, followed by an ethanol precipitation. Absorption spectra cannot differentiate between RNA and DNA. Contamination with DNA can result in an overestimation of total RNA amount and may lead to decreased sensitivity on the miRNA assay as less RNA than intended may be used in the assay.

It is strongly recommended that the total RNA be characterized for sample integrity and possible presence of contaminants. Characterization of the input total RNA using the Agilent 2100 bioanalyzer provides information on the sample quality prior to labeling and hybridization. You can use the RNA 6000 Nano or Pico kit to analyze total RNA. The small RNA assay characterizes the total RNA sample with an emphasis on the small RNA content of which a fraction is miRNA.

5. An ice-water bath must be used to ensure that the samples remain properly denatured. When using a thermocycler, the PCR plate must be transferred to an ice-water bath immediately after incubation at 100°C to prevent the RNA from re-annealing. Thermocyclers do not cool as quickly as necessary for this step.
6. The column purification step is an optional step that removes DMSO and much of the unincorporated Cy3-pCp from the labeling reaction. Inclusion of this optional step reduces the time needed for drying down the sample after labeling. In addition, this step reduces the amount of unincorporated dye in the hybridization sample. If disposal of unincorporated Cy3 dye is an issue at your institution, you may want to consider including this optional step. Use of the column purification does increase the hands on time required for the protocol and may be omitted, but speed vac times will be increased (see Note 8).
7. Optional step: The speed of the microcentrifuge must be accurately set (e.g., 1,000×*g*, not 1,000 rpm). If the spin speed is too low, sample can be lost. If the spin speed is too high, desalting may be ineffective and the column may break down.
8. The sample must be completely dried after labeling because residual DMSO in the sample will impact the hybridization efficiency and adversely affect the hybridization results. The type and age of the speed vac influences the length of time the sample requires to dry to completion.
9. Adjust your pipette to the sample final volume to avoid the introduction of air bubbles during application to the gasket well. This step will minimize sample leakage during the assembly and hybridization period.
10. When dispensing the hybridization solution do not to allow the pipette tip or the hybridization solution to touch the gasket as this can result in leakage of the hybridization solution during the hybridization.
11. The microarray slide should be placed on the gasket solution as parallel as possible to the gasket slide to avoid leaking of the hybridization solution. If you are not processing all of the arrays on a slide, apply the hybridization solution to the gasket well closest to the barcode. In the unused wells, pipette 45 μL of 1× Hybridization Buffer.

Refer to the Agilent Microarray Hybridization Chamber User Guide for in-depth instructions on how to assemble the arrays. This user guide is available with the Agilent Microarray Hybridization Chamber Kit and can also be downloaded from the Agilent Web site at http://www.agilent.com/chem/dnamanuals-protocols.

12. Be sure to balance the hybridization chambers on the oven rack so that there are an equal number of empty positions on each of the four rows of the hybridization rack.
13. Arrays should be hybridized for at least 20 h. Hybridization can occur for longer than 20 h but the actual hybridization time should be consistent if the results are to be compared. Failure to maintain consistent hybridization times may adversely affect the reproducibility of your data.
14. If the temperature of the laboratory in which this work is being performed is less than 25°C it may be necessary to heat the "Wash 2 dish" at a higher temperature to ensure that the dish does not cool too rapidly.
15. Some dish detergents may leave fluorescent residue on the dishes. Do not use any detergent (usually used for washing dishes) to wash the staining dishes since most detergents fluoresce and can impact the data quality if dried on the arrays. If detergent is used, all traces must be removed by copious rinsing with Milli-Q water.
16. Removing the slide rack with the microarray slides at an angle helps minimize residual droplets and encourages the wash buffer to sheet off the array surface more evenly.
17. The wash buffer in the "Disassembly dish" may appear pink due to the presence of unincorporated Cy3 dye, especially if the optional column purification step was not employed. Discard the used wash buffer according to appropriate institutional guidelines. Activated charcoal may be used to absorb the unincorporated Cy3 dye from the used wash buffer, if necessary. Extractors from Sigma can be used for this purpose.
18. Refer to the Agilent Microarray Scanner User Guide for more information on how to use the scanner. The User Guides can be found at http://www.agilent.com/chem/scanner. The Download Software tab at this site has information related to the Scan Control software and the Related Literature tab at this site has the related user guides.
19. Refer to the Agilent Feature Extraction Software *User Guide* or *Reference Guide* for more information to use the software. To get the most recent Feature Extraction software for miRNA expression, go to the Agilent Web site at http://www.agilent.com/chem/fe. To get the most recent Feature Extraction

protocols for miRNA, go to the Agilent Web site at http://www.agilent.com/chem/feprotocols. FE version 10.7.3 gives you the option to automatically download new grid templates, protocols, and QC metrics (QCM or QCMT) via eArray, and supports analysis of the miRNA spike-ins. Feature Extraction generates text data files in two different formats for analysis in downstream software. The "full text" file (e.g., file name US22502705_251911810010З_S01_miRNA_107_Sept09_1_1.txt) contains many data columns, array statistics, and feature extraction parameters, and is appropriate for downstream analysis in Agilent GeneSpring GX 10 or later. See the FE reference guide for a full description of data included in the full text file which could be found at http://www.agilent.com/chem/fe. An additional text file is also provided which includes a simplified view of the data. This "Gene View" file contains signal, error, and flag data at the miRNA level, and is suitable for analysis in Microsoft Excel or other spreadsheet applications.

20. GeneSpringGX version 11.5.1 or later can be used to profile the microRNAs expressed in a given sample and across samples by importing the Feature Extracted files. The microRNAs of interest can be selected and the potential mRNAs targeted can be determined by using the TargetScan function along with other ways of analyzing the data in GeneSpringGX.

References

1. Zamore, P. D. and Haley, B. (2005) Ribo-gnome: The big world of small RNAs. *Science* **309**, 1519–24.
2. Friedman, R. C., Farh, K. K., Burge, C. B. and Bartel, D. P (2009) Most mammalian mRNAs are conserved target of microRNAs. *Genome Res* **19**, 92–105.
3. Wang, H., Ach, R. A. and Curry, B. (2007) Direct and sensitive miRNA profiling from low-input total RNA. *RNA* **13**, 1–9.
4. Ach, R. A., Wang, H. and Curry, B. (2008) Measuring microRNAs: Comparisons of microarray and quantitative PCR measurements, and of different total RNA prep methods. *BMC Biotechnology* **8**, 69.
5. Git, A., Dvinge, H., Salmon-Divon, M., Osborne, M.., Kutter, C., Hadfield, J., et al. (2010) Systematic comparison of microarray profiling, real-time PCR, and next-generation sequencing technologies for measuring differential microRNA expression. *RNA* **16**, 991–1006.

Chapter 7

miRNA Expression Profiling Using Illumina Universal BeadChips

Jing Chen, Craig S. April, and Jian-Bing Fan

Abstract

We have developed a highly sensitive, specific, and reproducible method for microRNA (miRNA) expression profiling, using BeadArray technology. This method incorporates an enzyme-assisted specificity step, a solid-phase primer extension to distinguish between members of miRNA families. In addition, a universal PCR is used to amplify all targets prior to array hybridization. Using this method, highly reproducible miRNA expression profiles were generated with 100–200 ng total RNA input. The method has a 3.5–4 log (10^5–10^9 molecules) dynamic range and is able to detect 1.2- to 1.3-fold differences between samples. Expression profiles generated by this method are highly comparable to those obtained with RT-PCR ($R^2 = 0.85$–0.90) and direct sequencing ($R = 0.87$–0.89). This method should prove useful for high-throughput expression profiling of miRNAs in large numbers of tissue samples.

Key words: miRNA, Gene expression analysis, Microarray, BeadArray, DASL assay

1. Introduction

The Illumina® MicroRNA Expression Profiling Assay is an efficient and cost-effective system for high-throughput, multiplexed miRNA expression profiling (1). The MicroRNA assay is an adaptation of the DASL® (c*D*NA-mediated *A*nnealing, *S*election, Extension, and *L*igation) assay (2), incorporating its unique methodologies to make measurements of miRNAs. The system combines a novel, highly multiplexed expression assay, Universal BeadChips, and a precise confocal scanning system (the Illumina iScan™ System or the Illumina BeadArray™ Reader) to deliver unparalleled data quality and sample throughput.

1.1. BeadArray Technology

The BeadChip platform is composed of individual arrays manufactured on a microscope slide-shaped substrate (3). Each individual array on the BeadChip contains the same 1,536 IllumiCode

Jian-Bing Fan (ed.), *Next-Generation MicroRNA Expression Profiling Technology: Methods and Protocols*, Methods in Molecular Biology, vol. 822, DOI 10.1007/978-1-61779-427-8_7, © Springer Science+Business Media, LLC 2012

sequences attached to 3-μm beads. The beads are assembled into microwells etched into the slide. The BeadChip arrays feature multiple copies of each bead type. This built-in redundancy improves robustness and measurement precision. The BeadChip manufacturing process includes hybridization-based quality control for each array feature, allowing consistent production of high-quality, reproducible arrays. Each BeadChip has its own CD that contains a decode map (*.dmap) file for each array, which is necessary for analyzing the data.

1.2. miRNA Expression Profiling

Illumina has created standard panels of miRNA assays for human and mouse. The Human v2 and Mouse v2 panels contain 1,146 and 656 probes, respectively. This version updates the coverage of both products to greater than 95% of miRBase1 v12.0, to represent the majority of known miRNAs. The Human v1 miRNA panel contains 735 assay probes, corresponding to 470 well-annotated human microRNA sequences, and 265 potential microRNAs that were identified recently (4, 5). The Mouse miRNA v1 panel contains 380 assays for miRNAs.

1.3. GenomeStudio Integrated Informatics Platform

GenomeStudio, Illumina's new integrated data analysis software platform, provides a common environment for analyzing data obtained from microarray and sequencing technologies. Within this common environment, or framework, the GenomeStudio software modules allow one to perform application-specific analyses. The GenomeStudio Gene Expression Module, included with the Illumina MicroRNA Expression Profiling Assay, is an application for analyzing miRNA expression data from scanned microarray images collected from systems, such as the Illumina iScan System or BeadArray Reader. Experiment performance is based on built-in controls that accompany each experiment. Resulting GenomeStudio expression results can be exported and analyzed by most standard gene expression analysis programs. One can perform these analyses on individual arrays or on groups of arrays treated as replicates. For feature descriptions and instructions on using the GenomeStudio platform to visualize and analyze miRNA data, *see* the *GenomeStudio Framework User Guide* and the *GenomeStudio Gene Expression Module User Guide.*

2. Materials

Subheadings 2 and 3 assume that the user has access to either an Illumina iScan System or BeadArray Reader and either the Universal Starter Kit or both the GoldenGate and Gene Expression Option kits and associated reagents and equipment.

2.1. Recommended Kit for miRNA Enrichment

Although miRNA enrichment is an optional step in miRNA expression profiling assay, we recommend the PureLink™ miRNA Isolation Kit from Invitrogen for miRNA enrichment.

2.2. Illumina-Supplied Reagents for miRNA Assay

1. PAS (*Poly***a**denylation *S*ingle), reagent for 3′ Poly-A tail addition.
2. CSS (*c*DNA *S*ynthesis *S*ingle), reagent for cDNA synthesis
3. MAP (*M*icroRNA *A*ssay *P*ool), mixture of oligonucleotides designed to query cDNA target sequences.
4. OB1 (*O*ligo-*b*inding buffer *1*), oligo-annealing buffer which also contains paramagnetic particles to optimize washing, extension, and ligation steps of assay.
5. AM1 (*A*dd *M*EL buffer *1*), wash buffer used to remove excess mis- or unhybridized query oligonucleotides.
6. UB1 (*U*niversal wash *B*uffer *1*), wash buffer for several pre-PCR steps.
7. MEL (*M*aster mix for *E*xtension and *L*igation), optimized mixture of enzymes for extension/ligation step.
8. SCM (*S*ingle *C*olor *M*aster mix), PCR master mix which contains fluorescent and biotinylated common primers for multiplexed ligated oligonucleotide templates.
9. IP1 (*I*noculate *P*CR buffer *1*), elution buffer for inoculating PCR reaction with ligated templates.
10. MPB (*M*agnetic *P*article buffer *B*), suspension of paramagnetic particles used to bind PCR products.
11. UB2 (*U*niversal wash *B*uffer *2*), wash buffer used in post-PCR process.
12. MH1 (*M*ake *H*yb buffer *1*), buffer used to neutralize and prepare single-stranded sample for hybridization to BeadChip.
13. CHB (*C*hamber *H*umidification *B*uffer), buffer used to maintain humid environment in Hybridization (Hyb) Chamber.
14. PB1 (*P*repare *B*eadChips solution *1*), buffer used during staining process.
15. XC4 (*X*Stain Bead*C*hip solution *4*), solution used to coat BeadChips.

Please refer to Table 1 for reagent storage conditions.

2.3. Other Reagents Required for the Assay

1. 0.1N NaOH (Sigma-Aldrich).
2. TE buffer: 10 mM Tris–HCl, 1 mM EDTA, pH 8.0. (General lab supplier).
3. DNA polymerase (Clontech).
4. Uracil DNA glycosylase (UDG), 1 U/μL (Invitrogen).
5. Quant-iT RiboGreen RNA Assay Kit (Invitrogen).

Table 1
Reagent storage conditions

Reagent	Storage conditions	Shelf life	Comments
PAS	−25°C to −15°C	1 Year	Aliquot to refreeze
CSS	−25°C to −15°C	1 Year	Aliquot to refreeze
MAP	−25°C to −15°C	2 Years	Can be stored at 2–8°C up to 2 weeks
OB1	−25°C to −15°C	1 Year	Does not completely freeze
AM1	2–8°C	1 Year	
UB1	2–8°C	1 Year	
MEL	−25°C to −15°C	1 Year	Aliquot to refreeze
SCM	−25°C to −15°C	1 Year	Aliquot to refreeze after adding DNA polymerase
IP1	−25°C to −15°C	1 Year	
MPB	2–8°C	1 Year	Do not freeze
UB2	RT	1 Year	
MH1	RT	1 Year	Keep away from light
CHB	RT	1 Year	
PB1	RT	1 Year	
XC4	RT	1 Year	

6. Serological pipettes (10, 25, and 50 mL) (General lab supplier).
7. 96-Well, 0.2-mL skirted microtiter plates (Bio-Rad).
8. 96-Well, black, flat-bottom Fluotrac 200 plates (Greiner Bio One).
9. 96-Well cap mats, sealing mats, round cap, pierceable, nonautoclavable (ABgene).
10. Heat sealing foil sheets, thermo seal (ABgene).
11. Microtiter plate clear adhesive film, 2mil Sealplate Adhesive Film, Nonsterile (Phenix Research Products, LMT-SEAL-EX).
12. Microseal "A" Film, PCR plate sealing film (Bio-Rad).
13. Microseal "F" Film, aluminum adhesive film (Bio-Rad).
14. 96-Well V-bottom plates, Corning Costar Polypropylene (Fisher Scientific or VWR).
15. Multiscreen Filter plates, 0.45 μm, clear, Styrene (Millipore).

16. Multiscreen centrifuge alignment frames (Millipore).
17. Nonsterile solution basins (Labcor Products Inc or VWR).
18. Tweezers (General lab supplier).

2.4. Universal BeadChip

The manufacture of Illumina microarrays has been described elsewhere (2). Briefly, the miRNA Assay uses a multisample Universal-12 BeadChip as a readout, which is composed of 12 individual arrays manufactured on a single microscope slide-shaped silicon substrate.

2.5. Bead Array Imaging

The Illumina iScan System or BeadArray Reader is required for array imaging. Loading and unloading of BeadChips into the iScan System or BeadArray Reader can be automated with the optional AutoLoader2 or AutoLoader, respectively. Both AutoLoaders support unattended processing by placing BeadChips carriers in the imaging system's tray, so that it can scan the BeadChips. Both scanners use a laser to excite the fluor of the hybridized single-stranded product on the beads of the BeadChip sections. Light emissions from these fluors are then recorded in high-resolution images of the BeadChip sections. Data from these images are analyzed using Illumina's GenomeStudio Gene Expression Module. For Illumina iScan System or BeadArray Reader instructions, see the respective User Guides (http://www.illumina.com/support/documentation.ilmn).

3. Methods

3.1. RNA Enrichment

Please refer to the PureLink™ miRNA Isolation Kit protocol (Invitrogen).

3.2. miRNA Assay Protocols

These protocols are not intended to replace the miRNA Assay Guide supplied with Illumina systems but rather give a detailed overview of the process. Please consult the miRNA Assay Guide (http://www.illumina.com/support/documentation.ilmn) for the latest protocol updates before performing any experiments.

3.3. Assay Probe Design

The method is a modification of an assay that we developed previously for high-throughput gene expression profiling, the DASL Assay (c*D*NA-mediated *A*nnealing, *S*election, extension and *L*igation) (2). The miRNA method similarly targets specific sequences with sets of oligonucleotides that are extended, and then labeled during PCR amplification (1). As shown in Fig. 1, miRNAs are first polyadenylated using Poly-A Polymerase (PAP). The introduced poly-A tail is then used as a priming site for cDNA synthesis. The primer used for cDNA synthesis is biotinylated and contains a universal PCR primer sequence that is used later in the assay.

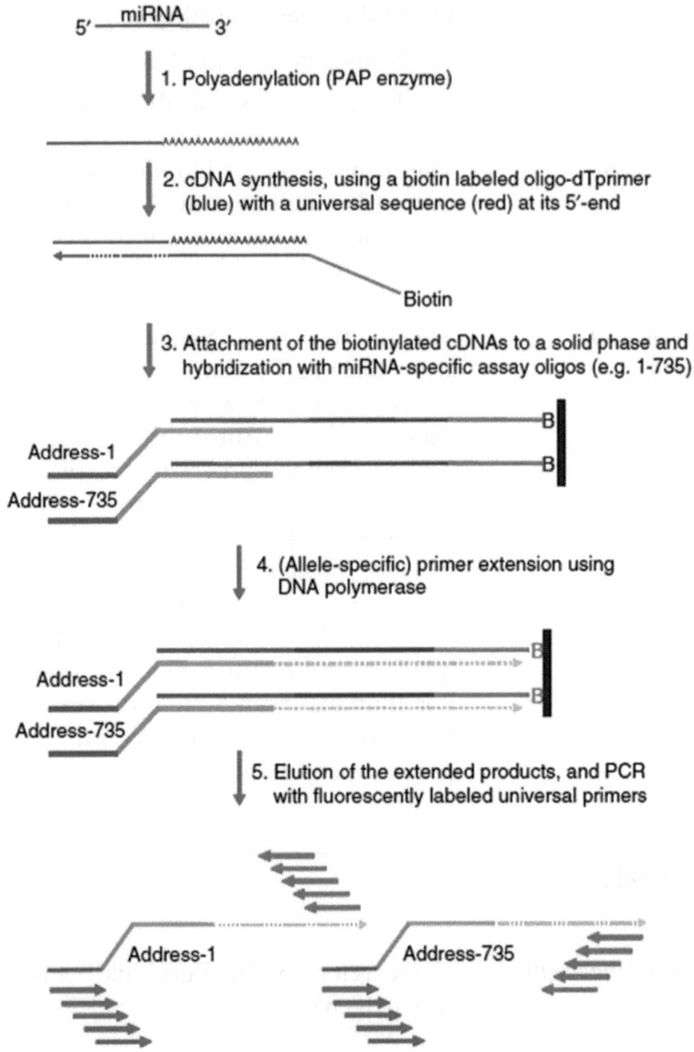

Fig. 1. miRNA assay scheme. (1) Polyadenylate RNA: add multiple A (>18 bases) to 3′-ends of total RNA or purified short RNA species, including microRNAs. (2) cDNA synthesis of microRNA: synthesize cDNA using a biotin-labeled oligo-dT primer with a universal sequence at its 5′-end. (3) Hybridize assay oligos to cDNA: attach biotinylated cDNA to a solid phase and hybridize with a pool of microRNA-specific oligos. (4) MicroRNA-specific primer extension: extend primers using DNA polymerase. (5) Universal PCR: elute the extended products and perform PCR with fluorescently labeled universal primers. Bind double-stranded PCR products to a solid phase and prepare the labeled, single-stranded PCR products for hybridization. (6) Hybridize ssDNA to arrays: hybridize PCR product to capture probes on the universal arrays.

After cDNA synthesis, miRNAs are individually interrogated using specific oligonucleotides. A single miRNA-specific oligo (MSO) is designed against each mature well-annotated miRNA sequence, which consists of three parts: the 5′-end contains another universal PCR priming site; the middle portion contains an address sequence used for capturing the product on the array; and the 3′-end contains a miRNA-specific sequence. The second universal PCR priming site is shared among all MSOs, and each address sequence is associated uniquely with each of the miRNA targets. Assay probes were designed with a T_m of 60 ± 8.6°C and a length of 17–21 nt (average 18 nt). To maximize assay specificity, candidate probes were examined collectively to minimize sequence similarity between probes, particularly at their 3′ ends.

3.4. miRNA Assay Protocol

3.4.1. The Make Poly-A Polymerase Process for 3′ Poly-A Tail Addition

1. Preheat two heat blocks to 37°C and 70°C and allow temperature to stabilize. Thaw PAS reagent to room temperature and vortex to fully mix tube contents.
2. Normalize intact RNA samples to 40–200 ng/μL with DEPC-treated H_2O. We recommend using 200 ng of miRNA containing total RNA for one assay (see Notes 1–5).
3. Add 5 μL of PAS to each well of microtiter plate labeled PAP. Transfer 5 μL of normalized RNA sample to each well of PAP plate, heat seal with foil, and vortex at 2,300 rpm for 20 s. Pulse-centrifuge sealed plate to 250 × *g* to prevent evaporation from wells during incubation.
4. Incubate PAP plate at 37°C for 60 min in preheated heat block.
5. When incubation is complete, remove PAP plate from heat block and pulse-centrifuge to 250 × *g* to collect any condensation.
6. Incubate PAP plate at 70°C for 10 min in preheated heat block.

3.4.2. The Make cDNA Synthesis Plate Process for cDNA Synthesis

1. Preheat heat block to 42°C and allow temperature to stabilize. Leave the other heat block at 70°C. Thaw CSS to room temperature and vortex to fully mix tube contents.
2. Add 8 μL CSS to each well of microtiter plate-labeled CSP, transfer 8 μL polyadenylated RNA sample from each well of the PAP plate to the corresponding well of the CSP plate. Heat seal with foil, and vortex at 2,300 rpm for 20 s. Pulse-centrifuge sealed plate to 250 × *g* to prevent evaporation from wells during incubation.
3. Incubate CSP plate at 37°C for 60 min in preheated heat block.
4. When incubation is complete, remove CSP plate from heat block and pulse-centrifuge to 250 × *g* to collect any condensation.
5. Incubate CSP plate at 70°C for 10 min in preheated heat block.

3.4.3. The Make Assay-Specific Extension Process for Annealing of Query Oligonucleotides to cDNA

1. Preheat heat block to 70°C and allow temperature to stabilize.
2. Remove MAP tube from freezer (if frozen, thaw, vortex, and then centrifuge). Thaw OB1 to room temperature and vortex. Do not centrifuge OB1 tube.
3. Dispense 5 μL of MAP to each well of a new, 96-well, 0.2 mL skirted microtiter plate-labeled "ASE."

4. Add 30 μL of well-resuspended OB1 to each well of *A*ssay-*S*pecific *E*xtension (ASE) plate.
5. Centrifuge CSP plate to 250 × *g* to collect samples at bottom of wells. Transfer 15 μL of biotinylated cDNA to ASE plate containing MAP and OB1 to bring final volume to 50 μL. Heat seal ASE plate and vortex briefly at 1,600 rpm to mix content of wells. Place ASE plate in 70°C heat block and immediately reduce temperature setting to 30°C. This will allow oligonucleotide annealing to cDNA targets by ramping temperature over approximately 2 h then holding at 30°C until next processing step.

3.4.4. The Add Master Mix for Extension and Ligation Process for Assay Oligonucleotide Extension and Ligation

1. Remove ASE plate from heat block, reset it to 45°C, and allow temperature to stabilize. Thaw MEL reagent to room temperature.
2. Place ASE plate with oligonucleotides annealed to cDNA templates on Illumina-supplied magnetic plate for at least 2 min, or until beads are completely captured. Washing beads removes excess and mis-hybridized oligonucleotides.
3. After paramagnetic particles are captured, remove heat seal from plate and remove and discard all liquid (~50 μL) from wells, retaining beads. Add 50 μL Add MEL buffer (AM1) to each well of assay plate. Seal plate with adhesive film and vortex at 1,600 rpm for 20 s or until all beads are resuspended.
4. Place ASE plate on magnet for at least 2 min, or until beads are completely captured. Remove all AM1 from each well, leaving beads in wells. Repeat addition of 50 μL AM1, vortexing, and removal of buffer.
5. Remove ASE plate from magnet and add 50 μL of UB1 to each well.
6. Place ASE plate on magnet for at least 2 min, or until beads are completely captured. Remove all UB1 from each well. Repeat addition of 50 μL UB1 and removal of buffer.
7. Add 37 μL of MEL to each well of ASE plate. Seal plate with adhesive film and vortex at 1,600 rpm for 1 min.
8. Incubate ASE plate on preheated 45°C heat block for 15 min (see Notes 6 and 7).

3.4.5. The Make PCR and Inoculate PCR Processes for Preparing the PCR Mix and Setting up the PCR Reaction

1. Prepare PCR master mix by adding 64 μL of DNA polymerase and 50 μL UDG to tube of SCM PCR reagent. Invert SCM tube several times to mix contents and aliquot 30 μL of mixture into each well of new 96-well 0.2 mL microtiter (PCR) plate.
2. Remove ASE plate from heat block after extension and ligation step and reset heat block to 95°C.

3. Place ASE plate on magnet for at least 2 min, or until beads are captured. Remove clear adhesive film from assay plate, and remove and discard supernatant (~50 μL) from all wells of ASE plate, leaving beads in wells. Leave ASE plate on magnet and add 50 μL UB1 to each well of plate.
4. Allow ASE plate to rest on magnet for at least 2 min to collect paramagnetic particles. Remove and discard all supernatant (~50 μL) from all wells of ASE plate, leaving beads in wells.
5. Add 35 μL of Inoculate (Inoc) PCR buffer 1 (IP1) to each well of assay plate and seal it with adhesive film. Vortex plate at 1,800 rpm for 1 min, or until all beads are resuspended. Place plate on 95°C heat block for 1 min.
6. Remove ASE plate from heat block and place it on magnet for at least 2 min, or until beads have been completely captured. Transfer 30 μL of supernatant from first column of ASE plate into first column of PCR plate. Repeat transfer for remaining columns, using new pipette tips for each column.
7. Seal PCR plate with Microseal "A" PCR plate sealing film.
8. Pulse-centrifuge PCR plate to 250 × *g* for 1 min. Immediately transfer PCR plate to thermal cycler and run following cycling program: 10 min at 37°C; 3 min at 95°C; 34 cycles (35 s at 95°C, 35 s at 56°C, 2 min at 72°C); 10 min at 72°C; and 4°C for 5 min.
9. Proceed immediately to preparation of single-stranded PCR products for precipitation or seal and store PCR plate at −15°C to −25°C.

3.4.6. The Make HYB Process to Prepare Samples for Array Hybridization

1. Vortex MPB reagent until beads are completely resuspended.
2. Dispense 20 μL of resuspended MPB into each well of PCR plate. Mix beads with PCR product by pipetting up and down, and then transfer mixed solution to filter plate. Cover filter plate with its cover and store at room temperature, protected from light, for 60 min.
3. Place filter plate containing bound PCR products onto new 96-well V-bottom waste plate using filter plate adapter. Centrifuge at 1,000 × *g* for 5 min at 25°C.
4. Remove filter plate lid. Add 50 μL of UB2 to each well of filter plate. Dispense slowly, so that beads are undisturbed. Replace lid of filter plate and centrifuge at 1,000 × *g* for 5 min at 25°C.
5. Prepare new 96-well, V-bottom plate and dispense 30 μL of MH1 to all wells of new intermediate plate. Place filter plate onto intermediate plate such that column A1 of filter plate matches column A1 of intermediate plate.
6. Dispense 30 μL of 0.1N NaOH to all wells of filter plate. Replace lid of filter plate and centrifuge immediately at 1,000 × *g*

for 5 min at 25°C. Gently mix contents of intermediate plate by moving it from side to side, without splashing. Proceed to next step for Beadchip Hyb process.

3.5. Hybridization to BeadChip and PostHyb Wash

3.5.1. Hybridization to BeadChip

1. Preheat hybridization oven to 60°C and allow temperature to equilibrate.
2. Prepare hybridization (Hyb) Chambers by pipetting 200 μL CHB into Hyb Chamber reservoirs, followed by sealing Hyb Chambers as described in miRNA Assay Guide.
3. Place multisample BeadChips into Hyb Chamber inserts.
4. Manually pipette 15 μL of sample into appropriate inlet ports on multisample BeadChips.
5. Place inserts with sample-laden BeadChips into Hyb Chamber, seal, and hybridize samples. Place the Beadchip containing Hyb Chamber into the preheated hybridization oven and incubate at 60°C for 30 min.
6. Change the temperature to 45°C. Hybridization then is conducted under a temperature gradient program from 60°C to 45°C over approximately 12 h.

3.5.2. Post-Hyb Wash

1. Prepare XC4 reagent. Add 335 mL 100% EtOH to XC4 bottle. Re-cap bottle, shake vigorously for 15 s by hand, place on rocker for 30–40 min to ensure that XC4 is completely resuspended. Each XC4 bottle contains enough to process up to eight BeadChips.
2. Remove Hyb Chamber from hybridization oven and place it on lab bench. Disassemble Hyb Chamber.
3. Using powder-free gloved hands, remove coverseal from first BeadChip. Using tweezers or powder-free gloved hands transfer BeadChip to slide rack submerged in dish containing 250 mL of PB1 solution.
4. Repeat steps 2 and 3 for all BeadChips from same Hyb Chamber.
5. Wash BeadChips in PB1 by moving slide rack up and down ten times. Incubate at RT static for 5 min.
6. Transfer the slide rack with BeadChips from PB1 dish to XC4 wash dish. Wash BeadChips in XC4 by moving slide rack up and down ten times. Incubate at RT static for 5 min.
7. Individually remove BeadChips from coating dish and place barcode side up on tube rack making sure BeadChips are not touching each other during drying step.
8. Place tube rack with horizontal BeadChips into vacuum desiccator, turn on vacuum to ~508 mmHg, and dry for 50–55 min at room temperature.

3.6. Image BeadChip on iScan System/ BeadArray Reader

The BeadChips are imaged using either the iScan System or the BeadArray Reader. Image processing and intensity data extraction are performed by the iScan Control Software (or BeadScan, for the BeadArray Reader). The iScan System incorporates advanced optics and sensors to support much higher throughput than the BeadArray Reader, while providing equally high-quality data.

3.7. Data Collection and Analysis

The GenomeStudio software package is included with the miRNA Assay product and is used as a tool for analyzing gene expression data from scanned microarray images collected from either the iScan System or the BeadArray Reader. Alternatively, GenomeStudio can be used to export the array intensity data for processing by most standard gene expression analysis programs. Specifically, GenomeStudio executes two types of data analysis: Gene analysis, which is to quantify gene expression levels and differential analysis, which is to determine whether gene expression levels are different between two experimental groups. Analyses can be performed on individual samples or groups of samples. GenomeStudio reports experiment performance based on built-in controls that accompany each experiment. In addition, GenomeStudio provides plotting (line plots, box plots, scatter plots) tools, able to generate dendrograms and heatmaps, and includes a genome viewer and chromosome browser all of which facilitate quick and visual means for exploratory analyses (see Note 8).

3.8. Examples of miRNA Assay Applications

Illumina's miRNA technology has been successfully used to profile a variety of samples types, such as human cell lines and tissues (1, 6–18), including human cord blood (19) and human fecal matter (20), mouse (21) and baboon (22) tissues, as well as formalin-fixed paraffin-embedded (FFPE) material (1).

4. Notes

1. Illumina does not recommend a specific RNA purification product. However, any product that yields pure, intact RNA of good quality that retains (at least) most of the small RNAs should work well with our miRNA assay. We have generated good data with RNAs extracted with several commercially available kits. For any given study though, it is ideal to isolate the RNAs using a single method.
2. Although we suggest starting inputs of 200 ng total RNA, we have compared the profiles generated with lower amount of starting material and have found that 200, 100, and 50 ng were highly correlated ($R^2 > 0.98$). Quite reproducible data were obtained in technically replicated experiments with as

little as 2 ng input total RNA ($R^2 > 0.94$), which makes it possible to assay small tissue samples, including archival tissue samples. Furthermore, about 90% of the miRNAs that were detected ($P < 0.05$) with 200 ng total RNA input were also detected when 2 ng total RNA was used (data not shown).

3. Excellent data have been generated using RNA extracted from FFPE tissue samples. Technical reproducibility is highly similar to intact RNA. In addition, Illumina has profiled artificially degraded RNA samples (95°C for 30 min). Excellent reproducibility was obtained with these samples. Further, the profiles generated with these samples are comparable to those generated with corresponding intact RNA samples.
4. The purification of small RNA species prior to sample labeling is unnecessary. This feature enables high-throughput miRNA analysis in a clinical setting where only limited amounts of biopsy material may be available. We obtained high concordance ($R^2 > 0.97$) between profiles generated with total RNA and low molecular weight (LMW) RNA enriched with Invitrogen's PureLink™ miRNA Isolation Kit. This result suggests that the assay is very specific, in which the presence of total RNA including mRNAs and ribosomal RNAs (rRNAs) background did not affect overall miRNA profiles.
5. RNA quantitation is an important step to ensure that sufficient material is used in the miRNA Assay to generate high-quality data. We recommend the Quant-iT RiboGreen RNA Assay Kit (Invitrogen) for quantitation of RNA samples. The RiboGreen assay measures RNA directly and can quantitate small RNA volumes. Other quantitation methods may be prone to measuring contaminants, such as small molecules and proteins. We also recommend using a fluorometer, as fluorometry provides RNA-specific quantitation, whereas spectrophotometry may be affected by DNA contamination, leading to artificially inflated amounts.
6. Specificity is achieved in the extension step (in the Add MEL step) and cycling. The extension step favors mature miRNAs because longer sequences will not achieve complete extension to the PCR primer portion of the poly-dT primer to the same degree as mature miRNAs.
7. Discrimination between miRNAs and total RNAs are achieved by a two-step process: (a) sequence hybridization – the specificity of the miRNA-specific probe which targets the pre and mature miRNA species and (b) enzymatic primer extension – to enhance the discrimination between members of miRNA families and between miRNAs and other similar sequences in the total RNA (e.g., mRNA targets). Illumina has obtained very similar expression profiles with total RNA and enriched

small RNA species, suggesting that cross-hybridization (if any) from the total RNA is minimal.

8. Another valuable approach to assessing data quality is to cluster the samples using the dendrogram tool in GenomeStudio. If there are replicate samples or known sample relationships, outlier samples can be readily identified by their failure to associate with replicate or related samples by hierarchical clustering.

Acknowledgments

We would like to thank Shawn Baker, Tanya Boyaniwsky, Kirt Haden, Mark Staebell, Christopher Streck, Scott Taylor, Joanne Yeakley, and John Stuelpnagel who are or were at Illumina; Louise Laurent, and Jeanne Loring at The Scripps Research Institute; Renee Rubio, Kristina Holton, and John Quackenbush at Dana-Farber Cancer Institute; Hua Gu at Columbia University and Guoping Fan at UCLA, for helpful discussions.

References

1. Chen, J., Lozach, J., Garcia, E. W., Barnes, B., Luo, S., Mikoulitch, I., et al (2008) Highly sensitive and specific microRNA expression profiling using BeadArray technology. *Nucleic Acids Res* **36**, e87.
2. Fan, J. B., Yeakley, J. M., Bibikova, M., Chudin, E., Wickham, E., Chen, J., et al (2004) A versatile assay for high-throughput gene expression profiling on universal array matrices *Genome Res* **14**, 878–85.
3. Barker, D. L., Theriault, G., Che, D., Dickinson, T., Shen, R., and Kain, R. (2003) Self-assembled random arrays: High-performance imaging and genomics applications on a high-density microarray platform. *Proc SPIE* **4966**, 1–11.
4. Berezikov, E., Thuemmler, F., van Laake, L. W., Kondova, I., Bontrop, R., Cuppen, E. and Plasterk, R.H. (2006) Diversity of microRNAs in human and chimpanzee brain. *Nat Genet* **38**, 1375–7.
5. Berezikov, E., van Tetering, G., Verheul, M., van de Belt, J.,van Laake, L., Vos, J., et al (2006) Many novel mammalian microRNA candidates identified by extensive cloning and RAKE analysis. *Genome Res* **16**, 1289–98.
6. Laurent, L. C., Chen, J., Ulitsky, I., Mueller, F. J., Lu, C., Shamir, R., et al (2008) Comprehensive microRNA profiling reveals a unique human embryonic stem cell signature dominated by a single seed sequence. *Stem Cells* **26**, 1506–16.
7. Siegrist, F., Singer, T. and Certa, U. (2009) MicroRNA Expression Profiling by Bead Array Technology in Human Tumor Cell Lines Treated with Interferon-Alpha-2a. *Biol Proced* **11**, 113–29.
8. Pallasch, C., P., Patz, M., Park, Y. J., Hagist, S., Eggle, D., Claus, R., et al (2009) miRNA deregulation by epigenetic silencing disrupts suppression of the oncogene PLAG1 in chronic lymphocytic leukemia. *Blood* **114**, 3255–64.
9. Cunningham, J. M., Oberg, A. L., Borralho, P. M., Kren, B. T., French, A. J., Wang, L. et al (2009) Evaluation of a new high-dimensional miRNA profiling platform. *BMC Med Genomics* **2**, 57.
10. Sarver, A. L., French, A. J., Borralho, P. M., Thayanithy, V., Oberg, A. L., Silverstein, K. A., et al (2009) Human colon cancer profiles show differential microRNA expression depending on mismatch repair status and are characteristic of undifferentiated proliferative states. *BMC Cancer* **18**, 401.
11. Wang, L., Tang, H., Thayanithy, V., Subramanian, S., Oberg, A. L., Cunningham, J. M., et al (2009) Gene networks and microRNAs implicated in aggressive prostate cancer. *Cancer Res* **69**, 9490–7.
12. Pradervand, S., Weber, J., Lemoine, F., Consales, F., Paillusson, A., Dupasquier, M., et al (2010) Concordance among digital gene expression, microarrays, and qPCR when

measuring differential expression of microRNAs. *Biotechniques* **48**, 219–22.

13. Git, A., Dvinge, H., Salmon-Divon, M., Osborne, M., Claudia Kutter, C., Hadfield, J., et al (2010) Systematic comparison of microarray profiling, real-time PCR, and next-generation sequencing technologies for measuring differential microRNA expression. *RNA* **16**, 991–1006.
14. Gaarz, A., Debey-Pascher, S., Classen, S., Eggle, D., Gathof, B., Chen, J., et al (2010) Bead array-based microrna expression profiling of peripheral blood and the impact of different RNA isolation approaches. *J Mol Diagn* **12**, 335–44.
15. Cox, M. B., Cairns, M. J., Gandhi, K. S., Carroll, A. P., Moscovis, S., Stewart, G. J. et al (2010) MicroRNAs miR-17 and miR-20a inhibit T cell activation genes and are under-expressed in MS whole blood. *PLoS One* **5**, e12132.
16. Kim, T. H., Kim, Y. K., Kwon, Y., Heo, J. H., Kang, H., Kim, G., et al (2010) Deregulation of miR-519a, 153, and 485-5p and its clinicopathological relevance in ovarian epithelial tumours. *Histopathology* **57**, 734–43
17. Smeets, A., Daemen, A., Vanden Bempt, I., Gevaert, O., Claes, B., Wildiers, H., et al (2010) Prediction of lymph node involvement in breast cancer from primary tumor tissue using gene expression profiling and miRNAs. *Breast Cancer Res Treat*, doi: 10.1007/s10549-010-1265-5.
18. Li, J. H., Xiao, X., Zhang, Y. N., Wang, Y. M., Feng, L. M., Wu, Y. M., et al (2011) MicroRNA miR-886-5p inhibits apoptosis by down-regulating Bax expression in human cervical carcinoma cells. *Gynecol Oncol* **120**, 145–51.
19. Iwaniuk, K. M., Schira, J., Weinhold, S., Jung, M., Adjaye, J., Müller, H. W., et al (2011) Network-Like Impact of MicroRNAs on Neuronal Lineage Differentiation of Unrestricted Somatic Stem Cells from Human Cord Blood. *Stem Cells Dev*, doi: 10.1089/scd.2010.0341.
20. Link, A., Balaguer, F., Shen, Y., Nagasaka, T., Lozano, J. J., Boland, C. R., et al (2010) Fecal MicroRNAs as novel biomarkers for colon cancer screening. *Cancer Epidemiol Biomarkers Prev* **19**, 1766–74.
21. Polikepahad, S., Knight, J. M., Naghavi, A. O., Oplt, T., Creighton, C. J., Shaw, C., et al (2010) Proinflammatory role for let-7 microRNAS in experimental asthma. *J Biol Chem* **285**, 30139–49.
22. Karere, G. M., Glenn, J. P., VandeBerg, J. L, and Cox, L.A. (2010) Identification of baboon microRNAs expressed in liver and lymphocytes. *J Biomed Sci* **17**, 54.

Chapter 8

MicroRNA Expression Analysis Using the Affymetrix Platform

Suzanne Dee and Robert C. Getts

Abstract

Microarrays have been used extensively for messenger RNA expression monitoring. Recently, microarrays have been designed to interrogate expression levels of noncoding RNAs. Here, we describe methods for RNA labeling and the use of a miRNA array to identify and measure microRNA present in RNA samples.

Key words: microRNA, snoRNA, scaRNA, Microarray, Gene expression, Dendrimer, Biotin, Regulation

1. Introduction

Low molecular weight RNA molecules such as microRNA (miRNA) have been shown to be involved in important biological processes such as mRNA degradation, transcriptional gene silencing (TGS), and translational repression (1–8). As a result, these noncoding RNAs are gaining the interest of the scientific community as possible new drug targets and for use in diagnostics. Microarrays are a useful tool for massively parallel assessment of messenger RNA (mRNA) expression levels (9, 10). GeneChip® miRNA microarrays provide comprehensive miRNA coverage, representing miRNA sequences from all organisms present in miRBase (http://www.mirbase.org) (11–13), as well as a number of homo sapiens small nucleolar RNAs (snoRNA) and small Cajal body-specific (sca) RNAs derived from snoRNABase (http://www-snorna.biotoul.fr/) (14) and Ensembl (http://www.ensembl.org) (15). Because miRNA sequences are very short (typically 21–23 nt), it is not possible to interrogate each miRNA with multiple probes differing in sequence. Instead, miRNA sequences are represented by probe

Jian-Bing Fan (ed.), *Next-Generation MicroRNA Expression Profiling Technology: Methods and Protocols*, Methods in Molecular Biology, vol. 822, DOI 10.1007/978-1-61779-427-8_8, © Springer Science+Business Media, LLC 2012

sets with identical probes complementary to the miRNA sequence on the arry. Longer RNAs, such as homo sapiens snoRNAs, scaRNAs, and 5.8 s rRNA have 9–11 probes per set. Also included are probe sets for hybridization controls and background subtraction, binned by GC content.

The miRNA is labeled and detected by fluorescent emission. Total RNA samples or total RNA samples enriched for low molecular weight (LMW) RNA may be used as starting material for the RNA labeling process. The Genisphere® FlashTag™ HSR labeling kit is used for microRNA labeling using a short two-step process that does not require any purification prior to array hybridization. Briefly, the process begins with a poly(A) tailing reaction followed by ligation of a biotinylated signal molecule to the target RNA sample. The labeled RNA sample is then added to a hybridization mix and hybridized to a miRNA microarray overnight, followed by a procedure to wash, stain, and scan the microarray to acquire the data. A free software tool (miRNA QC Tool) is available for data normalization, summarization, and quality control assessment.

2. Materials

2.1. RNA Isolation

The kits listed below for purification of total RNA or LMW RNA are compatible with FlashTag Biotin HSR. Elute or resuspend the RNA in nuclease-free water. Ensure that the purification method retains LMW species. Commercial products that have been tested successfully with FlashTag Biotin HSR include:

1. Applied Biosystems: mirVana™ miRNA Isolation Kit.
2. Qiagen: miRNeasy Mini Kit.
3. OriGene: Vantage™ microRNA Purification Kit.
4. Invitrogen: PureLink™ miRNA Isolation Kit.
5. Invitrogen: TRIzol® reagent (total RNA only) with additional overnight −20°C precipitation step during isopropanol precipitation.

When working with formalin fixed paraffin embedded (FFPE) samples, the RecoverALL™ Total Nucleic Acid Isolation Kit for FFPE (Applied Biosystems) is recommended to ensure complete recovery of both mRNA and microRNA and has been successfully used with FlashTag Biotin HSR and GeneChip miRNA Arrays.

2.2. RNA Labeling

1. Genisphere FlashTag Biotin HSR for Affymetrix miRNA Arrays (Genisphere). Store all components at −20°C.
2. Instructions for handling kit contents:

   Vials 1, 2, 5, 7, 9, 11, and 12: Thaw at room temperature, vortex, and briefly microfuge.

Vials 3, 8, and 10: Thaw on ice, microfuge if necessary, and *keep on ice at all times.*

Vials 4 and 6: Remove from freezer just prior to use, and briefly microfuge. Keep on ice at all times. Do not vortex.

3. Kit components

 Vial 1: 10× Reaction Buffer.

 Vial 2: 25 mM $MnCl_2$.

 Vial 3: ATP Mix.

 Vial 4: PAP Enzyme.

 Vial 5: **5**× FlashTag Biotin HSR Ligation Mix.

 Vial 6: T4 DNA Ligase.

 Vial 7: HSR Stop Solution.

 Vial 8: RNA Spike Control Oligos.

 Vial 9: Enzyme Linked Oligosorbent Assay (ELOSA) Spotting Oligos.

 Vial 10: ELOSA Positive Control.

 Vial 11: Nuclease-Free Water.

 Vial 12: 27.5% Formamide.

4. Nuclease-free water (Applied Biosystems cat. no. AM9932 or equivalent).
5. 1 mM Tris–HCl. Transfer 50 mL nuclease-free water to a 50 mL conical tube. Remove and discard 50 μL water. Add 50 μL of 1 M Tris–HCl, pH 8 (USB cat. no. 22638). After this dilution is made, do not take a pH reading. Store at room temperature for up to 3 months.

2.3. GeneChip miRNA Array Processing

1. GeneChip miRNA Array (Affymetrix).
2. GeneChip Hybridization, Wash and Stain Kit (Affymetrix).
3. GeneChip Hybridization Control Kit (Affymetrix).
4. Laser Tough-Spots® (3/8 in. diameter; 1/2 in. diameter; Diversified Biotech).
5. GeneChip Scanner 3000 7G System (Affymetrix).

3. Methods

Table 1 below describes general recommendations for RNA input. The amount of microRNA relative to total RNA is less in cultured cells than in tissues. Therefore, using more RNA isolated from cultured cells is recommended.

Table 1
RNA input recommendations

RNA sample	Input for FlashTag biotin HSR labeling
Total RNA containing LMW RNA	100–1,000 ng total RNA from tissues 500–1,000 ng total RNA from cultured cells
Enriched LMW RNA, quantitated	100–400 ng LMW RNA from tissues 200–400 ng LMW RNA from cultured cells
Enriched LMW RNA, not quantitated	Enriched from 100 to 1,000 ng total RNA

3.1. Poly(A) Tailing of RNA Sample

1. Adjust the volume of RNA to 8 μL with Nuclease-Free Water (Vial 11).
2. Transfer the 8 μL RNA to ice. Add 2 μL RNA Spike Control Oligos (Vial 8) and return to ice.
3. Dilute the ATP mix (Vial 3) in 1 mM Tris–HCl as follows:
   (a) For *total* RNA *samples*, dilute the ATP Mix 1:500.
   (b) For *enriched*, quantitated *samples*, calculate the dilution factor according to the following formula: 5,000/ng input LMW RNA (see Note 1).
   (c) For *enriched* samples *that are not quantitated*, calculate the dilution factor according to the following formula: 1,000/μg input total RNA (see Note 2).
4. Add the following components to the 10 μL RNA/Spike Control Oligos, for a volume of 15 μL:
   (d) 1.5 μL 10× Reaction Buffer (Vial 1).
   (e) 1.5 μL 25 mM $MnCl_2$ (Vial 2).
   (f) 1.0 μL diluted ATP Mix (Vial 3 dilution from step 3).
   (g) 1.0 μL PAP Enzyme (Vial 4) (see Note 3).
5. Mix gently (do not vortex) and microfuge.
6. Incubate in a 37°C heat block for 15 min. Discard any unused, diluted ATP Mix from step 2.

3.2. FlashTagBiotin HSR Ligation

1. Briefly microfuge the 15 μL of tailed RNA and place on ice.
2. Add 4 μL 5× FlashTag Biotin HSR Ligation Mix (Vial 5).
3. Add 2 μL of T4 DNA Ligase (Vial 6).
4. Mix gently (do not vortex) and microfuge.
5. Incubate at 25°C (room temperature) for 30 min.
6. Stop the reaction by adding 2.5 μL HSR Stop Solution (Vial 7). Mix and microfuge the 23.5 μL of ligated sample.

7. Remove 2 μL of the biotin-labeled sample for use in the ELOSA QC Assay. It is acceptable to store the 2 μL of biotin-labeled sample on ice for up to 6 h or at -20°C for up to 2 weeks, and run the ELOSA QC Assay at a convenient time.
8. The remaining 21.5 μL biotin-labeled sample may be stored on ice for up to 6 h or at -20°C for up to 2 weeks, prior to hybridization on Affymetrix GeneChip miRNA Arrays.

3.3. miRNA Microarray Hybridization

1. Install the miRNA array library files. This requires Affymetrix GeneChip® Command Console® (AGCC) software to be installed on the instrument control workstation. To install library files, download from the miRNA array product page on the Affymetrix Web site http://www.affymetrix.com. Use the Library File Importer to import the downloaded zip file from a local or network drive into AGCC (see *AGCC Installation Instructions User Guide*).
2. Set temperature of GeneChip Hybridization Oven 645 to 48°C. Set rotations per minute (rpm) to 60.
3. Remove miRNA array pouches from refrigerated storage, remove arrays from pouches, and allow to equilibrate to room temperature (10–15 min).
4. Bring the following reagents to room temperature. Completely thaw the 20× Hybridization Controls, then heat to 65°C for 5 min.
   (a) 2× Hybridization Mix (from Hybridization, Wash and Stain Kit).
   (b) 27.5% Formamide (Vial 12 from FlashTag Biotin HSR Kit).
   (c) DMSO (from Hybridization, Wash and Stain Kit).
   (d) Control Oligo B2, 3 nM (from Hybridization Control Kit).
   (e) 20× Hybridization Controls (from Hybridization Control Kit).
5. Register samples by selecting *Register* from the Samples menu in AGCC (*see AGCC User Manual*). After populating the array, barcode and attribute fields as desired for all arrays in the experiment, click *Save*.
6. Prepare hybridization mixes as follows, by adding each component in the order listed to the 21.5 μL of biotin-labeled sample for a total volume of 103.2 μL. A master mix of hybridization reagents may be prepared if multiple arrays will be processed.
   (a) 50 μL 2× Hybridization Mix.
   (b) 15 μL 27.5% Formamide (Vial 12).
   (c) 10 μL DMSO.

(d) 5 µL 20× Hybridization Controls.

(e) 1.7 µL Control Oligo B, 3 nM.

7. Heat the hybridization cocktail to 99°C for 5 min in a heat block, then 45°C for 5 min.

8. Each array has two septa. In order to fill the array with hybridization mix, first vent the array chamber by inserting a clean, unused pipette tip (20 µL or 200 µL unfiltered tips recommended) into the upper right septum. Transfer 100 µL of the hybridization mix that has been heated at 45°C, by inserting the pipette tip of a micropipettor into the remaining septum to fill.

9. Remove the pipette tip from the upper right septum. Cover both septa with 1/2 in. Tough-Spots to prevent leaking and/or minimize evaporation.

10. Place miRNA Arrays into the hybridization oven, arranging them in a balanced configuration around the axis to avoid stress to the motor. Hybridize at 48°C and 60 rpm for 16 h.

3.4. miRNA Microarray Wash/Stain/Scan

For more details around operation of the Fluidics Station 450 (FS450) and wash/stain protocols, see the user guides for the Hybridization, Wash and Stain (HWS) Kit and AGCC. The FS450 should be primed before use according to the HWS Kit User Guide. For more information regarding array scanning on the GeneChip Scanner 3000 7G, please refer to the AGCC User Manual (Chapter 6) or the GeneChip Expression Wash, Stain and Scan User Manual for Cartridge Arrays (Chapter 3). Turn on the GCS 3000 scanner and allow to warm up 10 min (when the green light is solid, it indicates the scanner is ready to scan).

1. Prior to the end of the 16 h hybridization, prepare the wash and stain reagents. Remove Stain Cocktail 1, Stain Cocktail 2, and Array Holding Buffer from the Stain Module, Box 1. Tap bottles gently to mix. Stain Cocktail 1 is light sensitive. Use amber microcentrifuge tubes and protect from light. For each array to be washed and stained, aliquot:

   (a) 600 µL of Stain Cocktail 1 into a 1.5-mL amber microcentrifuge vial.

   (b) 600 µL of Stain Cocktail 2 into a 1.5-mL (clear) microcentrifuge vial.

   (c) 800 µL of Array Holding Buffer into a 1.5-mL (clear) microcentrifuge vial.

2. Briefly centrifuge all vials to remove air bubbles.

3. After 16 h of hybridization, remove arrays from hybridization oven and remove Tough-Spots from arrays. Vent the array as described in Subheading 3.3, step 8 and extract the hybridization mix from each array and transfer to a new tube or well of a 96-well plate. Place at −80°C for long-term storage.

4. Fill each array completely with Array Holding Buffer. Allow arrays to equilibrate to room temperature prior to starting wash/stain procedure. Arrays containing Array Holding Buffer may be stored at 4°C up to 3 h before the wash/stain procedure.
5. Start AGCC Fluidics Control Software on the instrument workstation. Click the tab for the station to be used. In the barcode box, either type the barcode or enter by scanning arrays with a barcode scanner. Select appropriate fluidics script and enter for all fluidics modules used. Click *Run*.
6. Insert probe arrays into the appropriate modules on the fluidics station while cartridge lever is in the down or eject position. Return lever to up or engaged position. Ensure wash bottles and water reservoir have sufficient liquid for the run to complete and that the waste container is empty.
7. Remove any vials from sample holders on the fluidics station. Follow the LCD instructions on the fluidics station to place the three vials prepared in step 1, above as follows:
   (a) Place one vial containing 600 μL of Stain Cocktail 1 in sample holder 1.
   (b) Place one vial containing 600 μL of Stain Cocktail 2 in sample holder 2.
   (c) Place one vial containing 800 μL of Array Holding Buffer in sample holder 3.
8. Press down on the needle lever to snap needles into position and start the run. The LCD will indicate the run is complete by displaying the message *EJECT & INSPECT CARTRIDGE*. Press cartridge lever to eject position and remove arrays. If no bubbles are visible in probe array window, proceed to scanning on the GeneChip Scanner 3000 7G.
9. Remove Tough-Spots used during the wash and stain procedure. Clean excess fluid from around the septa. Clean glass surface with nonabrasive wipe and apply 3/8 in. Tough-Spots to septa to prevent leakage. Fill the Autoloader carousel with arrays (maximum capacity, 48 arrays). Start the AGCC Scan Control Software and click the Start button in the main tool bar.

3.5. Array Analysis

Use the free miRNA QC Tool software found on the Affymetrix Web site for data summarization, normalization, and quality control: http://www.affymetrix.com.

1. Edit the workflow in miRNA QC Tool to that illustrated in Fig. 1 below using the following steps:
   (a) Open miRNA QC Tool software.
   (b) Click *Cancel* in the Cel Files to process window (do not load any CEL files).
   (c) Select Tools → *Workflow Editor*.

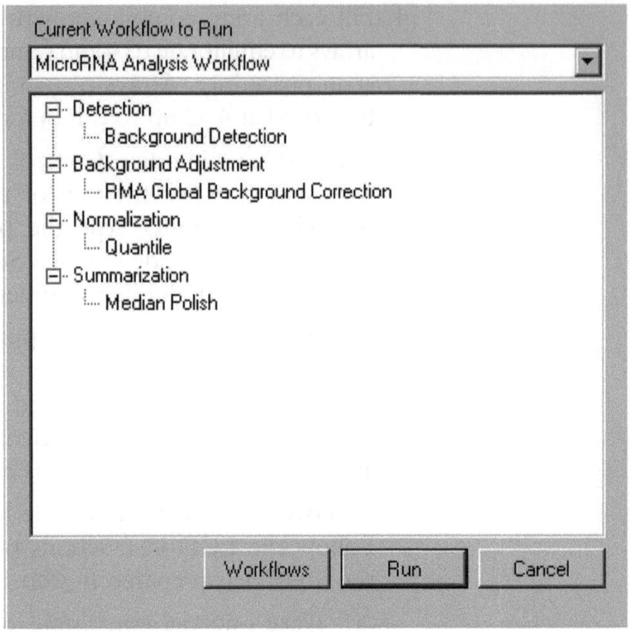

Fig. 1. Edited microRNA QC workflow for data analysis.

(d) Select *File* → *New*.

(e) Type a name for the new workflow.

(f) Add the procedures for Detection, Background Adjustment, Normalization, and Summarization (listed in Fig. 1) by selecting them on the left, and clicking the middle "arrow" button.

(g) Select *File* → *Save*.

(h) Click *OK*.

(i) Click *Close* to close the editor.

2. Ensure that library files match the CEL files being analyzed by selecting *Tools* → *Settings and Options…* from the menu bar. The Settings dialog box will appear. Click the *Browse* button to point the tool to the location of the following library files: CDF File, Annotation File, Background BGP File, List of Probes File, and Quality Control QCC file. Click OK when file locations have been entered.

3. To run an analysis with the settings described in step 1, select *File* → *Load CEL Files…* from the menu bar. The Cel Files to process window will open. Click the *Add* button to open a Windows input dialog box which allows the user to browse to the desired file location. Highlight the list of CEL files to be analyzed and click *Open*. When the CEL files appear in the window, click the *Load* button. When the loading is finished, the selection window will close and Analysis will be enabled. The Run Analysis dialog box will appear.

4. In the Run Analysis box, select the desired workflow from the *Current Workflow to Run* drop-down menu, then click the *Run* button. A Data Table will appear when analysis is complete. For each CEL file analyzed, three columns of data will be reported per probe set: robust multiarray analysis (RMA) summarized signal (log_2), *p*-value for the detection call and Detection call (TRUE/FALSE).
5. To evaluate the success of the target labeling and array hybridization, use the miRNA QC Tool software to evaluate the RNA Spike Control Oligos from Vial 8 in the list below. Select *Tables → Quality Control* to open the Quality Control Table. Each probe set should show >1,000 U (signal-background). The Affymetrix library file lists the following names for these probe sets:
   (a) spike_in-control-2_st
   (b) spike_in-control-23_st
   (c) spike_in-control-29_st
   (d) spike_in-control-31_st
   (e) spike_in-control-36_st

3.6. Array Performance

3.6.1. Sensitivity and Dynamic Range

Figure 2 summarizes the assessment of the sensitivity of the GeneChip microRNA array combined with FlashTag HSR labeling.

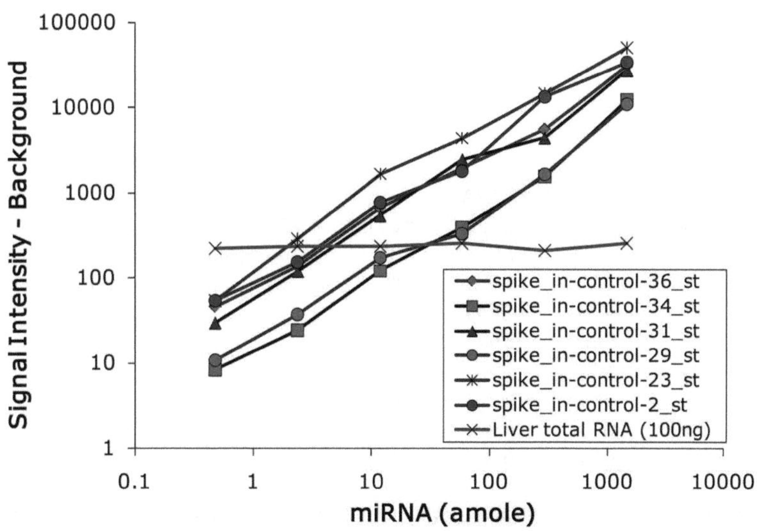

Fig. 2. Platform sensitivity: to determine sensitivity of the labeling and array, 0.5–7,400 amol of six synthetic RNA 22-mer oligos, complementary to features present on the GeneChip miRNA Array, were titrated in a background of 100 ng human liver total RNA (Ambion First Choice®), labeled with the FlashTag HSR kit and hybridized in duplicate. Average signal-background intensities were plotted for each miRNA spike input. A linear concentration vs. signal response was observed for all spiked RNAs for the concentration range tested. Less than 1 amol of each spike was detected significantly above background.

3.6.2. Specificity

Figure 3 summarizes the specificity of the GeneChip microRNA array combined with FlashTag HSR labeling to distinguish between miRNA transcripts with sequence differences of one or two nucleotides.

3.6.3. Reproducibility and Data Validity

Figure 4 demonstrates the reproducibility of the GeneChip microRNA array combined with FlashTag HSR labeling with respect to coefficient of variation (CV) and Pearson correlation of signal and fold change.

3.6.4. Data Validity

Cross platform validation of data is critical to reaching accurate conclusions for any microarray experiment. Figure 5 demonstrates the validation of experimental data observed on the GeneChip microRNA array combined with FlashTag HSR labeling compared to qPCR.

3.7. Labeling Reaction ELOSA QC Assay

The FlashTag HSR labeling procedure includes quality assessment procedure, ELOSA, designed to provide confirmation that the FlashTag HSR Labeling Kit has performed appropriately as a biotin labeling process. Specifically, the ELOSA is designed to detect the RNA Spike Control Oligos (Vial 8) included in all FlashTag HSR labeling reactions. Only 2 μL of the labeling reaction is required for the ELOSA. Successful biotin labeling is verified via a simple colorimetric ELOSA through the hybridization of the biotin-labeled RNA Spike Control Oligos (Vial 8) to complementary ELOSA Spotting Oligos (Vial 9) immobilized onto microtiter plate wells. This assay should be run prior to the use of any labeling reaction on microarrays to assure the FlashTag

FlashTag HSR

	Labeled MicroRNA			
	Let 7a	Let 7b	Let 7c	Let 7f
7a	100%	43%	37%	4%
7b	2%	100%	9%	0%
7c	8%	26%	100%	0%
7d	9%	5%	6%	0%
7e	1%	1%	1%	0%
7f	1%	2%	2%	100%
7g	0%	0%	0%	0%
7i	0%	0%	0%	0%

miRNA	Sequence
let-7a	UGAGGUAGUAGGUUGUAUAGUU
let-7b	UGAGGUAGUAGGUUGUGUGGUU
let-7c	UGAGGUAGUAGGUUGUAUGGUU
let-7d	AGAGGUAGUAGGUUGCAUAGUU
let-7e	UGAGGUAGGAGGUUGUAUAGUU
let-7f	UGAGGUAGUAGAUUGUAUAGUU
let-7g	UGAGGUAGUAGUUUGUACAGUU
let-7i	UGAGGUAGUAGUUUGUGCUGUU

Fig. 3. Platform specificity: to determine specificity, synthetic hsa-let7a, 7b, 7c, and 7f (0.02 ng each) were individually labeled with the FlashTag HSR kit in a background of 20 different synthetic miRNAs (0.4 ng each) and hybridized. The 20 synthetic miRNAs are known to have no cross-reactivity with the let-7 family arrayed probes. For analysis, array signals were normalized to the perfect match probe–target for each array. Some cross-hybridization was observed with base changes near the end of the transcript (hsa-let 7b compared to hsa-let-7a and hsa-let-7c; hsa-let-7c compared to hsa-let-7a), but the true target probes had the highest intensity compared to related RNAs.

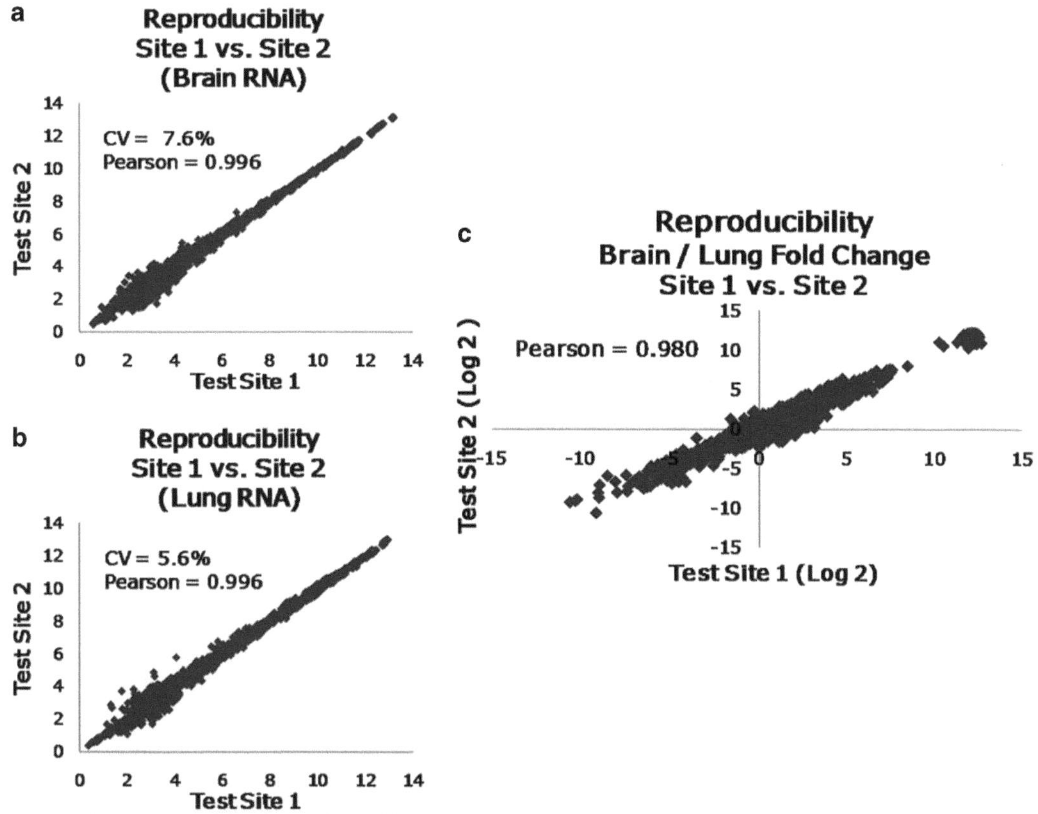

Fig. 4. Platform reproducibility: to determine reproducibility, 200 ng of Ambion Brain and Lung Total RNAs were labeled with the FlashTag HSR kit and hybridized in duplicate at two different sites. The % coefficient of variation (CV) and Pearson correlation coefficients were calculated using detected probe sets. (**a**) Signal correlation of Brain RNA test site 1 and 2. (**b**) Signal correlation of Lung RNA between test site 1 and 2. (**c**) Folder change correlation between site 1 and 2. A high level of correlation between test sites was observed for both signal ($R > 0.990$) and fold change ($R = 0.980$).

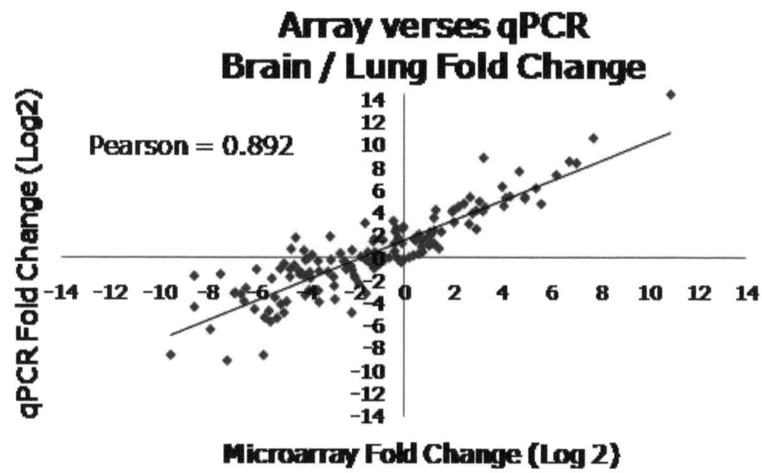

Fig. 5. Data validation: *Array*: 200 ng human brain and lung total RNAs (Ambion FirstChoice®) were labeled with FlashTag HSR in duplicate and hybridized to Affymetrix GeneChip miRNA Arrays. Replicate $\log_2$ signal intensities were averaged. To determine fold change, lung average $\log_2$ signal was subtracted from brain average $\log_2$ signal. *qRT-PCR*: Ct values were downloaded from Ambion TechNotes 14(2) – May 2007: miRNA Expression in FirstChoice® Human Brain Reference RNA (http://www.ambion.com/techlib/tn/142/5.html). To determine fold change, brain Ct values were subtracted from lung Ct values. Brain/lung fold changes were plotted for 149 miRNAs in common to both data sets. Despite the RNAs coming from different input total RNA, good correlation ($R = 0.892$) for fold change was observed for arrays compared to qRT-PCR data.

HSR labeling process worked appropriately with known controls. This procedure is not quantitatively predictive of the performance of any RNA sample on a microarray. To understand the validity of this ELOSA method, appropriate controls should be included in all ELOSAs. The ELOSA procedure includes the following general steps and is available on line and as part of the FlashTag HSR protocol. General steps for performing the ELOSA QC procedure:

(a) Coat an ELOSA plate with the ELOSA spotting oligos.

(b) Wash unbound oligos from the plate and block.

(c) Bind a portion of the FlashTag HSR labeling reaction to the spotted oligonucleotides on plate.

(d) After a brief wash, bind Streptavidin HRP to the bound biotin probes.

(e) Detect the signal.

4. Notes

1. Example: If using 100 ng of enriched LMW RNA, the dilution factor is 5,000 ÷ 100 = 50. Dilute the ATP Mix 1:50.
2. Example: If the sample was enriched from 500 ng total RNA, the dilution factor is 1,000 ÷ 0.5 = 2,000. Dilute the ATP Mix 1:2,000.
3. If at least five labeling reactions are simultaneously run, a master mix may be prepared at this step. Prepare one extra reaction's worth of reagents. For example, when five samples are run, prepare a master mix for six samples as shown below. Add 5 μL of master mix to the 10 μL RNA/Spike Control Oligos, for a volume of 15 μL.

   (a) 9 μL 10× Reaction Buffer (Vial 1).

   (b) 9 μL 25 mM $MnCl_2$ (Vial 2).

   (c) 6 μL diluted ATP Mix (Vial 3 dilution from step 3).

   (d) 6 μL PAP Enzyme (Vial 4).

References

1. Schembri, F., Sridhar, S., Perdomo, C., Gustafson, A. M., Zhang, X., Ergun, A., et al. (2009) MicroRNAs as modulators of smoking-induced gene expression changes in human airway epithelium. *Proc Natl Acad Sci USA* **106**, 2319–24.
2. Taylor, E. L. and Gant, T. W. (2008) Emerging fundamental roles for non-coding RNA species in toxicology. *Toxicol* **246**, 34–9.
3. Ronemus, M., Vaughn, M. W. and Martienssen, R. A. (2006) MicroRNA-Targeted and Small Interfering RNA–Mediated mRNA Degradation Is Regulated by Argonaute, Dicer, and RNA-Dependent RNA Polymerase in Arabidopsis. *Plant Cell* **18**, 1559–74.
4. Morel, J. B., Godon, C., Mourrain, P., Béclin, C., Boutet, S., Feuerbach, F., et al. (2002) Hypomorphic ARGONAUTE (ago1)

Mutants Impaired in Post-Transcriptional Gene Silencing and Virus Resistance. *Plant Cell* **14**, 629–39.

5. Krichevsky, A. M., King, K. S., Donahue, C. P., Khrapko, K., Kosik, K. S. (2003) A microRNA array reveals extensive regulation of microRNAs during brain development. *RNA* **9**, 1274–81.
6. Schmittgen, T. D., Jiang, J., Liu, Q. and Yang, L. (2004) A high-throughput method to monitor the expression of microRNA precursors. *Nucleic Acids Res* **32**, e43.
7. Thomson, J. M., Parker, J., Perou, C. M. and Hammond, S. M. (2004) A Custom Microarray Platform for Analysis of MicroRNA Gene Expression. *Nat Methods* **1**, 47–53.
8. Ambros, V. (2004) The functions of animal microRNAs. *Nature* **431**, 350–5.
9. Fodor, S. P., Read, J. L., Pirrung, M. C., Stryer, L., Lu, A. T. and Solas, D. (1991) Light-directed, spatially addressable parallel chemical synthesis. *Science* **251**, 767–73.
10. Lockhart, D. J., Dong, H., Byrne, M. C., Follettie, M. T., Gallo, M. V., Chee, M. S., et al. (1996) Expression monitoring by hybridization to high-density oligonucleotide arrays. *Nat Biotechnol* **14**, 1675–80.
11. Griffiths-Jones, S., Saini, H. K., van Dongen, S. and Enright, A. J. (2008) miRBase: tools for microRNA genomics. *Nucleic Acids Res* **36**, D154–8.
12. Griffiths-Jones, S., Grocock, R. J., van Dongen, S., Bateman, A. and Enright, A. J. (2006) miRBase: microRNA sequences, targets and gene nomenclature. *Nucleic Acids Res* **34**, D140–4.
13. Griffiths-Jones, S. (2004) The microRNA Registry. *Nucleic Acids Res* **32**, D109–11.
14. Lestrade, L. and Weber, M. J. (2006) snoRNA-LBME-db, a comprehensive database of human H/ACA and C/D box snoRNAs. *Nucleic Acids Res* **34**, D158–2.
15. Hubbard, T. J., Aken, B. L., Ayling, S., Ballester, B., Beal, K., Bragin, E., et al. (2009) Ensembl 2009. *Nucleic Acids Res* **37**, D690–7.

Chapter 9

Individualized miRNA Assay Panels Using Optically Encoded Beads

Keld Sorensen

Abstract

This chapter describes how to create a multiplexed miRNA profiling assay using readily available technology and reagents for midplex miRNA profiles (midplex used to denote mutliplexing up to about 100 miRNAs per well). The assay described is a direct hybridization assay, i.e., it involves no amplification. Further, the reagents are simple and an assay can be assembled in a few hours with simple design rules and very simple execution. The assay execution takes less than 5 h and involves only liquid addition, which makes it possible to create a multiplexed assay for a large number of miRNAs and run the entire profile for hundreds of samples in a matter of a couple of days. In addition, the profiles can be created for any short RNAs, i.e., they are not tied to human or any other species database for miRNAs, and the described assay works for both animal and plant miRNAs.

Key words: Bead-based assay, Encoded beads, miRNA, Nonamplified, Step-down hybridization, Array, Panels, Flow cytometry, Labeling

1. Introduction

Measurement of miRNA is an important element of understanding cell biology, in particular cancer – for a recent review, see Calin and Croce (1). A number of commercial methods exist for measuring levels of miRNA in a cell population. These methods can generally be divided into amplified and nonamplified methods. The amplified generally are based on PCR, whereas the nonamplified rely on capturing the miRNA and generate a signal that is proportional to the amount of miRNA present in the sample. Assay methods can also be classified as either singleplex or multiplex, where the term multiplex is used to denote assays where at least two results are simultaneously obtained from a single "well" of a reaction.

Jian-Bing Fan (ed.), *Next-Generation MicroRNA Expression Profiling Technology: Methods and Protocols*, Methods in Molecular Biology, vol. 822, DOI 10.1007/978-1-61779-427-8_9, © Springer Science+Business Media, LLC 2012

Singleplex is used to denote the assay, where a single well renders a single result. While the PCR-based methods have gained wide popularity (2), the inherent variability of extensive amplification has not been solved. This chapter describes a particular design of an assay that is a multiplexed nonamplified method that in addition has the advantage of not requiring labeling of the miRNA. The method can be used both for bead-based technologies and printed arrays. For sake of demonstration, the Luminex instrument is used for reading the assay, although other encoded beads and instruments can be utilized. Key elements in the assay are the optically encoded beads which are magnetic beads which have unique zip code sequences coupled to them. These sequences are universal array sequences that are not cross hybridizing with each other or with most known sequence in the biome, as established by BLAST searching. Although commercial beads are available, the literature describes how these can be made in the laboratory (3).

2. Materials

To create a customized miRNA assay for a panel of miRNAs, the following materials are needed.

2.1. Chimeric Probes

Chimeric probes are needed for each miRNA in the assay panel (see Note 1). These are oligo probes that are ordered from an oligo manufacturer or manufactured "in-house." Note that the probes are chimeric RNA–DNA probes (see Note 2) and are biotinylated. The probes are designed by the researcher as follows.

1. Obtain miRNA sequence (5′ to 3′).
2. Obtain zip-code sequence (5′ to 3′), for example, from Luminex (see Table 1 for examples) or other source of such universal array sequences (see Note 3).
3. Create reverse complement of miRNA sequence either manually or using software, such as that offered by oligo vendors.
4. Create reverse complement of zip-code sequence (as in step 3).
5. Concatenate the two sequences (RNA–DNA 5′ to 3′) and add biotin on the 5′ end.
6. Order this oligo.

   Design example

   miRNA sequence of interest mmu-miR-34b-5p: AGGCAGUGUAAUUAGCUGAUUGU.

   Create reverse complement: ACAAUCAGCUAAUUACACUGCCU (note that the sequence is RNA).

   Zip-code sequence on bead: GTTGTAAATTGTAGTAAAGAAGTA.

Table 1
This table demonstrates the first 14 zip-code entries in a list of the Luminex collection

Bead region	Sequence bound to each microsphere (zip code)
7	AAATTGTGAAAGATTGTTTGTGTA
8	TGTAAGTGAAATAGTGAGTTATTT
9	GAATTGTATAAAGTATTAGATGTG
12	AGTAGAAAGTTGAAATTGATTATG
13	AGTGAATGTAAGATTATGTATTTG
14	ATTGTGAAAGAAAGAGAAGAAATT
15	GTTGTAAATTGTAGTAAAGAAGTA
18	GTAATTGAATTGAAAGATAAGTGT
19	GTGTGTTATTTGTTTGTAAAGTAT
20	AAATTAGTTGAAAGTATGAGAAAG
21	ATTAAGTAAGAATTGAGAGTTTGA
22	GATTGATATTTGAATGTTTGTTTG
25	GTATGTTGTAATGTATTAAGAAAG
26	TTTGATTTAAGAGTGTTGAATGTA
…	…

Each sequence is unique and chemically attached to a bead "region" through an NH2 group and a spacer. Bead regions are used to identify the bead in the Luminex reader. For further information and complete list, please consult http://www.luminexcorp.com

Create reverse complement: TACTTCTTTACTACAA TTTACAAC (note that the sequence is DNA).

Chimeric oligo to order (concatenate of the above, plus biotin)

Biotin – ACAAUCAGCUAAUUACACUGCCUTACTTC TTTACTACAATTTACAAC.

Repeat this design process for each miRNA you wish to include in the panel. Software to create reverse complements is readily available at no cost or can easily be implemented in a spreadsheet. Note that the probes MUST be designed to be a perfect match to the miRNA under investigation, as the high specificity of the method (see Notes 4 and 5) allows discrimination between closely related miRNAs (see Fig. 1, for an example).

2.2. Beads with DNA Capture Probes (Zip Codes)

If the user wishes to use commercially available capture probes, these can be purchased, although manufacture of these is simple. A set of sequences obtained from one vendor is listed in Table 1.

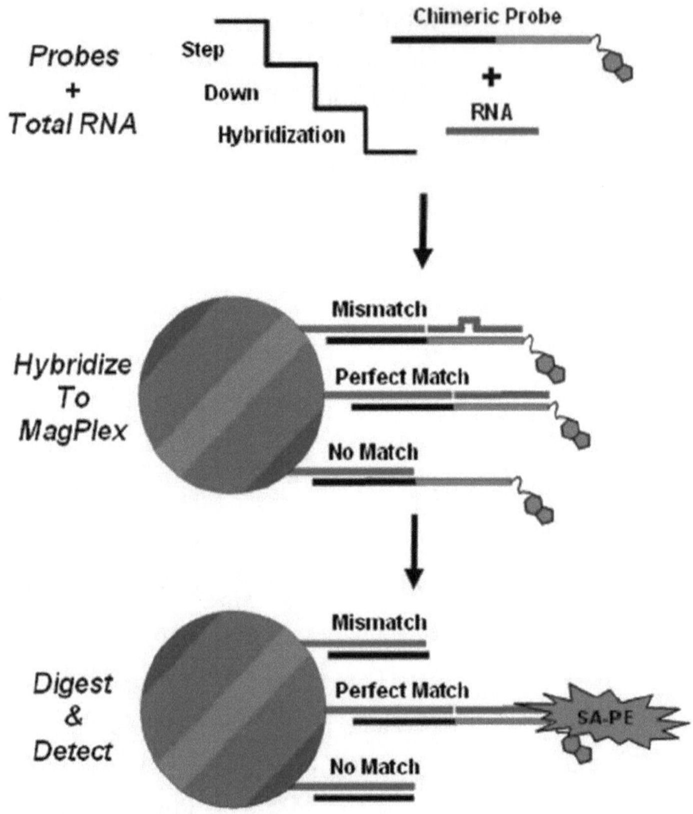

Fig. 1. Process of assay. For each chimeric probe intended to capture a given miRNA, a total of three situations may occur: a perfect match (signal is desired), a mismatch (signal is not desired), and the absence of the target miRNA (signal should not be generated). The assay utilizes a step-down hybridization protocol and an enzymatic treatment to remove mismatches and "no matches," i.e., chimeric probes that do not capture an miRNA target are removed by the RNase One treatment. This ensures that signal is only generated when the correct target is hybridized to the chimeric probe and captured on the beads. Unwanted reactions are absent or minimized by the step-down protocol which favors perfect matches and the enzymatic treatment which removes the unwanted reactions.

These describe the commercial bead sequences from Luminex. It is recommended that the user verifies sequence information on commercial products prior to design of the assay, as these sequences may change.

2.3. Miscellaneous Reagents

1. RNase ONE (Promega).
2. Stock streptavidin–phycoerythrin conjugate (SAPE) – can be ordered from a vendor, such as Moss Substrates. About 1 mg/mL stock solution.
3. Buffer: A single buffer is used for both wash and hybridization buffer. The composition is 10 mM Tris, 200 mM sodium acetate, 5 mM EDTA, and 0.05% Tween 20 – adjust pH to 7.7.

4. Instrument capable of reading beads. There are several models on the market. If the beads are purchased from a vendor, these vendors usually have recommendations for the hardware.
5. Samples to be tested. The sample should be "total RNA" – it is NOT recommended to purify miRNA or "small RNA" as most purification methods to purify only miRNAs may introduce unwanted bias by selectively purifying some species over others, and often have losses that may be universal or specific. Total RNA has the advantage that the RNA is the ideal "carrier" for the miRNA as it is chemically identical. Traditional methods, such as phenol/chloroform, may be used or alternatively bind and elute methods that are known to purify total RNA including short RNA, such as miRNA. Some older bind and elute methods are not suitable as they do not recover miRNAs. Please verify that you are using a suitable method.

2.4. Instruments and Supplies Required

1. PCR cycler for "step-down" hybridization protocol.
2. Luminex instrument for reading of the assay (Luminex 100, 200 or similar flow cytometric reader) (see Note 6).
3. 96-Well PCR plate non-skirted (ThermoFisher).
4. 8-Well flat cap strips, product (ThermoFisher).
5. 1.5-mL tubes for mixing of reagents.
6. Pipette tips (barrier type highly recommended).
7. Bath sonicator.
8. Vortex mixer.
9. Magnetic separation plate, such as Dynal MPC 96 S or Luminex equivalent.
10. Plate shaker for microwell plates. The shaker should ideally accommodate a PCR plate and be able to form a visible vortex in each well.

3. Methods

3.1. Preparation

1. Pick the number of zip-code-carrying beads equal to the number of miRNAs to be measured. This selection can be done by simply starting at the top of a table, such as Table 1, and simply take the number needed for the assay (number of beads = number of miRNAs).
2. Order the biotinylated chimeric DNA/RNA probes from an oligo vendor.
3. Program PCR cycler as follows:
   a. Start with temperature 90°C for 3 min.
   b. Drop temperature to 80°C for 6 min.

c. Drop temperature 1°C every 6 min until 60°C is achieved, i.e., steps are 80°C for 6 min, then 79°C for 6 min, 78°C for 6 min, etc. until 60°C is reached.

d. Drop temperature to 37°C and HOLD until user intervention.

e. Keep temperature at 37°C for 30 min after user intervention.

f. Drop temperature to 30°C and HOLD for user intervention.

g. Keep temperature at 30°C for 30 min.

h. END of profile.

4. Create chimeric probe mix: Each incoming chimeric probe is dissolved to 100 μM in water or TE for plexes up to 100. If higher plex numbers are needed, the concentration should be proportionally higher. The probes can be aliquoted and stored frozen at this point (as concentrates). To create an equimolar mix of chimeric probes, calculate the amount of total probe mix needed (1.25 μL for each sample), and then calculate the total volume needed to create a 1:100 dilution of each probe. Example: A 5-plex miRNA profile is to be analyzed on 96 samples. The total number of samples require 96 × 1.25 μL = 120 μL total probe mix needed. To ensure overage for pipetting, etc., a slightly higher number is recommended, such as 150 μL. Since each of the chimeric probes are at 100 μM and need to be diluted 1:100, each probe should be added at 1.5 μL – the combined probe addition into an empty tube is, thus, 6 × 1.5 = 9 μL. Addition of 141 μL hybridization buffer creates 150 μL of probe mix, where each of the chimeric probes is present at 1 μM.

5. Create bead mix: The incoming MagPlex-Tag bead solution should be adjusted to 50,000 beads/μL (dilute with hybridization buffer as required). Since each region should contribute 1,000 beads per reaction and since 4 μL bead mix is added to the reaction, a mixture consisting of each region is created as follows: Multiply the number of samples by 4 μL. If 100 samples are to be tested, this would be 400 μL. Add about 20% overage for a total of 480 μL. Since each region should be diluted 200-fold to achieve 1,000 beads per region, 2.4 μL of each region should be added to an empty tube and the volume made up to 480 μL with hybridization buffer. This achieves a mix, where 4 μL delivers 1,000 beads per region to each well. Shortly before use, vortex and sonicate the prepared bead mix for 10–15 s. At the 37°C HOLD, resuspend the bead mixture by vortexing and add 4 μL to each well.

6. Prepare sample: Adjust the sample total RNA to the desired input amount – 250–500 ng per sample is recommended. A volume of 2.5 μL is recommended. Example: If the sample RNA concentration is 1 mg/mL, then combine 0.5 μL RNA with 2.0 μL of hybridization buffer to achieve 500 ng in the 2.5 μL recommended sample volume (see Note 7).

3.2. Execution of Assay (Workflow)

1. Biotinylated chimeric probes and samples are mixed and a step-down protocol which allows target miRNAs to hybridize to capture probes (2 h, step-down thermal cycler).
2. Bead hybridization: The probes with their cargo of miRNAs hybridize to beads (30 min, 37°C).
3. Enzyme treatment: Any excess probes and nontarget RNA are digested (30 min, 30°C).
4. Wash to remove excess nontarget RNA, etc. (15 min).
5. SAPE incubation (signal generation on the biotin) (30 min, ambient).
6. Wash to remove excess SAPE.
7. Read on instrument.

3.3. Detailed Protocol

1. Create the master mix in nuclease-free PCR strips (96-well format) – with caps. Add the reagents in the order shown (multiply by number of samples plus a few extra).

   Hybridization buffer 17.25 μL.

   Chimeric probe mix 1.25 μL.

   Total volume 18.50 μL.

   To obtain the total volume of master mix needed, multiply the above number by the total number of samples. If a total of ten samples (as an example) are to be tested, multiply these numbers with 10 and add at least 20% overage, i.e., make a total of 12 × 18.5 μL master mix. This allows for the pipetting losses. Mix the master mix well.
2. Pipette 18.5 μL of the master mix into each well. Make sure that the reagent is in the tip of the well, not the walls.
3. Pipette 2.5 μL of the total RNA sample into the bottom of each well. The RNA sample should be "total RNA" and should contain 250–500 ng total amount of RNA in the 2.5 μL.
4. Close PCR plate/tubes. Vortex for 5 s. Flick to remove bubbles.
5. Quick spin the PCR plate/tubes to ensure that all reagents are at the bottom of the well. 1–2 s is appropriate.
6. Place samples in thermal cycler programmed with the step-down profile described earlier.
7. At the 37°C HOLD step, pause the thermal cycler and add 4 μL of the bead mix to each well. Mix well by pipetting up and down or remove the resealed plate, vortex for 10–15 s, and quick spin for 1–2 s.
8. Resume step-down program (37°C for 30 min).
9. Enzyme digestion: 5 min prior to the completion of the step-down program, prepare a 1:500 dilution of the stock enzyme using the hybridization buffer as the diluent. At the 30°C

HOLD, pause the thermal cycler and add 2.5 μL of the diluted enzyme to each well while the plate remains in the cycler. It is important that the enzyme is pipetted into the bottom of the tube, not onto the walls.

10. Remove the plate briefly from the thermal cycler, and mix the content well, followed by a brief spin (1–2 s) to bring the entire content down into the tip of the tube.
11. Return plate to the thermal cycler for the final step of 30°C for 30 min.
12. Five minutes prior to the end of the 30°C step, prepare a 1:500 dilution of the fluorescent reporter (SAPE). Calculate the volume by using 75 μL per well with an average of about 10–20%. If ten samples are tested, make 12 × 75 μL of reporter solution.
13. Remove the reaction supernatant prior to the addition of SAPE as follows: Place the plate on a magnetic separator. Let the magnetic beads migrate for 2 min.
14. With a multichannel pipette, remove the fluid gently from the wells without disturbing the bead pellet.
15. Add 200 μL of wash buffer to each well and resuspend the beads by pipetting up and down three or four times.
16. Return the plate to the magnetic separator and again allow the microspheres to migrate and form a pellet (if you are using a low plex number, the pellet may not be readily visible).
17. Again, remove the supernatant gently and carefully.
18. Add 75 μL of the diluted SAPE solution. Cap tubes and mix by vortexing.
19. Place plate on a plate shaker able to form a vortex in each well.
20. Remove plate from plate shaker and place it on the magnetic separator, allowing the microspheres to migrate for 2 min.
21. Remove the SAPE solution by pipetting carefully and gently without disturbing the pellet.
22. Add 200 μL of wash buffer to each well and resuspend the beads by pipetting up and down 3–4 times.
23. Return the plate to the magnetic separator and allow beads to migrate, and remove the supernatant.
24. Repeat the above two steps.
25. Add 100 μL wash buffer for the final resuspension of the beads into this volume. The beads should be totally suspended in this volume by pipetting up and down 3–4 times.
26. Read the plate in the Luminex instrument which has been adjusted for the type of plate. If you prefer, you may transfer the 100 μL of bead suspension to a standard ELISA plate if your instrument is set up for that type of plate.

27. Read the assay on a LUMINEX reader (flow cytometry) which has been set up for the regions of beads you have chosen. See instrument instruction manual.

List of tag sequences that are available from Luminex can be viewed on the Luminex Web site http://www.luminexcorp.com; other sequences can be used, but the user must couple them to beads. Such sequences are generally known as zip-code sequences.

4. Notes

1. A key element in the assay is the chimeric probes which have one RNA sequence, which is a 100% complement to the miRNA to be measured, and one DNA moiety, which is 100% complementary to the zip-code sequence on the bead selected for that particular miRNA. It is imperative that each probe is designed for a specific miRNA as the inherent specificity of the assay does not allow for mismatches in the design.
2. The probes must be chimeric, i.e., RNA–DNA chimers, since the enzyme treatment that affords the specificity is single-strand RNA specific.
3. Zip-code sequences: The zip-code sequences can be obtained in a number of places, including commercially. One such source is Luminex which is using the trademark xTAG for these sequences. The length of zip codes is commonly 24 bases and usually the sets of zip codes are "isothermal," i.e., they have a similar Tm to each other within the set. Zip-code sequences are also known as universal array sequences. End users may design their own sets as long as these are not cross-reacting. Different designs of zip codes may require a different temperature profile for the hybridization steps.
4. *Sensitivity*: The lowest amount of an miRNA that gives a signal is the definition of the assay sensitivity. This may be different for each miRNA under measurement. This is often measured as the lower limit of detection (LOD), and can be estimated by measuring the "background" signal (in the absence of a miRNA), and from that BG measurement establish the mean plus 3 SDs which with great certainty is the signal that can be differentiated from the background "noise." In other words, any signal above background mean plus 3 SDs is a true signal. Anything less must be considered noise. This signal can then be related to the amount of miRNA that generates such a signal and that concentration is the LLD. The sensitivity of an miRNA assay is easily measured by creating a dilution series of miRNA and measuring the signal from each sample.

This should always be done when setting up an assay for the first time. Note that each miRNA in a multiplex has its own sensitivity (limit of detection).

5. *Specificity*: The ability to measure a given miRNA in the presence of other related miRNAs. In particular, "families" of miRNAs should be investigated, since these families may differ only by a few or even a single base. Depending on the nature of the assay, it may be better at differentiating differences in the middle of the molecule, toward the ends, or even at the very end. It is important for the researcher to understand the inherent technology used in an assay and how this affects specificity. The assay demonstrated here utilizes a dual approach to enhance specificity, a step-down hybridization protocol, and an enzymatic treatment of the captured duplex to remove mismatches. The specificity obtained with the assay described here is demonstrated in Fig. 2.

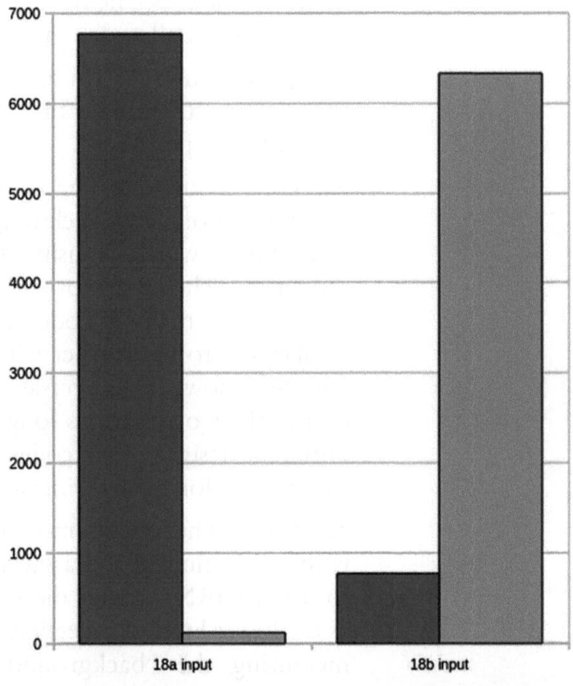

Fig. 2. Specificity of assay shown with miR18a and miR18b at the same time to establish specificity of the assay. At 1 fmol input of each in a mixture of total RNA, the signals plotted in the figure were obtained. At this input level which is in the dynamic range of the method, there is less than about 10% cross hybridization. It is noteworthy that hybridization events are not symmetrical, i.e., the fact that one miRNA gives slightly higher cross hybridization to a different target than the inverse. In the case demonstrated here, miR18a has a cross hybridization to the miR18b capture probe of less than 2%, whereas the inverse cross hybridization, i.e., miR18b cross hybridizing to the miR18a capture probe, is about 10%. At lower input amounts, the cross hybridization is much lower.

6. Flow cytometric readers, such as the Luminex instrument, have limits of plexing, often in the range of 100, i.e., no more than 100 miRNAs can be measured in a single panel on the most popular Luminex instrument.
7. It is highly recommended that novice users first create a sample with synthetic miRNAs added to a complex RNA sample. This allows the user to become comfortable with the method and to establish performance characteristics, such as LOD, CV values, etc. These performance characteristics are dependent on the reagents and instruments used. These synthetic miRNAs can be chosen either as mimics of known biological RNAs or as designed short RNAs that have no relationship to biological material.

Acknowledgments

The method was conceptualized and developed by Ramin Saberi, Brian Schrader, and Cora Lahey, all scientists at Luminex, Austin, TX. My gratitude goes to them for being team players and excellent scientists during my time with Luminex; without them, this assay would not exist, nor would this chapter.

References

1. Calin, G. A. and Croce, C. M. (2006) MicroRNA signatures in human cancers. *Nat Rev Cancer* **6**, 857–66.
2. Chen, C., Ridzon, D. A., Broomer, A. J., Zhou, Z., Lee, D. H., Nguyen, J. T. et al., (2005) Real-time quantification of microRNAs by stem–loop RT–PCR. *Nucleic Acids Res* **33**, e179.
3. Lu, J, Getz, G., Miska, E. A., Alvarez-Saavedra, E., Lamb, J., Peck, D. et al., (2005) MicroRNA expression profiles classify human cancers. *Nature* **435**, 834–8.

Chapter 10

Microfluidic Primer Extension Assay

Markus Beier and Valesca Boisguérin

Abstract

MicroRNAs (miRNAs) are a new class of biomarkers. They represent a group of small, noncoding RNAs that regulate gene expression at the posttranslational level by degrading or blocking translation of messenger RNA (mRNA) targets. miRNAs are important players when it comes to regulating cellular functions and in several diseases, including cancer (Cancer Res 66:7390–7394, 2006; Nature 435:834–838, 2005). So far, miRNAs have been extensively studied in tissue material. Only recently, it was found that miRNAs also exist in a broad range of body fluids (Clin Chem 56:1733–1741, 2010). A major challenge still is the efficient and specific detection of miRNAs. The short length of miRNAs, with only 17–27 base pairs, comes with technical difficulties for analysis. Furthermore, individual miRNAs, especially members of a miRNA family (e.g., the let-7 family), show high sequence homology, with sequences differing by as little as a single base pair. Although miRNAs are abundant in higher copy numbers compared to mRNAs, miRNAs lack a common feature like a poly-A tail that eases detection in a complex background of other RNA species. Besides qPCR, in situ hybridization, and next-generation sequencing, microarrays are versatile tools for high-throughput analysis of already known miRNAs (PLoS One 12:e9685, 2010; Nat Genet 38:S2–S7, 2006; Nature Methods 50:298–301, 2010). Different assay formats have been proposed for expression analysis of miRNAs on microarrays, of which most employed prelabeled RNA molecules. As a modification, the so-called RAKE assay was developed that combined the use of unlabeled RNA with on-chip enzymatic labeling by exonuclease cleavage and polymerase primer extension (RNA 12:187–191, 2006; Nature Methods 1:155–161, 2004; Genome Res 16:1289–1298, 2006).

Here, we describe a simple method for detection of miRNAs based on a combination of stringent hybridization and enzymatic primer extension on a microfluidic microarray starting from total RNA material, without the need for enrichment, amplification, or labeling of the native RNA samples (N Biotechnol 25:142–149, 2008). This assay can be used with starting material as low as 30 ng of total RNA. We have used this technique extensively for identifying specific sets of miRNAs (miRNA signatures) for diagnosis of cancer and cardiovascular or inflammatory diseases from blood samples of patients (Br J Cancer 103:693–700, 2010; BMC Cancer 9:353, 2009; PLoS One 4:e7440, 2009; BMC Cancer 10:262, 2010; Basic Res Cardiol 106(1):13–23, 2011).

Key words: microRNA, Biomarker, Microarray, Hybridization, Primer extension, Klenow

Jian-Bing Fan (ed.), *Next-Generation MicroRNA Expression Profiling Technology: Methods and Protocols*, Methods in Molecular Biology, vol. 822, DOI 10.1007/978-1-61779-427-8_10, © Springer Science+Business Media, LLC 2012

1. Introduction

MicroRNA (miRNA) expression analysis is a powerful tool for biomarker development. Today, biomarkers play a key role in early diagnosis, risk stratification, and therapeutic management of various diseases. Disease-specific expression of miRNAs has been reported in many human cancers employing tissue material as the primary miRNA source. In this context, miRNAs' expression profiles were found to be useful in identifying the tissue of origin for cancers of unknown primary origin (11). Various miRNA biomarkers found in tissue material have been proposed to be correlated with certain diseases, e.g., cancer (1–2). However, there is still a need for novel miRNAs as biomarkers for the detection and/or prediction of cancer and other types of diseases. Especially desirable are noninvasive biomarkers that allow for quick, easy, and cost-effective diagnosis/prognosis which cause only minimal stress for the patient, eliminating the need for surgical intervention.

Since recently, it became clear that miRNAs are not only present in tissues but also in various body fluid samples (e.g., serum, plasma, tears, mother milk, sputum, urine), including human blood (3, 4–10, 12–15). Nevertheless, the mechanism why miRNAs are found in body fluids, especially in blood, or their function in these body fluids is not understood yet. The method described here (5) allows for simple and efficient miRNA analysis starting directly from total RNA. We have used the described method extensively for identification of miRNA biomarkers that are differentially regulated in blood samples between diseased (e.g., subjects suffering from cancer) and healthy individuals.

2. Materials

2.1. Blood Collection and RNA Extraction

1. PaxGene Blood RNA tubes (Beckton Dickinson), miRNeasy Mini Kit (Qiagen, Germany), chloroform, ethanol.
2. Nanodrop 1000 (Fisher Scientific), Bioanalyzer 2100 (Agilent), RNA 6000 Pico and Nano Kit (Agilent).

2.2. Expression Analysis Employing Microfluidic Primer Extension Assay

1. Hybridization buffer: 20× SSPE buffer concentrate, formamide, 100× TE buffer, BSA, Tween-20, DEPC water. Mix 66 μL of SSPE (20×), 22 μL of formamide, 22 μL of TE buffer (1×), 4.4 μL of BSA (50 mg/mL), 2.2 μL of Tween-20 (1%), and 103.4 μL of dd water.
2. Stringent washing buffer (0.5× SSPE): 20× SSPE buffer concentrate, DEPC water. Mix 25 mL of 20× SSPE with 975 mL of ultrapure water.

3. Nonstringent washing buffer (6× SSPE): 20× SSPE buffer concentrate, DEPC water. Mix 300 mL of 20× SSPE with 700 mL of ultrapure water.
4. Streptavidin–phycoerythrin (SAPE) solution: 6× SSPE buffer concentrate, SAPE, BSA. Add 44 μL of SAPE (1 mg/mL) to a solution of 9,000 μL of SSPE (6×) and 360 μL BSA (50 mg/mL). Mix 1,750 μL of stain buffer (2×), 140 μL BSA (50 mg/mL), 35 μL goat IgG solution (10 mg/mL), 21 μL biotinylated anti-streptavidin antibody solution; mix well by inverting.
5. 12× MES buffer: MES-free acid monohydrate, MES sodium salt. Dissolve 64.61 MES-free acid monohydrate and 193.3 MES sodium salt in 1,000 mL of dd water. Adjust pH to 6.5–6.7 and filter through 0.2-μm filter. Mix by inverting, and store at 4°C; protect from light.
6. Antibody solution: MES hydrate, MES sodium salt, NaCl, Tween-20, BSA, goat IgG, biotinylated anti-streptavidin antibody, DEPC water. Mix 41.7 mL of MES buffer (12×), 92.5 mL of NaCl (5 M), 25 mL of Tween-20 (1%), and 90.8 mL of dd water.
7. Prepare stain buffer (2×): Mix 41.7 mL of MES buffer (12×), 92.5 mL of NaCl (5 M), 25 mL Tween-20 (1%), 90.8 mL dd water; store at 4°C. Prepare goat IgG solution (10 mg/mL): Dilute the goat IgG from manufacturer with 5 mL NaCl (150 mM); mix by inverting; store away aliquots of 37 μL at 4°C.
8. Prepare biotinylated anti-streptavidin antibody solution (0.5 mg/mL): Dilute the biotinylated anti-streptavidin antibody with 1 mL of dd water; mix by inverting; store away aliquots of 22 μL at −20°C.
9. Equilibration solution: 10× NEBuffer 2, DEPC water. Mix 160 μL of 10× NEBuffer 2 with 1,440 μL of DEPC water to obtain 1,600 μL of 1× NEBuffer 2. Pipette 200 μL of 1× NEBuffer 2 into each of eight 1.5-mL Eppendorf tubes.
10. Primer extension mix (enzymatic elongation and labeling solution): 10× NEBuffer 2, DEPC water, Biotin-11-dATP, Klenow exo- (50,000 U/mL). Mix 44 μL of NEBuffer 2 (10×), 44 μL of Biotin-11-dATP (40 μM), 2.9 μL Klenow exo- (50,000 U/mL), and 349.1 μL dd water. Ensure that all components have been mixed by pipetting up and down. Pipette 50 μL of primer extension mix into eight 1.5-mL Eppendorf tubes and store on ice until the "equilibration program" is finished. Prepare the stop solution. Pipette 200 μL of SSPE (6×) in each of eight 1.5-mL tubes.

3. Methods

The microfluidic primer extension assay described here (5) (see Fig. 1) makes use of total RNA as starting material. There is no need for enrichment of the small RNA fraction out of the complex pool of different RNA species (e.g., mRNA, tRNA, rRNA) contained in total RNA. No amplification step is required that may introduce bias to the miRNA expression levels. Furthermore, the total RNA is directly employed for hybridization to a microfluidic microarray without prior labeling. This reduces cost and time and adds significant sensitivity to the assay since potential background signals arising from nonspecific hybridization are reduced.

The microfluidic microarrays employed here are fabricated in a microfluidic structure by light-directed in situ synthesis (16). The DNA probes are chemically grown in a high parallel fashion on the solid support, anchoring the probes with the 3′-end to the support. One microstructure contains eight microfluidic microarrays with 15,000 probes each. Each microfluidic microarray comprises a channel-like structure with seven meanders (see Fig. 2). The DNA probes on the microfluidic microarrays are complementary to all known miRNAs, as annotated in the miRBase (17, 18) at the Sanger Institute (www.mirbase.org). Each miRNA is represented by 11 replicates within the microfluidic microarray. The part of the DNA probe that is complementary to miRNAs (M-element in Fig. 1) is attached to the solid surface. On the 5′-end of the M-element, the probes extend to the so-called elongation element

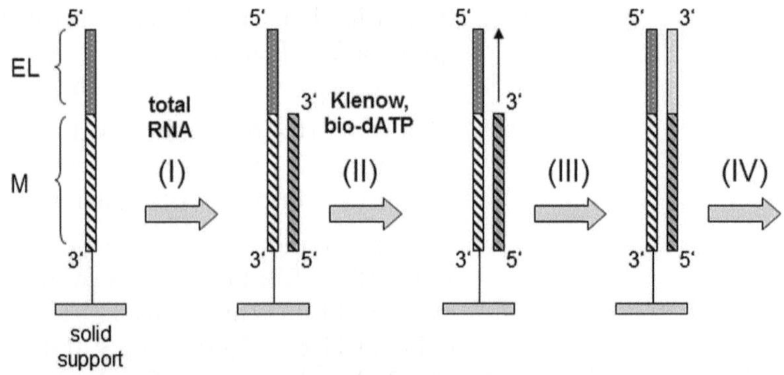

Fig. 1. Microfluidic primer extension assay (MPEA): The solid support-bound probes contain a first sequence stretch complementary to known microRNAs (M-element) and an elongation sequence (EL element) that contains preferably five thymidine nucleotides. (I) stringent hybridization with total RNA to the M-element of the probe; (II) the DNA–microRNA duplex is reacted with Klenow polymerase and biotin-dATP; (III) primer extension using the bound microRNA as primer and the EL element as template, thereby extending the microRNA and simultaneously incorporating labels by use of biotin-labeled dATP; (IV) subsequent staining and detection.

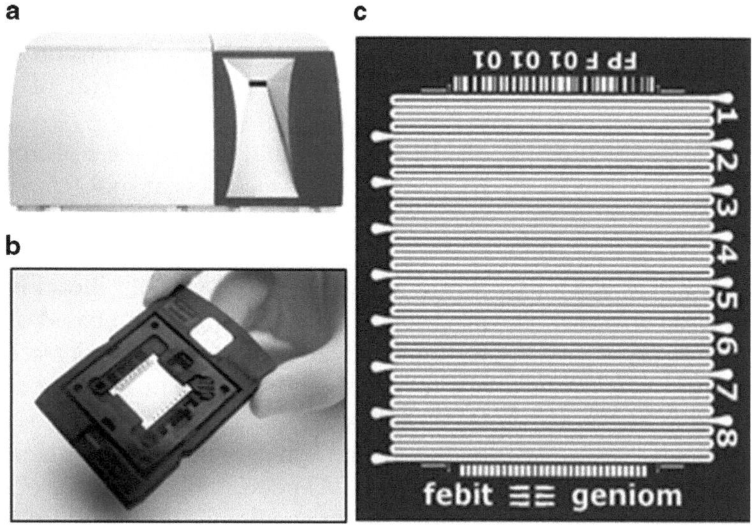

Fig. 2. Instrumentation for running the microfluidic primer extension assay: (**a**) Geniom RT Analyzer instrument, (**b**) biochip cartridge, (**c**) microfluidic biochip containing eight microfluidic microarrays.

(EL element). This element comprises a homomeric nucleotide stretch (e.g., 5 thymidine units) that is not complementary to miRNA sequences, but represents the template for a Klenow polymerase-based primer extension reaction to elongate miRNA molecules that previously were bound to the microfluidic microarray by stringent hybridization.

In a first step, the total RNA is hybridized under stringent condition to the miRNA element (M-element) of the probes on the microfluidic microarray (see Fig. 1). By applying stringent buffers together with elevated temperatures, it is assured that a high fraction of perfect-matched sequences bind to the support-bound DNA probes. These conditions allow for high-specificity recognition of targets, especially at the central position of the probe. Afterward, unbound and nonspecific RNA molecules are washed away from the microarray. The duplexes bound to the microfluidic microarray are DNA–RNA chimeras with the 3′-end of the miRNAs pointing toward the elongation element of the DNA probe. In the next step, a template-oriented primer extension reaction is carried out, employing the support-bound miRNA molecules as primers and the elongation element as template. By using the Klenow polymerase together with biotin-labeled triphosphate-nucleotides, not only the miRNA primers are elongated, but simultaneously biotin labels are incorporated into the support-bound miRNAs. These biotin moieties can be stained via the SAPE system in the subsequent detection steps.

Altogether, the described procedure comprises two consecutive specificity steps, namely, a first stringent hybridization step with a high discrimination power at the central position of the miRNA and a second primer extension step that ensures high discrimination at the terminal 3′-end of the miRNA by use of a polymerase.

In the following, we provide a protocol that allows for identification of diagnostic relevant miRNAs starting from whole blood as the source of total RNA.

3.1. Collection and Extraction of Total RNA from Blood Samples

Blood was drawn in PAXgene™ Blood RNA Tubes that contain a proprietary reagent composition based on a patented RNA stabilization technology that stabilizes intracellular RNA. By that, ex vivo changes of expression profiles are avoided. The RNA can be stabilized by PAXgene™ system for 3 days at 18–25°C, for 5 days at 2–8°C, and over a longer period (at least 6 months) at –20°C to 70°C. The PAXgene™ tubes contain 6.9 mL of RNA-stabilizing solution and are suitable for the sampling of 2.5-mL blood per tube and patient.

1. Collect 2.5 mL of blood directly into a PAXgene tube.
2. Shake tube well.
3. Store away at 4°C overnight or at –20°C for a period of several days before proceeding to RNA isolation.

For extraction of total RNA from blood collected in PAXgene tubes, we employ the miRNeasy Mini Kit (Qiagen).

1. When frozen, equilibrate the PAXgene tube for at least 2 h at room temperature.
2. Pellet the blood cell fraction by centrifugation at 5,000×*g* for 10 min at room temperature.
3. Resuspend the pellet in 10 mL RNase-free water and centrifuge again (5,000×*g*, 10 min, room temperature).
4. Resuspend the pellet in 700 μL of QIAzol lysis reagent employing the miRNeasy mini Kit (Qiagen) and incubate for 5 min at room temperature.
5. Add 140 μL of chloroform, vortex for 15 s, and incubate for 2–3 min at room temperature.
6. Centrifuge for 15 min with 10,000×*g* at room temperature.
7. Remove the upper aqueous phase containing the RNA fraction and mix it with 1.5 volumes of absolute ethanol to precipitate the RNA.
8. Make aliquots of 700 μL each and place each aliquot on a single column provided with the miRNeasy Kit.
9. Centrifuge for 15 s at 10,000×*g* at room temperature. Discard the flow through.
10. Add 700 μL of RWT buffer to each column and centrifuge for 15 s at 10,000×*g* at room temperature. Discard the flow through.

11. Add 500 μL of RPE buffer to each column and centrifuge for 2 min at 10,000 × *g* at room temperature. Discard the flow through.
12. Dry the columns by centrifugation at 10,000 × *g* for 1 min.
13. Elute the total RNA by adding 40 μL of RNase-free water to each of the columns and centrifugation at 10,000 × *g* for 1 min at room temperature.
14. Store the total RNA at −70°C until use.

3.2. Quality Control

Due to the omnipresence of RNases and the instability of RNA, integrity checks and sample quantitation are essential steps before any RNA-dependent application.

1. Check the RNA concentration using the Nanodrop ND-1000 photometer (see Nanodrop manufacturers' instructions).
2. Depending on the RNA concentration, use either the Agilent RNA 6000 Nano Kit (for concentration: 5–500 ng/μL) or the Agilent RNA 6000 Pico Kit (for concentration: 50–5,000 pg/μL).
3. Employ the Agilent 2100 Bioanalyzer instrument for measuring the RNA integrity number (RIN) of the total RNA. Follow the manufacturer's protocol for handling the Agilent 2100 Bioanalyzer instrument (see Note 1).
4. Any sample with an RIN below 7 has failed the QC. Continue only with samples that have passed the QC.

3.3. MicroRNA Expression Analysis Employing the Microfluidic Primer Extension Assay

Total RNA samples were analyzed with the Geniom RT Analyzer instrumentation (febit group, Germany) using the Geniom Biochip miRNA *Homo sapiens*. Each biochip contains eight microfluidic microarrays for analyzing eight samples in parallel. The following protocol is set up for processing eight miRNA analyses in parallel on one biochip.

3.3.1. Stringent Hybridization of Total RNA

1. Carefully resuspend 300 ng of total RNA in 25 μL of hybridization buffer (see Note 2).
2. Denature the RNA sample for 3 min at 95 C, place the sample immediately on ice, centrifuge to collect little droplets from the lids of the tubes, and keep on ice until start of the hybridization.
3. Loading of the Geniom RT Analyzer instrument with reagents:
   (a) Make sure that the following solutions have been loaded to the instrument.
   (b) Stringent washing buffer (0.5× SSPE, 1,000 mL)
   (c) Nonstringent washing buffer (6× SSPE, 1,000 mL)

4. Place eight total RNA samples, each resuspended in 25 μL of hybridization buffer within 1.5-mL Eppendorf tubes, into the sample rack and start the program for automatic loading of the eight samples to each of the eight microfluidic microarrays.
5. On the Geniom RT Analyzer, start the "stringent hybridization program."
6. The stringent hybridization is carried out at 42°C for 14 h with the sample being agitated by forward and backward pumping. After hybridization, the microarray is washed with stringent hybridization buffer (0.5× SSPE) at 45°C, followed by a nonstringent washing step (6× SSPE) at 25°C.
7. Pipette 50 μL of primer extension mix into eight 1.5-mL Eppendorf tubes and store on ice until the "equilibration program" is finished.
8. Pipette 200 μL of SSPE (6×) in each of eight 1.5-mL tubes.
9. Loading of the Geniom RT Analyzer instrument with reagents:
   - Nonstringent washing buffer (6× SSPE, 1,000 mL)
   - SAPE solution (9 mL)
   - Antibody solution (3.5 mL) (see Note 3)
10. Equilibrate the microarray for primer extension.
    - To prepare the microarray for the primer extension reaction, the nonstringent washing buffer, which remains in the microarrays after the "stringent hybridization of total RNA" is finished, is replaced.
    - Place the eight tubes containing the equilibration buffer into the sample rack of the Geniom RT Analyzer instrument and start the "equilibration program" (see Note 4).
11. Running the primer extension reaction:
    - Place the eight tubes with primer extension mix (see Note 5) into the sample rack of the Geniom RT Analyzer instrument and start the "primer extension program."
    - The primer extension is carried out at 37°C for 30 min with the solution not being agitated.
12. Stopping the primer extension reaction:
    - While the primer extension reaction is running, replace the eight tubes of the sample rack that contained the primer extension mix with the eight tubes containing the stop solution.

3.3.2. Detection

1. After the primer extension reaction, the Geniom RT Analyzer automatically starts the staining and detection procedures. All reagents required have been already loaded to the instrument before.

2. The instrument automatically runs a first incubation of the microarrays for 15 min at 25°C with SAPE conjugate, followed by nonstringent washing at 25°C. The CCD detection module of the Geniom RT Analyzer acquires a first detection image.

4. Notes

1. When using the Agilent Bioanalyzer: Do not leave any wells empty otherwise the chip does not run properly. Add 6 μL of the RNA marker (green lid) to each unused sample well.
2. Do not include any genomic DNA into the primer extension reaction since this causes unspecific elongation and high background signals.
3. Discard any thawed antibody solutions.
4. After having finished the equilibration reaction, proceed as soon as possible with the primer extension reaction and detection/staining, as the microarrays should not be left dry too long.
5. Make sure that the Klenow enzyme employed is the exo^{-} version; otherwise, the probes might be partially degraded.

References

1. Calin, G.A. and Croce, C.M. (2006) MicroRNA-cancer connection: the beginning of a new tale. *Cancer Res* **66**, 7390–4.
2. Lu J, Getz G, Miska EA, Alvarez-Saavedra E, Lamb J, Peck D., et al. (2005) MicroRNA expression profiles classify human cancers. *Nature* **435**, 834–8.
3. Weber JA, Baxter DH, Zhang S, Huang DY, Huang KH, Lee MJ., et al. (2010) The MicroRNA Spectrum in 12 Body Fluids. *Clin Chem* **56**, 1733–41.
4. Kroh EM, Parkin RK, Mitchell PS, and Tewari M. (2010) Analysis of circulating microRNA biomarkers in plasma and serum using quantitative reverse transcription-PCR (qRT-PCR). *Methods* **50**, 298–301.
5. Vorwerk S, Ganter K, Cheng Y, Hoheisel J, Stähler PF, and Beier M. (2008) Microfluidic-based enzymatic on-chip labeling of miRNAs. *N Biotechnol* **25**, 142–9.
6. Häusler SF, Keller A, Chandran PA, Ziegler K, Zipp K, Heuer S. et al. (2010): Whole blood-derived miRNA profiles as potential new tools for ovarian cancer screening. *Brit J Cancer* **103**, 693–700.
7. Keller A, Leidinger P, Borries A, Wendschlag A, Wucherpfennig F, Scheffler M et al. (2009) miRNAs in lung cancer - studying complex fingerprints in patient's blood cells by microarray experiments. *BMC Cancer* **9**, 353.
8. Keller A, Leidinger P, Lange J, Borries A, Schroers H, Scheffler M et al. (2009) Multiple sclerosis microRNA expression profiles accurately differentiate patients with relapsing-remitting disease from healthy controls. *PLoS One* **4**, e7440.
9. Leidinger P, Keller A, Borries A, Reichrath J, Rass K, Jager SU et al. (2010) High-throughput miRNA profiling of human melanoma blood samples. *BMC Cancer* **10**, 262.
10. Meder B, Keller A, Vogel B, Haas J, Sedaghat-Hamedani F, Kayvanpour E et al. (2011) MicroRNA signatures in total peripheral blood as novel biomarkers for acute myocardial infarction. *Basic Res Cardiol* **106**, 1:13–23.
11. Rosenfeld N, Aharonov R, Meiri E, Rosenwald S, Spector Y, Zepeniuk M et al. (2008) MicroRNAs accurately identify cancer tissue origin. *Nat Biotechnol* **26**, 462–9.
12. Mitchell PS, Parkin RK, Kroh EM, Fritz BR, Wyman SK, Pogosova-Agadjanyan EL et al. (2008) Circulating microRNAs as stable blood-based markers for cancer detection. *Proc Natl Acad Sci USA* **105**, 10513–8.

13. Kosaka, N., Iguchi, H. and Ochiya, T. (2010) Circulating microRNA in body fluid: a new potential biomarker for cancer diagnosis and prognosis. *Cancer Sci* **101**, 2087–92.
14. Gilad S, Meiri E, Yogev Y, Benjamin S, Lebanony D, Yerushalmi N, et al. (2008) Serum microRNAs are promising novel biomarkers. *PLoS One* **3**, e3148.
15. Brase JC, Wuttig D, Kuner R, Sültmann H. (2010) Serum microRNAs as non-invasive biomarkers for cancer. *Mol Cancer* **9**, 306.
16. Baum, M. (2003) Validation of a novel, fully integrated and flexible microarray benchtop facility for gene expression profiling. *Nucleic Acids Res* **31**, e151.
17. Griffiths-Jones, S. (2004): The microRNA Registry. *Nucleic Acids Res* **32**, D109–11.
18. Griffiths-Jones S, Grocock RJ, van Dongen S, Bateman A, Enright AJ. (2006) miRBase: microRNA sequences, targets and gene nomenclature. *Nucleic Acids Res* **34**, D140–4.

Chapter 11

MicroRNA Profiling Using μParaflo Microfluidic Array Technology

Xiaochuan Zhou, Qi Zhu, Christoph Eicken, Nijing Sheng, Xiaolin Zhang, Litao Yang, and Xiaolian Gao

Abstract

The diverse functions of microRNA (miRNA) molecules have drawn broad and intensive interest in various biological fields, biomedical applications, and technology development. Which are endogeneous cellular short RNA molecules found in the cytoplasm as well as in various serum fluids. miRNAs are transcriptional and translational regulatory molecules active in cell division, growth, and apoptosis (1). Dysregulated expression of miRNAs has been implicated in various disease states and has been tested as biomarker candidates (2–4). miRNAs are endogeneous cellular short RNA molecules found in the cytoplasm as well as in various serum fluids. miRNAs are transcriptional and translational regulatory molecules active in cell division, growth, and apoptosis (Bartel, Cell 116:281–97, 2004). Dysregulated expression of miRNAs has been implicated in various disease states and has been tested as biomarker candidates (He et al., Nature 435:828–833, 2005; Lu et al., Nature 435:834–838, 2005; O'Donnell, et al., Nature 435:839–843, 2005). In this chapter, we describe the methods using μParaflo® microfluidic oligonucleotide microarray technology for applications in miRNA profiling. One unique feature of this technology is the flexibility that provides users with the freedom to select sequence content either for focused studies wherein only the most relevant sequences are included or for discovery studies wherein the most updated sequence content such as those newly derived from deep sequencing. This chapter provides detailed information from experimental design to sample preparation, as well as data analysis for a miRNA array experiment.

Key words: MicroRNA profiling, miRNA detection, Oligonucleotide microarray, Microfluidics, Microfluidic biochip, Photogenerated acid, Parallel synthesis, Digital photolithography

1. Introduction

In the last decade, in the wake of rapid accumulation of abundant genomic information, there has been an increasing need for miniaturized biological assay systems for inexpensive high-throughput experiments requiring minimal sample input. The emergence of deep sequencing technology (or Next Generation Sequencing

Jian-Bing Fan (ed.), *Next-Generation MicroRNA Expression Profiling Technology: Methods and Protocols*, Methods in Molecular Biology, vol. 822, DOI 10.1007/978-1-61779-427-8_11, © Springer Science+Business Media, LLC 2012

(NGS)) in recent years has led to an unprecedentedly rapid growth of sequence content information for array-based profiling studies. Today's array technology must be simple to use, flexible for implementing custom content, and high quality in terms of data uniformity and sequence specificity so as to meet diverse requirements in the biological, biomedical, and clinical research fields as well as other applications.

In this chapter, we describe the application of μParaflo® microfluidic microarray (array) chips for expression profile analysis of microRNAs (miRNAs) (13, 27, 31). μParaflo® microfluidic technology encompasses a novel high-throughput biopolymer synthesis chemistry carried out using a new class of microfluidic reaction devices and an advanced digital light synthesizer apparatus and is developed for carrying out picoliter scale chemical and/or biochemical reactions. Figure 1 shows images of μParaflo® microfluidic chips and the associated internal fluidic structures. The functionalized μParaflo® chips are particularly suited for applications where small sample consumption, contamination-free, and performance-reproducibility are primary concerns. This technology enables massively parallel synthesis of high-quality DNA and RNA oligonucleotides as well as peptides and peptidomimetics in picoliter-scale reaction chambers inside μParaflo® chips. Addressable arrays are now widely accepted as efficient assay forms for analyzing nucleic acids and proteins (5–12). A critical component of an array platform is the content, i.e., the molecules, such as oligonucleotides or peptides, immobilized on the surface for specific detection or identification of analytes. In the last decade, various technologies have been developed for creating such arrays. Among these, in situ synthesis technologies have been particularly successful for the preparation of DNA oligonucleotides (13–27) and peptide arrays (14, 28–39), which

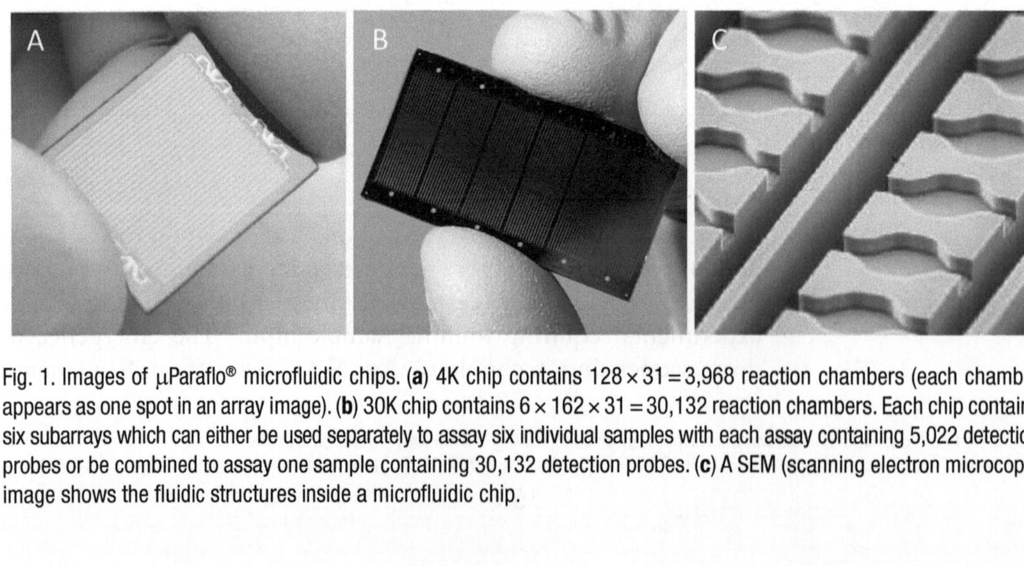

Fig. 1. Images of μParaflo® microfluidic chips. (**a**) 4K chip contains 128 × 31 = 3,968 reaction chambers (each chamber appears as one spot in an array image). (**b**) 30K chip contains 6 × 162 × 31 = 30,132 reaction chambers. Each chip contains six subarrays which can either be used separately to assay six individual samples with each assay containing 5,022 detection probes or be combined to assay one sample containing 30,132 detection probes. (**c**) A SEM (scanning electron microcope) image shows the fluidic structures inside a microfluidic chip.

are now the major tools for genome-scale analysis of nucleic acids and proteins. In situ synthesis methods are capable of simultaneously synthesizing thousands or more biopolymer sequences directly on well-controlled substrate surfaces. Several chemistries have been developed for the synthesis. One major difference is in the use of conventional chemistry (13, 16, 21, 25, 28, 31) versus specialized chemistry involving photolabile group protected monomer building blocks (14, 19, 34). Compared to the specialized chemistry, the conventional chemistry has the advantage of higher step-wise synthesis yield, wide range of off-shelf synthesis monomer compounds including varieties of modified nucleotides, and lower material cost. Flexible in situ synthesis methods have the added advantage of making arrays of different sequence designs with no need for hardware changes, therefore, significantly lowering the cost of custom arrays. By comparison, a spotting-based array making method requires presynthesis of large numbers of oligos or other probe sequences and suffers from spotting tip, surface quality, and environmental variation. The method is often time and cost prohibitive for exploratory research projects and tends to produce arrays of variable quality. This chapter focuses on the use of μParaflo® microfluidic chips for miRNA profiling applications. The chips are made by a flexible, in situ synthesis method using conventional chemistry.

2. Materials

In the following method and protocol descriptions, unless specifically noted, μParaflo® 4K chips (LC Sciences) are used. These chips contain 3,968 isolated reaction sites in a 1.4-cm^2 area and the total internal volume of the chip is 9.6 μL. For other chip formats, such as 30K or larger chips, the protocols may be used with appropriate adjustments.

1. Oligonucleotide arrays containing probes of 16–25 nt long, designed for detecting mature human and/or other species miRNA sequences (miRBase database (40)), and as an option, customized sequences, such as those unique to the pre-miRNAs, deep sequencing detected small RNAs, and predicted sequences, are also present in the array. The arrays also contain control probes for chip quality analysis, detection of spike-in RNA sequences, and internal positive controls. Standard internal positive controls include 5–6 probes targeted to different regions of 5S ribosomal RNA of the corresponding species. Most probes are repeated three times or more. The probes are synthesized using natural DNA and modified nucleotide residues for producing enhanced sensitivity and specificity in miRNA detection.

2. Total RNA sample: 0.5 μg per assay without fractionation or 5 μg per assay with fractionation. The total RNA sample must be prepared by a procedure or a kit specifically developed for miRNA and/or small RNA studies. When multiple samples are involved in a differential expression study, all samples must be prepared using the same procedure. Each sample must pass quality control: UV absorption 260:280 ratio≥2.0, UV absorption 260:230 ratio≥1.8, Bioanalyzer (Agilent) RIN≥8 for eukaryotic samples (see Subheading 3.2 for details on RIN).
3. Genepix™ 4000B microarray scanner (Molecular Devices/Axon).
4. Fluid Station (LC Sciences).
5. Chip adapter for liquid circulation (LC Sciences).
6. Chip adapter for image scanning (LC Sciences).
7. Water: filtered water (US Filter, Purelab Plus or Milli-Q, Millipore).
8. Nuclease-free water (Ambion).
9. T4 RNA ligation reaction buffer (Promega).
10. Hybridization buffer: 6× SSPE, 25% formamide, pH 6.8 (LC Sciences).
11. Blocking buffer: 6× SPPE, 25% formamide, 0.01% BSA, pH 6.8 (LC Sciences).
12. Stripping buffer: 0.01× SSPE, 50% formamide, 0.5% SDS (LC Sciences).
13. Wash buffer: 1:1 hybridization solution and water, 0.2% SDS (LC Sciences).
14. Spike-in control (synthetic) RNA mixes I, II, and III (LC Sciences). Spike-in control RNA mix II contains 11 sequences of different concentrations ranging across 4 orders of magnitude.

Other materials are used in specific methods and are described in the following corresponding sections.

3. Methods

The discovery of miRNA as transcriptional and translational regulators is an important event in recent years. miRNAs are small, non-coding RNA sequences recently found across various organisms and species (40). MiRNAs are generated from pre-miRNA by endogenous processing enzymes, such as DROSHA, DICER, and DGCR8 in human. The length of pre-miRNA known thus far is mostly ~70 residues or longer and that of its product, mature miRNA, is 17–26 residues (40). Although miRNAs have only recently appeared in the literature (41, 42), the ubiquitous existence of this family of molecules has been established in many eukaryotic species and miRNA expression can be associated with cell development, tumorigesis, and a growing list of important cellular activities. It is therefore of great interest to understand the expression of miRNA associated with different cell- or tissue-types and at different cell states or disease stages. As of January 2011, miRBase (a comprehensive database of miRNAs (40)) contains 15,172 entries of miRNA from living species including 1,048 entries of miRNA from humans.

Microarrays have proven to be powerful tools for the rapid profiling and analysis of a large number of RNA transcripts. However, array applications for miRNA (pri-miRNA, pre-miRNA, and mature-miRNA) profiling and analysis are not straightforward extensions of those of mRNA expression profiling. A few unique properties are associated with miRNA detection: (a) mature miRNAs do not have a 3′-poly-A (polyadenylated) tail; (b) mature miRNAs are short (~20 residues) and therefore limited sequence choices are available for hybridization probe designs; (c) terminal isoforms are present in most mature miRNA sequences at both 3′ and 5′ ends while variations at the 3′ end are generally larger (43, 44); (d) in developed cells, many mature miRNAs have relatively large copy numbers as compared to average mRNAs, while together mature miRNAs represent a small weight percentage (less than 0.01%) of total RNA mass. As a result, these small RNA molecules cannot be labeled and/or amplified using oligothymine-based reverse transcription and dye-NTP (or modified NTP) incorporation as commonly used for mRNA detection. Detection probes on the chip cannot be optimized as in the case of probe design for mRNA detection, where probes of different sequences and variant lengths can be selected from the target mRNA for optimal hybridization. The recent reorganization of mature miRNAs end variation as revealed by deep sequencing data has led to a rethinking regarding using end-specific probe designs and assay strategies as in some applications (46, 57). The relatively large copy numbers of individual miRNAs make it possible to use direct labeling detection (without transcript amplification) for most applications.

A number of oligonucleotide array-based miRNA profiling methods have been reported (45–57). In general, the miRNA profiling steps consist of (a) experimental design, (b) array and/or probe selection, (c) total RNA extraction, (d) sample quality analysis and control, (e) fractionation of small RNA from total RNA, (f) labeling, (g) hybridization, and (h) data acquisition and analysis. Some of these steps may be altered and/or removed depending on specific applications which will be described in detail in the following sections. Validation of array results is an important part of postarray miRNA profiling studies. The most commonly used validation methods include real-time PCR and Northern blot hybridization. Readers should refer to the relevant literature for details (58, 59).

3.1. Experimental Design

An effective as well as efficient experimental design is important for deriving statistically sound conclusions from miRNA array profiling data at a minimal cost. An experimental design aims to determine how many biologically categorized groups of samples will be used in a project, how many biological replicate samples will be included in each group, how many technical replicates will be used for each sample, and whether single sample or dual sample array assays will be used. Design rules are made based on the understanding of data analysis of differential expression and the ranking of the various sources of experimental variation. In a differential expression study involving two sample groups, statistical significance is determined by comparing the average expression level difference between the two groups to the expression level variation within the two groups. The degree of significance of a specific miRNA sequence can be mathematically assessed in a *T*-test using the *T* function

$$T = \frac{\overline{S_A} - \overline{S_B}}{\sqrt{\frac{\sigma_A^2}{n_A} + \frac{\sigma_B^2}{n_B}}} \quad (1)$$

where $\overline{S_A}$ and $\overline{S_B}$ are the average values of groups A and B; σ_A and σ_B are the standard deviations of groups A and B; n_A and n_B are the number of samples in groups A and B. The higher the T function value, the higher the probability of groups A and B being statistically different. A corresponding *p*-value may be derived using a *t*-distribution table (available through computer programs, e.g. R and Excel). Equation 1 reveals two mathematical facts. First, a statistical test can be performed only when each group contains two or more samples. Second, accurate assessments of standard deviations among samples within each group are as important as an assessment of differences between the groups. While the first fact is obvious, the second fact is not fully recognized by many biological experimentalists.

The standard deviations in Eq. 1 reflect the cumulative result of the various sources of experimental variation, which are generally ranked as following:

Biological variation among biological replicates $>$
Run-to-run variation due to RNA extraction $>$
Labeling dye induced variation or bias $\geq$
Run-to-run variation due to sample labeling $>$
Chip-to-chip variation due to chip fabrication $\approx$
Chip-to-chip variation due to hybridization $>$
Spot-to-spot variation within a chip

This ranking is valid for assays using in situ synthesized array chips (e.g., the ones from LC Sciences). Biological variation among biological replicates of various types of samples is generally ranked as following:

human samples > lab animal samples > cell lines

Based on the above experimental variation rankings and the analysis of large number of experimental data (not presented here), we suggest the following rules as general guidance for experimental designs.

1. Biological replicates versus technical replicates
   For most applications biological replicates are used and technical replicates are not needed. Biological replicates are defined as samples derived from specimens of individual human or animal subjects or individual cell line growth batches of the same testing group. Technical replicates are defined as samples originated from a single biological specimen but are separately processed at a certain step of the assay process. In the case where in situ synthesized array chips are used, assay-induced variation is generally less than biological-replicate variation. The use of technical replicates alone will not be able to assess the biological variation within the same groups, will underestimate the standard deviation (σ_A and σ_B) in Eq. 1, and therefore lead to false-positive calls.
2. Number of samples per group
   - Human: 10 or more
   - Lab animal: 3 or more
   - Cell line: 3
3. Pooling samples
   Pooling samples before performing an array assay has been used by some scientists to reduce array assay costs while used by others to supplement limited sample inputs. When samples are pooled, critical information on sample-to-sample variation (σ_A and σ_B in Eq. 1) within the same groups is lost and therefore identification of biologically significant differentially expressed miRNAs may no longer be possible. Therefore, in general, the use of pooled samples is not recommended. However, if pooled samples have to be used due to limited sample amounts from individual specimens, the number of samples in each pool should be minimized, pooled specimens should be selected among closely related biological specimens, and multiple pools should be used as substitutes for biological replicates.
4. Single-sample versus dual-sample assays
   In single-sample assays, all samples are labeled with the same dye and each labeled sample is hybridized to one chip (or to one subarray of a chip). In a dual-sample assay, one sample is labeled with Cy3 (or Alexa Fluor 546, Oyster-550, or any other equivalent dye); the other sample is labeled with Cy5 (or Alexa Fluor 647, Oyster-650, or any other equivalent dye); and the two samples are combined and co-hybridized to one

chip (or to one sub-array of a chip). The main advantage of the single-sample assay is the absence of dye-related bias. The method is good for all applications.

The dual-sample assay has the advantage of lower per-sample cost as compared to that of the single-sample assay. During the early days of microarray development, when chip-to-chip consistency was an issue, dual-sample assay was touted as a preferred method for differential expression studies since the method cancels chip-to-chip variation between two co-hybridized samples. However, the development of in situ array synthesis processes has significantly improved chip qualities to the extent that chip-to-chip variation is now less than experimental variation from sample preparation and labeling steps. Dual sample assays should be selected only when large numbers of samples are involved in each testing group and assay cost is the determining factor in the assay method selection.

A good dual sample design should have half of the members of each group labeled with Cy3 and the other half labeled with Cy5. This is called dye swap, which makes it possible to cancel dye biases by combining signal data from the same group. Sample pairing for hybridization may be done between two different groups or within the same group. Intergroup pairing will produce array ratio images directly revealing differences between two corresponding groups (see Subheading 3.8) and is more commonly used. However, for the purpose of canceling dye biases the two pairing strategies serve equally well. Tables 1 and 2 list two example pairing designs for a 12-sample experiment involving three testing groups of untreated (A1–A4), treated at dosage 1 (B1–B4), and treated at dosage 2 (C1–C4).

3.2. Probe Content Selection

μParaflo® technology, with its flexible (or programmable) in situ synthesis process, removes technical as well as cost barriers in probe content selection. The following list provides several commonly used methods for selecting biological content.

1. Using standard arrays as provided by LC Sciences. MiRNA content of these arrays is selected from miRBase database (40). Currently, 33 standard arrays are listed at http://www.lcsciences.com, including 15 covering single species and 18 covering multiple species. Multiple species arrays are useful when miRNA sequence information of the study subject is not complete or not even available. Standard arrays are mostly used by medical researchers who want to focus their studies to validated and/or annotated miRNA sequences.
2. Adding custom sequences to standard arrays. Many researchers collect sequences from the most updated literature, computational predictions, and/or sequencing experiments and add them as custom sequences to standard arrays of the corresponding species. The custom sequences are not limited to miRNAs.

Table 1
Intergroup pairing

	Untreated (A)	Dosage 1 (B)	Dosage 2 (C)
Chip 1	A1–Cy3	B1–Cy5	
Chip 2	A2–Cy5	B2–Cy3	
Chip 3		B3–Cy5	C1–Cy3
Chip 4		B4–Cy3	C2–Cy5
Chip 5	A3–Cy5		C3–Cy3
Chip 6	A4–Cy3		C4–Cy5

Table 2
Intragroup pairing

	Untreated (A)		Dosage 1 (B)		Dosage 2 (C)	
Chip 1	A1–Cy3	A2–Cy5				
Chip 2	A3–Cy5	A4–Cy3				
Chip 3			B1–Cy5	B2–Cy3		
Chip 4			B4–Cy3	B3–Cy5		
Chip 5					C1–Cy3	C2–Cy5
Chip 6					C4–Cy5	C3–Cy3

For example, piRNAs, snoRNAs, tRNAs, and other small RNAs have been incorporated (60). This approach is very useful for discovery studies.

3. Using deep sequencing results for array content. Deep sequencing yields result that cover genome-wide miRNAs without the need for any prior sequence knowledge. However, it costs substantially more and takes a longer time to complete than a corresponding array assay. More importantly, its sample preparation involves significantly more steps than the array assay and therefore results in additional profiling distortion (61, 62). Integration of the two complementary technologies makes a comprehensive, effective as well as efficient tool for performing profiling studies on large number of samples without worrying about missing any sequencing information. For most applications, sequencing data are produced in a single sequencing run (e.g., using one lane on Illumina GAIIx Genome Analyzer) using a pooled sample combining all representative sample types involved in a corresponding project. Sequencing data are

analyzed and array probe sequences are produced through Seq-Array^{SM} software pipeline (LC Sciences).

In addition to the aforementioned biological content, an array also contains control probes for chip quality analysis, spike-in controls, internal positive controls, and sometimes negative controls as well. More than 12 spike-in controls are used in each assay. Corresponding perfect matched and single-base mismatched detection probes are implemented for hybridization specificity assessment. Standard internal positive controls include 5–6 probes targeting different regions of the 5S ribosomal RNA of the corresponding species. Most probes are repeated three times or more. Synthesized probes consist of natural DNA and modified nucleotide residues to produce uniform binding coefficients across all probes including short and/or GC-poor sequences.

3.3. RNA Extraction

Total RNA samples are prepared using total RNA extraction kits that are specifically designed for retaining small RNAs. Commercial kits include miRNeasy from Qiagen, miRVana from Ambion, Total RNA Purification Kit from Norgen Biotek, etc. For most applications, small RNA enrichment during total RNA extraction is not recommended. If needed, a separate small RNA fractionation process is added in a latter step (see Subheading 3.5).

As mentioned in Subheading 3.1, run-to-run variation due to total RNA extraction ranks high among all sources of experimental variation. Use care to maintain consistency across all samples for each project. For example, use only one type of extraction kit for all the samples in one project, use exactly the same extraction conditions for all samples, and if possible have the same person perform the extractions on all samples.

Many laboratories have obtained excellent results from total RNA samples extracted using TRIzol-based methods. However, skill, experience, and sometime sample types may become critical factors in obtaining consistently good sample quality. The conventional TRIzol method involves isopropanol precipitation at room temperature. This step needs to be modified by doubling the usual isopropanol volume and leaving the RNA at −80°C for 15 min so as to ensure the precipitation of small nucleic acids.

3.4. RNA Sample Quality Control

The quality of the total RNA samples is of utmost importance to the success of any miRNA assay. Three quality controls are performed including quantity, purity, and integrity. The quantity and purity are evaluated using NanoDrop 1000 spectrometer (Thermo Scientific). UV absorption at 260 nm is used to measure RNA quantity. Absorption 260/280 ratio $\geq$ 2.0 and 260/230 ratio $\geq$ 1.8 are used as indicators of acceptable purity. Low 260/280 ratios are often attributed to phenol and/or protein contamination. Low 260/230 ratios are usually attributed to salt (e.g., guanidine isothiocyanate) and/or phenol contamination. Contaminants may cause serious downstream problems by inhibiting enzymatic labeling reactions. "High-salt," seen as 260/230 ratio < 1.0, is an often occurring contamination problem that causes a significantly increased absorption at 260 nm and results in the overestimation of RNA quantity. Grossly erroneous quantitative measurements will lead to significant RNA aliquot variation, increase experiment-induced variation, and reduce high confidence calls in profiling data (see Subheading 3.9).

RNA sample integrity is evaluated using the Bioanalyzer 2100 (Agilent) or denaturing agarose gel electrophoresis. Bioanalyzer analysis is preferred due to its low sample input requirement and automated quality assessment. For total RNA quality analysis, the RNA 6000 Pico LabChip kit is used by following the corresponding Agilent User Manual. A high-quality eukaryotic RNA sample should have a band intensity ratio of 28S rRNA to 18S rRNA of two. Electrophoretic traces from Bioanalyzer contain additional information relating to RNA integrity. Agilent introduced an analysis algorithm, RIN (RNA Integrity Number) to standardize RNA integrity interpretation. RIN ranges from 1 to 10, with 1 being the most degraded profile and 10 being the most intact. A total RNA with RIN ≥ 8 is generally considered of acceptable integrity for mRNA as well as miRNA profiling analysis. As an added note, the RIN reflects largely integrity of large RNAs which count for 95% by weight of total RNA and which dominates the electrophoretic trace signals. It is reasonable to expect cases when RINs do not reflect the integrity of miRNAs in total RNA samples (see Note 1).

3.5. Fractionation of Small RNAs

Small RNAs are optionally fractionated from total RNAs to reduce the probability of cross-hybridization due to larger RNAs. The following protocol describes fractionation using a size-filtration column, Microcon YM-100 device (Millipore).

1. Combine the following in a 1.5 mL microcentrifuge tube: 300 μL of 0.1 mM EDTA at pH 8.0, 4 μL of spike-in control RNA mix II (LC Sciences), 5 μg of total RNA sample, and nuclease-free water (Ambion) to make up a total volume of 350 μL. Mix the solution gently.
2. Heat the solution at 95°C for 7 min, let the solution cool down to room temperature, and then briefly spin the tube.
3. Transfer the entire contents of the tube into a Microcon YM-100 filer tube without touching the membrane, place the filter device into a collection tube (included in the Microcon device), and then centrifuge the YM-100 device at 5,000 × *g* in a microcentrifuge for 7 min.
4. Measure the RNA concentration of the collected solution using a Nanodrop spectrometer, measure the volume of the collected solution using pipette, and then calculate the fractionation yield as the ratio of the collected small RNA amount over the starting total RNA amount.

5. Flag any total RNA sample that has a fractionation yield greater than 15%. An exceedingly high fractionation yield indicates possible degradation or a high level of impurity of the total RNA sample. Such samples often produce poor assay results.
6. Concentrate the small RNA fraction by transferring the entire collected volume into a Microcon YM-3 filer device, placing the filter device into a collection tube (included in the Microcon device), and then centrifuging the YM-3 device at 13,000 × *g* for 50 min.
7. Add 150 μL nuclease-free water to the YM-3 filter device and centrifuge the YM-3 device at 13,000 × *g* for 27 min.
8. Invert the YM-3 filter device in a new collection tube. Centrifuge for 2 min at 13,000 × *g* to collect the concentrated small RNA.
9. Measure the quantity of the concentrated small RNA using a Nanodrop spectrometer and then store the sample at −80°C for subsequent applications. One of the limitations of the fractionation process is the relatively large sample quantity requirement of 5 μg total RNA. Without fractionation, the requirement is reduced to 0.5 μg. The direct use of total RNA has produced satisfactory results in most applications. First, detection probes for miRNAs are short at about 22 nt long. The short probes are capable of producing high sequence specificity yielding single-base discrimination (especially for mismatches at mid regions of the detection probes) under optimized hybridization conditions. Therefore, off-target cross-hybridization contributions to most of the miRNA probe signals are small. Second, in our miRNA assay native RNA sequences are labeled directly at the 3′ terminus. No modification is made to midnucleotides and therefore there is no disruption to RNA folding. While most mature miRNAs do not have a stable secondary structure, long native RNA sequences including miRNA precursors tend to form stable secondary structures or folding. The stable intramolecular folding reduces binding between long RNA sequences and short array detection probes (see Note 2). Additional advantages of using total RNA include the simplification of the sample preparation process, the removal of fractionation related process variation, and therefore the improvement of assay robustness.

3.6. Methods of Labeling

Two labeling methods, direct ligation and poly-A tailing plus ligation, are described here (Fig. 2). The direct ligation method is robust, tolerant to various sample conditions, and is recommended for single-sample assay use. The poly-A tailing method produces minimum dye-related variation and is suitable for dual-sample assay use.

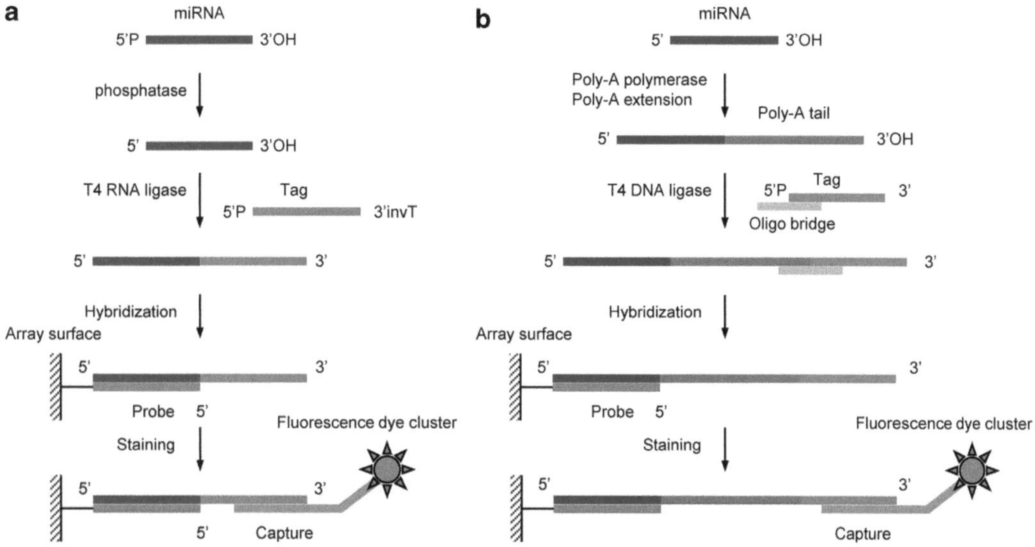

Fig. 2. Process flows of two labeling methods. (**a**) In the direct ligation method, a 5′ modified DNA tag sequence is added to the 3′ end of a miRNA sequence by T4 RNA ligation to form a tagged sequence. (**b**) In the poly-A tailing plus ligation method, a poly-A tail is first added to the 3′ end of a miRNA sequence using a PAP enzyme. An oligonucleotide tag sequence is then ligated to the poly-A tail using a bridge sequence and T4 DNA ligase. Both methods utilize signal amplification staining for miRNA target sequence detection.

3.6.1. Direct RNA Ligation

The direct ligation method is schematically shown in Fig. 2a. In this method, the 5′ phosphate group of a miRNA sequence is first removed to avoid self-ligation and circularization in the subsequent ligation step. Then a 5′ modified DNA tag sequence is added to the 3′ end of the miRNA sequence by T4 RNA ligation to form a tagged sequence. After the tagged sequence is hybridized to an array chip, the chip is stained with a capture sequence containing a fluorescence dye cluster. The capture sequence is complementary to the tag sequence and hybridizes to the immobilized tagged sequence. Each fluorescence dye cluster (from Genisphere or Invitrogen) contains about 900 dye molecules (cyanine or Alexa Fluor dyes) to produce signal amplification.

Dephosphorylation

1. On ice, combine the following components in a 1.5-mL microcentrifuge tube, 3.5 μL of 10× Antarctic Phosphatase Buffer (New England BioLabs), 1 μL of RNasin® Plus RNase Inhibitor (Promega), 1.5 μL of Antarctic Phosphatase (New England BioLabs), 0.5 μg of total RNA sample (or fractionated small RNA derived from 5 μg of total RNA sample), 5 μL of spike-in control RNA mix II (LC Sciences, when using fractionated small RNA this is replaced with 1 μL of spike-in control RNA mix III), and nuclease-free water (Ambion) to make up a total volume of 35 μL. Mix the reaction solution gently by pipetting up and down several times.

2. Incubate the reaction solution in a 37°C water bath for 2 h.
3. Briefly spin the tube.

Purification

4. Add 175 μL of RLT Buffer from an RNeasy MinElute™ Cleanup Kit (Qiagen) and 525 μL of 100% ethanol (EMD Chemicals) to each sample and mix well by pipetting.
5. Transfer the mixture above to the MinElute™ column (Qiagen) from the kit. Spin the column in a microcentrifuge (Eppendorf) at 845×*g* for 1 min and discard the flow through.
6. Place the column in a new collection tube. Add 600 μL RPE Buffer (ethanol added) to the column. Spin for 15 s at 9400×*g* and discard the flow through.
7. Add another 600 μL RPE Buffer (ethanol added) to the column. Spin for 15 s at 9400×*g* and discard the flow through.
8. Spin the column again for 1 min at 9400×*g*.
9. Place the column into a new 1.5 mL microtube. Add 30 μL of nuclease-free water to the center of the column, wait for 1 min, and then spin at 16000×*g* for 2 min to collect the eluted solution. Add another 30 μL of nuclease-free water to the column and repeat the elution process.
10. Measure the RNA concentration of the eluted solution using a Nanodrop spectrometer (Thermo Scientific). A typical concentration is about 8 ng/μL.
11. Reduce the total volume to 6 μL or less using a Vacufuge Concentrator (Eppendorf) at 45°C for about 25 min.
    Ligation
12. Combine the following components in a 1.5-mL microcentrifuge tube, 15 μL of ligation buffer (LC Sciences), 1 μL of RNasin® Plus RNase Inhibitor (Promega), 2 μL of 5′ modified DNA tag sequence solution (LC Sciences), dephosphorylated and purified RNA sample, 1 μL of T4 RNA Ligase (Promega), and nuclease-free water (Ambion) to make up a total volume of 25 μL. Mix the reaction solution gently by pipetting up and down several times.
13. Incubate the sample solution at 37°C water bath for 30 min.
14. Briefly spin the microtube containing the RNA sample solution after incubation.
15. Add 25 μL of hybridization buffer (LC Sciences) and store the mixture at −80°C.

3.6.2. Poly-A Tailing Plus Ligation

The method of poly-A tailing plus ligation is schematically shown in Fig. 2b. In this method, a poly-A tail is first added to the 3′ end of the miRNA sequence using a PAP enzyme. An oligonucleotide tag sequence is then ligated to the poly-A tail. An oligo bridge joining the poly-A tail and the tag sequence and T4 DNA ligase are used to facilitate the ligation reaction. For dual-sample assay applications, two distinct tags of different sequences are used to tag two samples. After the tagged sequences are hybridized to an array chip, the chip is stained with one or two capture sequences containing fluorescence dye clusters. Each capture sequence is complementary to one tag sequence. Each fluorescence dye cluster (from Genisphere or Invitrogen) contains about 900 dye molecules (cyanine or Alexa Fluor dyes) to produce signal amplification.

Poly-A Tailing

1. Prepare 1 mM ATP solution by adding 1 μL of 10 mM ATP (Invitrogen) to 9 μL of 1 mM Tris, pH 8.0.
2. On ice, combine the following components in a 1.5-mL microcentrifuge tube: 2.5 μL of Poly-A Polymerase 10× Reaction Buffer (Epicentre), 1 μL of 1 mM ATP solution, 1 μL of Poly-A Polymerase (Epicentre), 0.5 μg of total RNA sample (or fractionated small RNA derived from 5 μg of total RNA sample), 5 μL of spike-in control RNA mix II (LC Sciences, when using fractionated small RNA this is replaced with 1 μL of spike-in control RNA mix III), and nuclease-free water (Ambion) to make up a total volume of 25 μL. Mix the reaction solution gently by pipetting up and down several times.
3. Incubate the reaction solution in a 37°C water bath for 15 min.
4. Briefly spin the tube.

Ligation

5. Add the following to the above tailed RNA sample: 6 μL of 6× Cy3 or Cy5 Ligation Mix (Genisphere) 2 μL of T4 DNA Ligase (Invitrogen), and 3 μL of nuclease-free water (Ambion). Total volume is 36 μL (see Note 3).
6. Incubate the reaction solution in a 25°C water bath for 50 min.
7. Stop the ligation reaction by adding 4 μL of 0.5 M EDTA at room temperature. Briefly vortex and then spin the tube.

Purification

8. Add 700 μL of the Binding Buffer from a MinElute™ PCR Purification Kit (Qiagen) to each microtube and mix well by pipetting. For a dual sample assay, add 350 μL of Binding Buffer to each of the two differently tagged samples, then combine these two samples into one microtube.
9. Transfer the entire volume onto a spin cartridge from the MinElute™ kit (Qiagen). Spin the cartridge in a microcentrifuge (Eppendorf) at 845×*g* for 3 min and discard the flow through.

10. Add 600 μL of ethanol added Wash Buffer (Qiagen) to the cartridge. Spin for 0.5 min at 16000×*g* and discard the flow through.
11. Spin the cartridge again for 1 min at 16000×*g*.
12. Place the cartridge into a new microtube. Add 22.5 μL of nuclease-free water to the center of the cartridge, wait for 7 min.
13. Spin the cartridge at 16000×*g* for 2 min to collect the eluted solution.
14. Measure the RNA concentration of the eluted solution using a Nanodrop spectrometer (Thermo Scientific). A typical concentration is about 10 ng/μL for single sample and about 20 ng/μL for dual samples.

3.7. Hybridization

Hybridization is performed on a Fluidic Station (LC Sciences), which contains micropumps for sending sample and buffer solutions through μParaflo® microfluidic chips at predetermined flow rates and a temperature control module for maintaining the chips at predetermined temperatures. The station may be set up to flush a solution through the chips in an open-loop directly into a waste bottle or to circulate a solution through the chips in a closed-loop for an extended period of time.

System wash

1. Place a blank chip (LC Sciences) in a Fluidic Station and then flush the station sequentially with 1 mL of hot 1% SDS solution (preheated to 95°C), 1 mL of hot DI water (preheated to 95°C), and 3 mL of water.

Preparation for hybridization

2. Place a miRNA array chip (LC Sciences) in the Fluidic Station. Flush (open-loop) 200 μL of the stripping buffer through the chip and then circulate (closed-loop) 800 μL of the buffer through the chip at a flow rate of 500 μL/min and a chip temperature of 50°C for 15 min.
3. Flush (open-loop) 200 μL of hybridization buffer (LC Sciences) through the chip and then circulate (closed-loop) 800 μL of the buffer through the chip at a flow rate of 500 μL/min and a chip temperature of 32°C for 10 min.
4. Circulate (closed-loop) 50 μL of blocking buffer (LC Sciences) through the chip at a flow rate of 500 μL/min and a chip temperature of 32°C for 10 min.

Hybridization

5. Add 50 μL hybridization buffer into each tagged sample and then heat the sample to 95°C for 5 min. Spin the sample tube.

6. Circulate (closed-loop) the sample solution through the chip at a flow rate of 500 μL/min and a chip temperature of 40°C for 16 h.
7. Flush (open-loop) 200 μL of hybridization buffer (LC Sciences) through the chip at a flow rate of 500 μL/min and a chip temperature of 32°C.

Staining

8. Prepare staining solution in a 0.65-mL microcentrifuge tube by mixing following components: 50 μL of hybridization buffer (LC Sciences), 2 μL of 10% SDS, and 2.5 μL of each capture sequence (Cy3/Cy5 from Genisphere or AF-3/AF-5 from Invitrogen; in case of single sample assay, only one capture sequence is added). Mix the solution gently by pipetting up and down several times. Heat the solution in a water batch at 53°C for 10 min and then spin down.
9. Circulate (closed-loop) the staining solution through the chip at a flow rate of 500 μL/min and a chip temperature of 32°C for 2 h.
10. Flush (open-loop) 500 μL of hybridization buffer (LC Sciences) through the chip at a flow rate of 500 μL/min and a chip temperature of 32°C.
11. Flush (open-loop) 200 μL of wash buffer (LC Sciences) through the chip at a flow rate of 100 μL/min and a chip temperature of 32°C.
12. Circulate (closed-loop) 1 mL of the wash buffer through the chip at a flow rate of 100 μL/min and a chip temperature of 50°C for 20 min. While maintaining the circulation, lower the chip temperature to 32°C.

3.8. Image Acquisition

Mount a hybridized and wash buffer filled chip to a 3″ × 1″ slide adapter (LC Sciences) and place the adapter on the support platform of GenPix 4000B scanner with the chip facing down. Set the scanning pixel size at 10 or 5 μm for the 4K chip or 30K chip, respectively. Adjust the scanner optimal focal distance by maximizing and balancing the signal intensities of the control replicates around the four corners of the chip. In the case of dual channel data collection, adjust only the focal distance for the 532 nm channel. Cy3 and Cy5 signals are collected using 532 and 635 nm channels, respectively. In each channel, adjust the PMT (Photo-Multiplier Tube) gain such that the strongest signal within the whole imaging area is just below the saturation level of 65,535. After the completion of scanning, save the scan image in TIFF format. Refer to the manufacturer's manual for additional details on scanner and scanner operations.

Figure 3 shows example array images, in pseudo color, from a dye swap experiment involving two different samples S1 and S2.

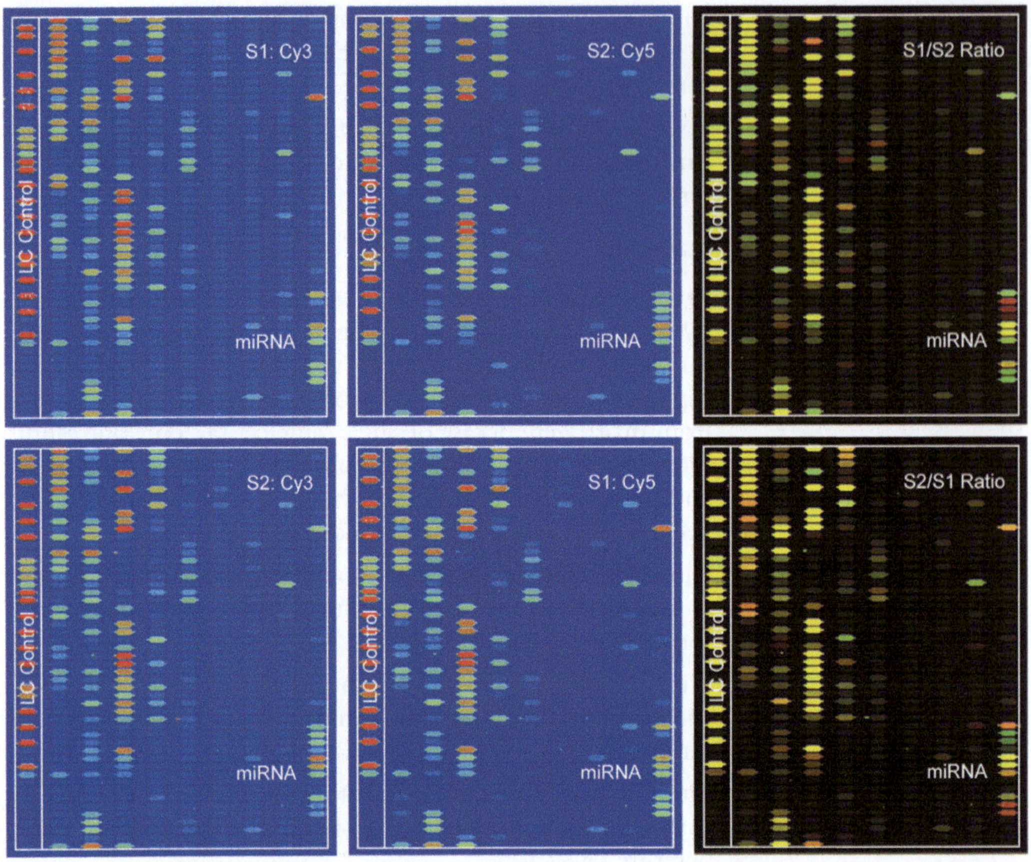

Fig. 3. Example array images of dual sample assays, in pseudo color, from a dye swap experiment involving two different samples S1 and S2. The *upper row* images were obtained from a chip hybridized with S1 and S2 labeled with Cy3 and Cy5, respectively. The *lower* three images were obtained from a reverse labeled chip with S2 and S1 labeled with Cy3 and Cy5, respectively.

The upper three images were obtained from a chip hybridized with S1 and S2 labeled with Cy3 and Cy5, respectively. The lower three images were obtained from a reverse labeled chip with S2 and S1 labeled with Cy3 and Cy5, respectively. The two images on the left are Cy3 fluorescence intensity images and the two images in the middle column are Cy5 fluorescence intensity images. As fluorescence intensity increases from 1 to 65,535 the corresponding color changes from blue to green, to yellow, and to red. The two images on the right are ratio images in which the green, yellow, and red colors indicate higher Cy3 signal, equivalent Cy3 and Cy5 signals, and higher Cy5 signals, respectively, for the corresponding probes, implying different relative expression levels of the corresponding sequences in the two samples. One may easily observe the results of dye-swapping by comparing the two ratio plots. A red (or green) colored spot in the upper right image would become a green (or red) colored spot at the same position in the lower right image.

3.9. Data Analysis

Data analysis includes image digitization, background subtraction, signal significance analysis, normalization, and differential analysis. The analysis is carried out using miRNA array analysis software (LC Sciences).

1. Image digitization
Image digitization is performed using the program "Array-Pro Analyzer" (from MediaCybernetics). Digitized data is extracted by overlaying an appropriate grid on top of a scan image. Each element of the grid covers one spot (or feature) area which is composed of multiple image pixels. A pixel is the smallest addressable digital image element that has a finite value between 0 and 65,535. Digitization calculates the mean, median, and standard deviation values of the pixels within each spot area. No background subtraction is performed during the digitization. A table is generated to list attributes of all spots including, row index, column index, mean value, median value, and the standard deviation value. The table is saved as a text file which is considered "Raw Data." Refer to the User Manual for Array-Pro Analyzer or other array image processing programs for details of the digitization procedure.
2. Background subtraction
For an array image, the fluorescence signal of each spot can usually be attributed to two different sources, background signal and hybridization target signal (63). The background signal derives from various sources including nonspecific dye staining, auto-fluorescence of probe molecules, and auto-fluorescence of glass substrate material. The background signal intensity is largely independent on the probe sequences. The background signal follows a normal distribution and its characteristics are derived by analyzing whole array signals. Figure 4a shows a ranked raw signal intensity profile which is obtained by sorting signal intensities of nonblank spots. An average (or mean) background level can be estimated by identifying an inflection point of the induction section of the profile. Figure 4b shows a corresponding histogram. As a first approximation, the left side of the histogram peak (Fig. 4b) is attributed to background. The background mean value is determined by peak position. The background standard deviation is calculated by

$$\sigma_{\mathrm{bkg}} = \sqrt{\frac{2}{2N-1}\sum_{i-1}^{N}\left(I_i - \overline{I_{\mathrm{bkg}}}\right)^2} \qquad (2)$$

where $\overline{I_{\mathrm{bkg}}}$ is the mean value, I_i is the signal (median) value of individual spots that are below or equal to the background mean value, N is the number of spots below or equal to the background mean value. When more than one block or subarray is processed, background calculation and subtraction are performed within individual blocks.

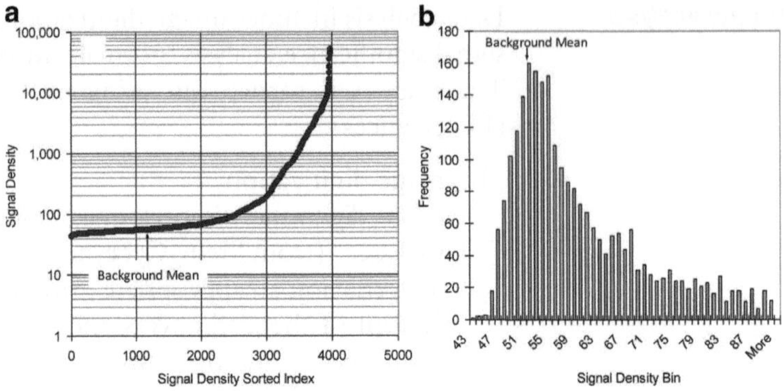

Fig. 4. Raw data signal distributions are presented in the form of a (**a**) ranked raw signal intensity profile which is obtained by sorting signal intensities of all nonblank spots and (**b**) histogram of the raw signals included in (**a**).

3. Signal significance analysis

   A signal is considered detectable when it is significantly above background. The degree of significance is evaluated using the following statistic equations (64)

$$\Upsilon = \frac{\overline{I - I_{bkg}}}{(\sigma + \sigma_{bkg})\sqrt{2}} \quad (3)$$

$$p = \mathrm{erfc}(\Upsilon) = \frac{2}{\sqrt{\pi}} \int_{\Upsilon}^{\infty} e^{-t^2} \mathrm{d}t \quad (4)$$

   Equation 3 calculates a normalized difference between the signal mean (of row data) and the background mean, where $\overline{I}$ is the signal mean value, $\overline{I_{\mathrm{bkg}}}$ is the background mean value, σ is the signal standard deviation, and σ_{bkg} is the background standard deviation. When replicate spots are used, the signal standard deviation σ includes both spot-to-spot and in-spot pixel-to-pixel variation. Equation 11.4 is the complementary error function which calculates the p-value of the corresponding statistics. The lower the p-value the more significant the signal is above background. Generally, $p \le 0.01$ is used as a threshold for reporting detectable signals.

   The signal standard deviation σ in Eq. 3 is derived by considering all pixel intensities covered by replicate spots and is expressed as

$$\sigma \approx \sqrt{\frac{1}{n_{\mathrm{spot}}} \left[\sum_{i=1}^{n_{\mathrm{spot}}} \sigma_i^2 + (n_{\mathrm{spot}} - 1)\sigma_{\mathrm{spot}}^2 \right]} \quad (5)$$

   where σ_i is the standard deviation among all pixels within spot i, σ_{spot} is the spot-to-spot standard deviation among replicate spots, n_{spot} is the number of replicate spots. Detailed derivation of σ is beyond the scope of this chapter.

4. Normalization

Normalization is carried out using the LOWESS (locally weighted scatterplot smoothing) method (65–67) on background-subtracted data. The normalization removes system-related variation, such as sample amount variation, labeling dye differences, signal gain differences of scanners, and nonlinear signal intensity responses to target concentrations so that biological variation is faithfully revealed. For miRNA data analysis, the LOWESS method is preferred over the Quantile method (67), which is popular for messenger RNA gene expression data analysis but has a tendency to cause unjustifiable data manipulation when small data sets such as miRNA data sets are processed.

LOWESS is a pair-wise regression method. For spot k of two arrays i and j, intensities I_{ki} and I_{kj} are first transformed into intensity A_{kij} and ratio M_{kij}.

$$A_{kij} = \frac{1}{2}\log_2(I_{ki}I_{kj}) \quad (6)$$

$$M_{kij} = \log_2\left(\frac{I_{ki}}{I_{kj}}\right) \quad (7)$$

Then M_{kij} is plotted against A_{kij} (MA plot), where $k = 1\ldots n$ represent all spots to be used in the regression fitting. A normalization curve, M''_{kij} against A_{kij}, is fitted to an MA plot with each point being obtained by linear regression of a subset of the data (68). The selection of the subset is determined by the following weight function.

$$w(A) = \begin{cases} \{1, A_i - \frac{W}{2} \le A \le A + \frac{W}{2} \\ 0, A < A_i + \frac{W}{2}, A > A_i + \frac{W}{2} \end{cases} \quad (8)$$

where W is a user defined bandwidth. A wider bandwidth will produce a smoother fitted curve while a narrower bandwidth will produce a fitted curve that better conforms to the scatter data. Equation 8 suggests that the number of data points in each subset may vary while the intensity bandwidth W is fixed across the fitted regions. This is different from the original LOWESS method in which a tricube weight function is applied to a subset of a fixed number of data points (65). Regular miRNAs arrays contain only several hundred to at most a couple of thousands unique probe sequences. The corresponding data points are often distributed less densely than cRNA or mRNA array data that generally contain thousands to tens of thousands unique probe sequences for which LOWESS was

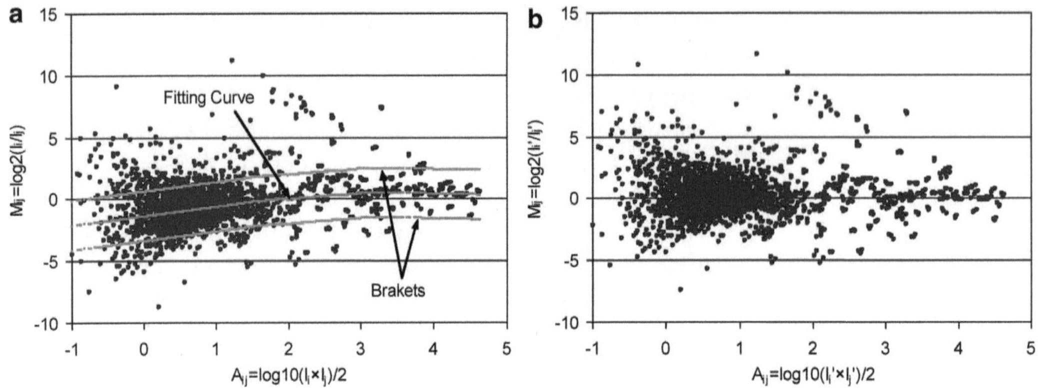

Fig. 5. MA scatter plots of intensity data (**a**) before normalization correction and (**b**) after normalization correction. Axis A_{ij} is plotted in $\log_{10}$ instead of $\log_2$ scale for easy conversion to the decimal scale.

originally applied (69, 70). Equation 11.8 is selected specifically for achieving an overall smooth and conforming fit for the full dynamic range of miRNA data. Figure 5a shows a typical MA plot with a fitted curve. The normalized (or adjusted) intensities, I'_{ki} and I'_{kj}, are obtained by

$$\mathrm{M}'_{kij} = \mathrm{M}_{kij} - \mathrm{M}''_{kij} \tag{9}$$

$$I'_{ki} = 2^{\mathrm{A}_{kij} + \frac{\mathrm{M}'_{kij}}{2}}, \quad I'_{kj} = 2^{\mathrm{A}_{kij} - \frac{\mathrm{M}'_{kij}}{2}} \tag{10}$$

where M'_{kij} is the normalization adjustment. Figure 5b shows the MA scatter plot of the normalized intensities.

To normalize more than two arrays (N arrays) cyclic LOWESS is used. First, normalization adjustments of all distinct pair-wise combinations (total $N(N-1)/2$ combinations) are calculated. For each array, $N-1$ sets of adjusted intensities are calculated by pairing with all other arrays and then an averaged set of adjusted intensities is obtained by combining the $N-1$ sets of adjusted intensities. The process is reiterated. In most cases, five iterations are sufficient to produce stabilized intensities.

The standard LOWESS method is applicable under an assumption that averaged gene expression levels are the same between any two samples and within a moving intensity bandwidth. This assumption is valid in most experimental conditions such as normal versus disease tissues of the same tissue type and treated versus untreated animals. However, this assumption is sometimes not valid. The LC Sciences software contains several options to handle various conditions.

In some experiments, a few miRNAs or other transcripts are highly differentiated as compared to the rest of the array content. In some other experiments, arrays contain a large number of probes for (often custom selected) transcripts that have similar responses to the experimental conditions. In the first case, "outlier" data points that significantly deviate from the main body of data points are formed in a MA scatter plot. In the second case, densely packed data clusters may form outside of the main body data points on a MA scatter plot. Both cases would cause distortions to corresponding regression fitting curves. To solve the problem an algorithm has been implemented to identify the flow of main body data points and to exclude outlier data points from regression calculations. While discussion of the algorithm is beyond the scope of this chapter, Fig. 5a illustrates the calculation result. A pair of brackets separate highly differentiated data points from the main body data points. A smooth and conforming fitted curve is obtained on a widely scattered data set.

In some experiments, overall miRNA expression levels may be different in different samples. The standard LOWESS method uses the main body data points as an internal reference, cancels the overall miRNA level differences, and therefore is not adequate for these experiments. To solve the problem, an option is available for the use of spike-in controls as a (external) normalization reference. MA transformations are applied to all data points while the LOWESS regression calculation is applied only on the spike-in control data points to produce a fitted curve. Then the fitted curve is used to perform normalization adjustments of all the data points. This method assumes that the measured amount of total RNA is an accurate reflection of the number of cells. For this method to be used correctly, one must make sure that the ratio of the spike-in controls to total RNA is precise across all the samples. This is sometimes challenging especially when the total RNA concentration is low or the sample solution contains impurities which lead to an inaccurate RNA concentration measurement (see Note 4).

Another available option to handle the overall miRNA expression level changes is the use of endogenous controls as an internal normalization reference. The calculation method is similar to that of option (2). An advantage of this method is the elimination of using sometimes inaccurate external normalization references. Proper use of the method requires the selection of suitable endogenous control sequences that are small, are each expressed at a constant level across all samples, can be labeled along with miRNAs, and hybridize effectively to detection probes under the same conditions as that of miRNAs. For adequate normalization across the full miRNA expression range, at least ten endogenous controls of different expression levels are required. Candidates for the endogenous controls include small noncoding RNAs, miRNAs that are known to be invariant in the corresponding experimental conditions, and tRNAs. A number of small noncoding RNAs have been established as endogenous controls in real-time PCR measurements (71) (see Note 5).

5. Differential analysis

Differential analysis reveals significantly differentiated miRNAs between individual groups of samples. This chapter is limited to two basic but effective analysis methods, ANOVA (Analysis of Variance) and *T*-test, while interested readers may refer to the literature for a wide range of statistic methods and tools (72). As mentioned earlier in Subheading 3.1, the degree of significance is evaluated by comparing between-group variation to within-group variation, the principle shared by both ANOVA and *T*-test. In general, for an experiment involving more than two sample groups ANOVA is first applied to produce a miRNA expression profile overview across all samples; then *T*-tests are performed to identify significantly differentiated miRNAs among all interested combinations of the two groups.

Figure 6 depicts clustered heat maps (also called clustergraphs) of miRNA differential expression profiles across three groups with each group containing four samples. The clustered heat map is a visual tool to examine expression profiles of large number of genes or miRNAs over multiple numbers of samples in a naturally intuitive manner (73). Closely clustered miRNAs have a likelihood of having similar functions. To generate the heat maps of Fig. 6, a *Z* transformation is first applied to signal intensities across all samples for each specific miRNA

$$Z_i = \frac{\Upsilon_i - S_\Upsilon}{\sigma_\Upsilon} \quad (11)$$

where Z_i is the transformed intensity of the sample *i*, $\Upsilon_i = \log I_i$, I_i is the signal intensity of sample *i*, S_Υ is the average value of Υ_i over all samples, and σ_Υ is the standard deviation of Υ_i over all samples. In heat maps of Fig. 6, *Z* values are presented in color-coded blocks with green being a negative value, black being zero, and red being a positive value. A negative *Z* indicates down regulation and a positive *Z* indicates up regulation. The brighter the color the higher the absolute *Z* value. The miRNA rows in Fig. 6 are arranged by gene clustering, which groups together miRNAs of similar changes across all samples (73). Within cluster V, miRNAs are consistently expressed at lower levels in group A than in groups B and C. Expression level variation within the three groups is less than either between groups A and B or between A and C. In contrast, within cluster W, miRNA expression levels vary at similar levels within each individual group and between any pairs of groups. ANOVA is used to identify miRNAs that are consistently differentially expressed by evaluating the statistic significance of the changes. The degree of significance is measured by *p*-value, which is the probability of subject groups

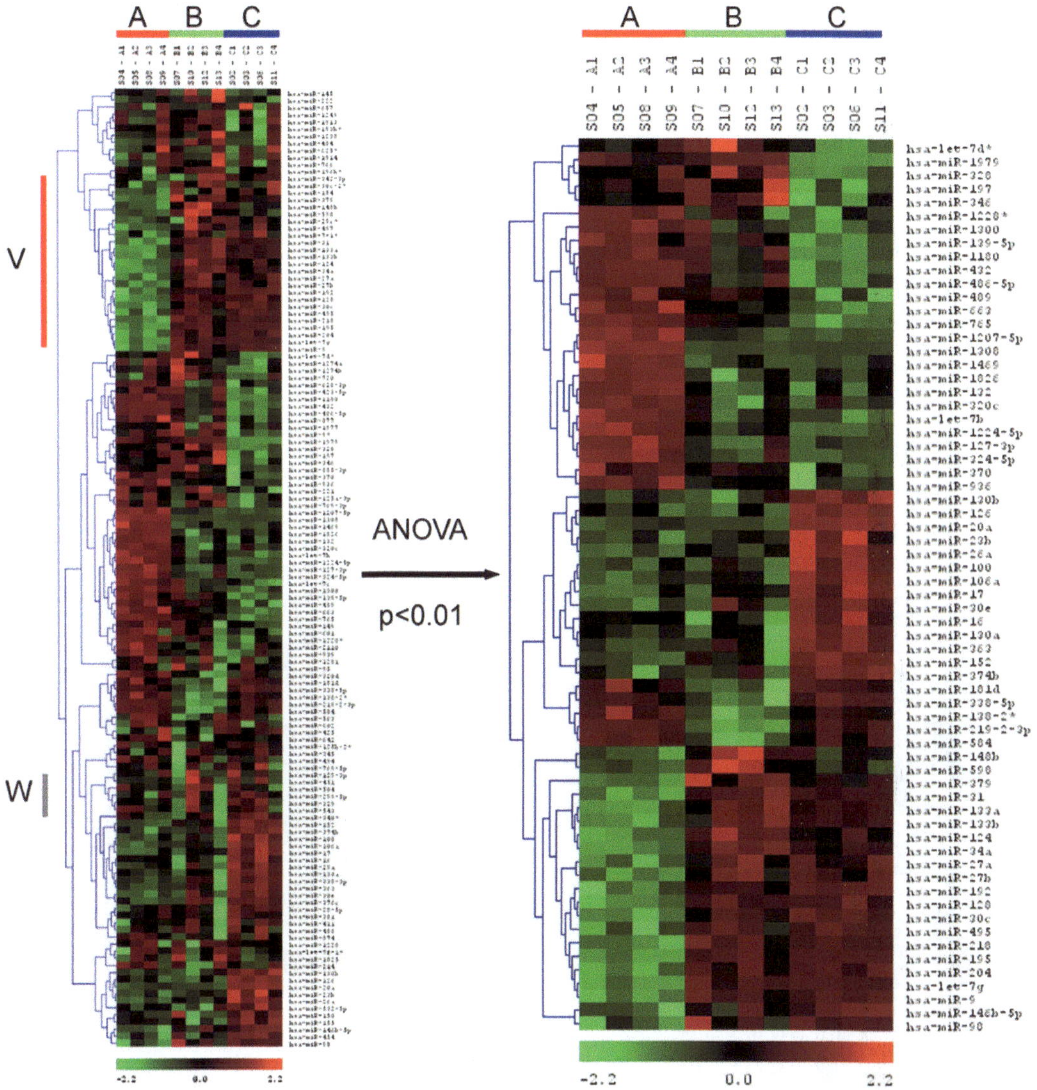

Fig. 6. Clustered heat maps illustrate miRNA differential expression profiles across 12 samples for three experimental conditions, A, B, and C with each condition applied to four biological replicate samples. The *left* heat map includes a number of arbitrarily selected miRNAs. The *right* heat map includes ANOVA selected miRNAs with *p*-values less than 0.01.

not being differentiated. The right-side clustered heat map of Fig. 6 shows statistically significant miRNAs with a *p*-value less than 0.01. To reveal differentially expressed miRNAs between two sample groups a *T*-test is used. Figure 7 shows clustered heat maps of miRNA expression profiles between groups A and B before and after a *T*-test analysis.

In some cases such as time-course experiments or low-budget preliminary studies, only one sample is collected and assayed at each time point or experimental condition. In these experiments, replicate detection probes on individual arrays may be used to form "groups" in ANOVA and *T*-test analysis. However, one must keep

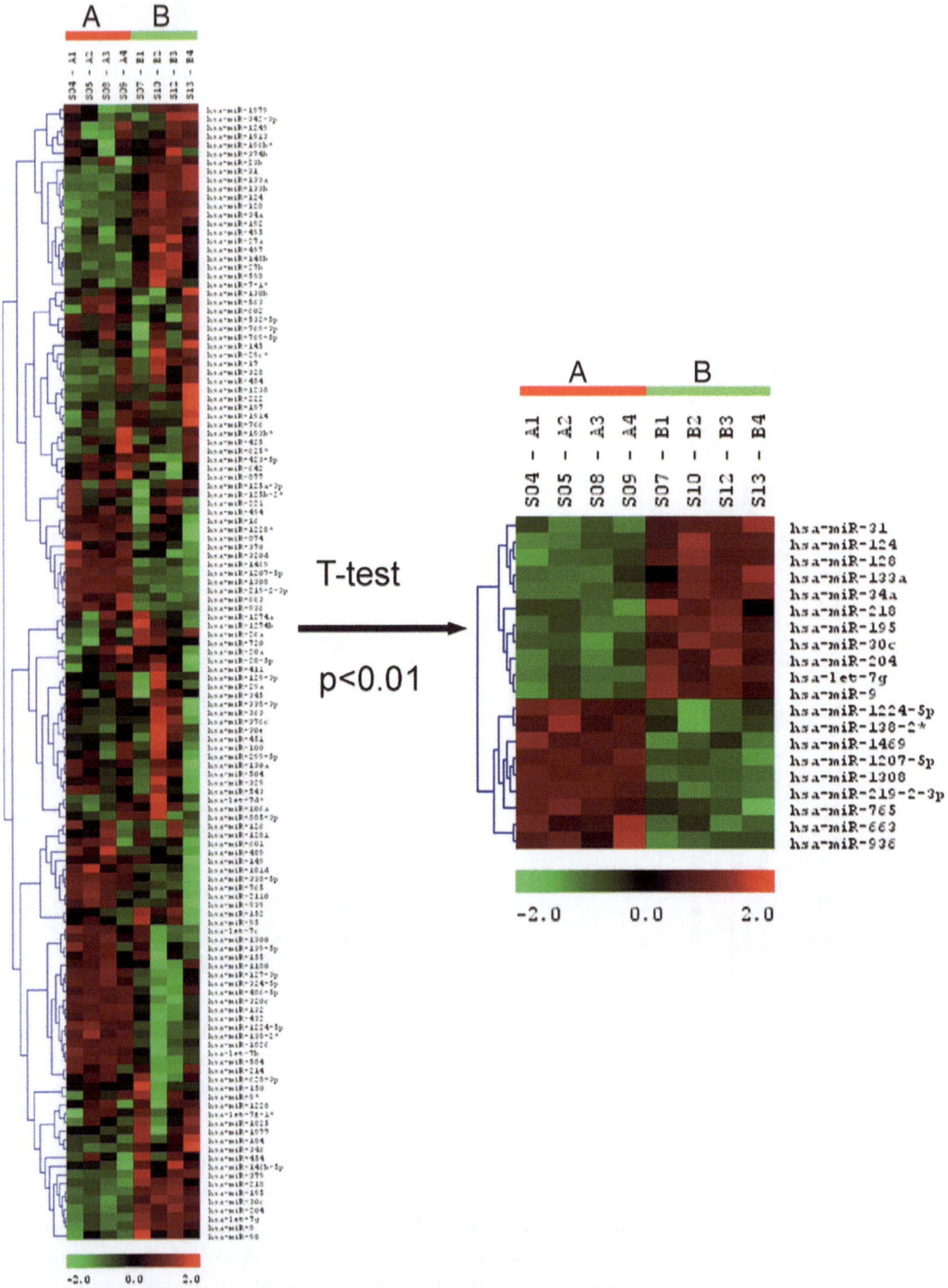

Fig. 7. Clustered heat maps illustrate miRNA differential expression profiles between two experimental conditions, A, and B with each condition applied to four biological replicate samples. The left heat map includes a number of arbitrarily selected miRNAs. The right heat map includes *T*-test selected miRNAs with *p*-values less than 0.01.

in mind that since in-group biological variation and sample preparation related variation are not included in the statistic analysis, the analysis will likely report some statistically significant miRNAs that may not be of biological significance.

4. Notes

1. We indeed have observed a few cases in which normal miRNA profiles were obtained from total RNA samples of low RINs (less than 5). In this case, the specific nucleases that effectively cut larger RNAs were not so effective for degrading miRNAs. A caution to new miRNA researchers is not to expect all samples of high RIN to produce strong miRNA array signals. A total RNA sample of very high RIN (larger than 9) does not guarantee the retaining or good integrity of miRNAs in the sample. Small RNAs could have been lost during RNA extraction or due to certain type of degradation reactions in a seemingly good total RNA sample. We have encountered a few cases of very weak miRNA signals being produced from samples of high RINs (larger than 9). A Bioanalyzer Small RNA kit is available from Agilent for the quantitative analysis of small RNAs. However, due to low mass, miRNAs usually appear as a slowly climbing baseline in electrophoretic trace and are difficult to be reliably evaluated. For a project involving a large number of samples, it is prudent to have a pilot run by using array assay as the final qualifier for the RNA extraction process.
2. Experiments have been performed to compare hybridization signals of synthetic mature miRNAs and corresponding miRNA precursors of same concentration. The mature miRNAs produced significantly higher signal intensities than corresponding precursors did (unpublished data).
3. For a dual sample assay, prepare Cy3 and Cy5 reaction solutions separately for the two corresponding RNA samples. The cyanine ligation mix may be replaced with 6× Alexa Fluor®-3 (AF-3) and/or Alexa Fluor®-5 (AF-5) Ligation Mix from Invitrogen.
4. Always add the spike-in controls into total RNA instead of fractionated or enriched small RNA samples from which it is often difficult to back-track the amount of starting total RNA.
5. Most of the small noncoding RNAs are longer than 50 nt. Many of these not so small RNAs have stable secondary structures which affect their hybridization with array probes. Therefore, care must be taken to optimize probe design and perform validation before these small RNAs are used as array normalization references.

Acknowledgments

This research is supported by grants from NIH (R44GM076941, R33CA126209, and R41DA029169).

References

1. Bartel, D. P., (2004) MicroRNAs: Genomics, Biogenesis, Mechanism, and Function. *Cell* **116**, 281–97.
2. He, L., Thomson, J. M., Hemann, M. T., Hernando-Monge, E., Mu, D., and Goodson, S. (2005) A microRNA polycistron as a potential human oncogene. *Nature* **435**, 828–33.
3. Lu, J., Getz, G., Miska, E. A., Alvarez-Saavedra, E., Lamb, J., Peck, D., et al (2005) MicroRNA expression profiles classify human cancers. *Nature* **435**, 834–8.
4. O'Donnell, K. A., Wentzel, E. A., Zeller, K. I., Dang, C. V., and Mendell, J. T. (2005) c-Myc-regulated microRNAs modulate E2F1 expression. *Nature* **435**, 839–43.
5. Hinds, D. A., Stuve, L. L., Nilsen, G. B., Halperin, E., Eskin, E., Ballinger, D. G. et al. (2005) Whole-genome patterns of common DNA variation in three human populations. *Science* **307**, 1072–9.
6. Ekins, R. P. (1989) Multi-analyte immunoassay. *J Pharm Biomed Anal* 7, 155–68.
7. Mirzabekov, A. and Kolchinsky, A. (2002) Emerging array-based technologies in proteomics. *Curr Opin Chem Biol* **6**, 70–5.
8. Schena, M., Shalon, D. D., Davis, R. W., and Brown, P. O. (1995) Quantitative monitoring of gene expression patterns with a complementary DNA microarray. *Science* **270**, 467–0.
9. Lockhart, D. J., Dong, H., Byrne, M. C., Follettie, M. T., Gallo, M. V., Chee, M. S., et al (1996) Expression monitoring by hybridization to high-density oligonucleotide arrays. *Nat Biotechnol* **14**, 1675–80.
10. Ekins, R. P. (1998) Ligand assays: from electrophoresis to miniaturized microarrays. *Clin Chem* **44**, 2015–30.
11. Stoll, D., Templin, M. F., Bachmann, J. and Joos, T. O. (2005) Protein microarrays: applications and future challenges. *Curr Opin Drug Discov Devel* **8**, 239–52.
12. The chipping forecast. (1999) *Nat Genet* **Supplement 21**, 3–60.
13. Gao, X., Gulari, E., and Zhou, X. (2004) In situ synthesis of oligonucleotide microarrays. *Biopolymers* **73**, 579–96.
14. Fodor, S. P., Leighton, P. A. J., Pirrung, M. C., Stryer, L, and Solas, D. (1991) Light-directed spatially addressable parallel chemical synthesis. *Science* **251**, 767–73.
15. Maskos, U., and Southern, E. M. (1992) Parallel analysis of oligodeoxyribonucleotide (oligonucleotide) interactions. I. Analysis of factors influencing oligonucleotide duplex formation. *Nucleic Acids Res* **20**, 1675–8.
16. Blanchard, A. P., Kaiser, R. J., and Hood, L. E. (1996) High-density oligonucleotide arrays. *Biosens Bioelectron* **11**, 687–90.
17. Blanchard, A. P., and Hood, L. E. (1996) Sequence to array: probing the genome's secrets. *Nat Biotechnol* **14**, 1649.
18. Gao, X., Yu, P. Y., LeProust, E., Sonigo, L., Pellois, J. P., and Zhang, H. (1998) Oligonucleotide synthesis using solution photogenerated acids. *J Am Chem Soc* **120**, 12698–9.
19. Singh-Gasson, S., Green, R. D., Yue, Y., Nelson, C., Blattner, F., Sussman, M. R., et al (1999) Maskless fabrication of light-directed oligonucleotide microarrays using a digital micromirror array. *Nat Biotechnol* **17**, 974–8.
20. Gao, X., LeProust, E., Zhang, H., Srivannavit, O., Gulari, E., Yu, P., et al (2001) Flexible DNA chip synthesis gated by deprotection using solution photogenerated acids. *Nucleic Acids Res* **29**, 4744–50.
21. Hughes, T. R., Mao, M., Jones, A. R., Burchard, J., Matthew, J. B., Marton, M.J., et al (2001) Expression profiling using microarrays fabricated by an ink-jet oligonucleotide synthesizer. *Nat Biotechnol* **19**, 342–7.
22. Bulter, J. H., Cronin, M., Anderson, K. M., Biddison, G. M., Chatelain, F., Cummer, M., et al (2001) In situ synthesis of oligonucleotide arrays by using surface tension. *J Am Chem Soc* **123**, 8887–94.
23. McGall, G. H., and Fidanza, J. A. (2001) Photolithographic synthesis of high-density oligonucleotide arrays. *Methods Mol Biol* **170**, 71–101.
24. Luebkel, K. J., Balog, R. P., and Garner, H. R. (2003) Prioritized selection of oligodeoxyribonucleotide probes for efficient hybridization to RNA transcripts. *Nucleic Acids Res* **31**, 750–8.
25. Tesfu, E., Maurer, K., Ragsdale, S. R. and Moeller, K. D. (2004) Building addressable

libraries: the use of electrochemistry for generating reactive Pd(II) reagents at preselected sites on a chip. *J Am Chem Soc* **126**, 6212–3.

26. Srivannavit, O., Gulari, M., Gulari, E., LeProust, E., Pellois, J. P., Gao, X., et al (2004) Design and fabrication of microwell array chips for a solution-based, photogenerated acid-catalyzed parallel oligonuclotide DNA synthesis. *Sensors and Actuators A* **116**, 150–60.
27. Zhou, X., Cai, S., Hong, A., You, Q., Yu, P., Sheng, N., et al (2004) Microfluidic PicoArray synthesis of oligodeoxynucleotides and simultaneous assembling of multiple DNA sequences. *Nucleic Acids Res* **32**, 5409–17.
28. Frank, R. (2002) The SPOT-synthesis technique. Synthetic peptide arrays on membrane supports - principles and applications. *J Immunol Methods* **267**, 13–26.
29. Reimer, U., Reineke, U. and Schneider-Mergener, J. (2002) Peptide arrays: from macro to micro. *Curr Opin Biotechnol* **13**, 315–20.
30. Lam, K. S. and Renil, M. (2002) From combinatorial chemistry to chemical microarray. *Curr Opin Chem Biol* **6**, 353–8.
31. Gao, X., Pellois, J. P., Na, Y., Kim, Y., Gulari, E. and Zhou, X. (2004) High density peptide microarrays. In situ synthesis and applications. *Mol Divers* **8**, 177–87.
32. Panse, S., Dong, L., Burian, A., Carus, R., Schutkowski, M., Reimer, U. et al (2004) Profiling of generic anti-phosphopeptide antibodies and kinases with peptide microarrays using radioactive and fluorescence-based assays. *Mol Divers* **8**, 291–9.
33. Pease, A. C., Solas, D., Sullivan, E. J., Cronin, M. T., Holmes, C. P. and Fodor, S. P. (1994) Light-generated oligonucleotide arrays for rapid DNA sequence analysis. *Proc Natl Acad Sci USA* **91**, 5022–6.
34. Holmes, C. P., Adams, C. L., Kochersperger, L. M., Mortensen, R. B. and Aldwin, L. A. (1995) The use of light-directed combinatorial peptide synthesis in epitope mapping. *Biopolymers* **37**, 199–211.
35. Pellois, J. P., Wang, W. and Gao, X. (2000) Peptide synthesis based on t-Boc chemistry and solution photogenerated acids. *J Comb Chem* **2**, 355–60.
36. Pellois, J. P., Zhou, X., Srivannavit, O., Zhou, T., Gulari, E. and Gao, X. (2002) Individually addressable parallel peptide synthesis on microchips. *Nat Biotechnol* **20**, 922–6.
37. Komolpis, K., Srivannavit, O. and Gulari, E. (2002) Light-directed simultaneous synthesis of oligopeptides on microarray substrate using a photogenerated acid. *Biotechnol Prog* **18**, 641–6.
38. Li, S., Bowerman, D., Marthandan, N., Klyza, S., Luebke, K. J., Garner, H. R. et al (2004) Photolithographic synthesis of peptoids. *J Am Chem Soc* **126**, 4088–9.
39. Li, S., Marthandan, N., Bowerman, D., Garner, H. R. and Kodadek, T. (2005) Photolithographic synthesis of cyclic peptide arrays using a differential deprotection strategy. *Chem Commun (Camb)* 7, 581–3.
40. miRbase: Integrating miRNA annotation and deep sefuencing data. Kozomara A and Griffiths- Jones S (2011). Nucleic Acid Research 39 (dabase issue): D152–D157 http://www.mirbase.org/.
41. Lagos-Quintana, M., Rauhut, R., Lendeckel, W. and Tuschl, T. (2001) Identification of novel genes coding for small expressed RNAs. *Science* **294**, 853–8.
42. Lee, R. C. and Ambros, V. (2001) An extensive class of small RNAs in Caenorhabditis elegans. *Science* **294**, 862–4.
43. Lee, L. W., Zhang, S., Etheridge, A., Ma, L., Martin, D., Galas, D., et al (2010) Complexity of the microRNA repertoire revealed by next generation sequencing. *RNA* **16**, 1–11.
44. Kozomara, A. and Griffiths-Jones, S. (2010) *Nucleic Acids Res* **39**, D152-7.
45. Thomson, J. M., Parker, J., Perou, C. M. and Hammond, S. M. (2004) A custom microarray platform for analysis of microRNA gene expression. *Nat Methods* **1**, 47–53.
46. Nelson, P. T., Baldwin, D. A., Scearce, L. M., Oberholtzer, J. C., Tobias, J. W. and Mourelatos, Z. (2004) Microarray-based, high-throughput gene expression profiling of microRNAs. *Nat Methods* **1**, 155–61.
47. Liang, R. Q., Li, W., Li, Y., Tan, C. Y., Li, J. X., Jin, Y. X., et al (2005) An oligonucleotide microarray for microRNA expression analysis based on labeling RNA with quantum dot and nanogold probe. *Nucleic Acids Res* **33**, e17.
48. Shingara, J., Keiger, K., Shelton, J., Laosinchai-Wolf, W., Powers, P., Conrad, R., et al (2005) An optimized isolation and labeling platform for accurate microRNA expression profiling. *RNA* **11**, 1461–70.
49. Barad, O., Meiri, E., Avniel, A., Aharonov, R., Barzilai, A., Bentwich, I., et al (2004) MicroRNA expression detected by oligonucleotide microarrays: system establishment and expression profiling in human tissues. *Genome Res* **14**, 2486–94.
50. Liu, C. G., Calin, G. A., Meloon, B., Gamliel, N., Sevignani, C., Ferracin, M., et al (2004) An oligonucleotide microchip for genome-wide microRNA profiling in human and mouse tissues. *Proc Natl Acad Sci USA* **101**, 9740–4.
51. Babak, T., Zhang, W., Morris, Q., Blencowe, B. J. and Hughes, T. R. (2004) Probing microRNAs

with microarrays: tissue specificity and functional inference. *RNA* **10**, 1813–9.

52. Baskerville, S. and Bartel, D. P. (2005) Microarray profiling of microRNAs reveals frequent coexpression with neighboring miRNAs and host genes. *RNA* **11**, 241–7.
53. Miska, E. A., Alvarez-Saavedra, E., Townsend, M., Yoshii, A., Sestan, N., Rakic, P., et al (2004) Microarray analysis of microRNA expression in the developing mammalian brain. *Genome Biol* **5**, R68.
54. Sioud, M. and Rosok, O. (2004) Profiling microRNA expression using sensitive cDNA probes and filter arrays. *Biotechniques* **37**, 574–6, 578–80.
55. Sun, Y., Koo, S., White, N., Peralta, E., Esau, C., Dean, N. M., et al (2004) Development of a micro-array to detect human and mouse microRNAs and characterization of expression in human organs. *Nucleic Acids Res* **32**, e188.
56. Krichevsky, A. M., King, K. S., Donahue, C. P., Khrapko, K. and Kosik, K. S. (2003) A microRNA array reveals extensive regulation of microRNAs during brain development. *RNA* **9**, 1274–81.
57. Wang, H., Ach, R. A., and Curry, B. (2007) Direct and sensitive miRNA profiling from low-input total RNA. *RNA* **13**, 1–9.
58. Chen, C., Ridzon, D. A., Broomer, A. J., Zhou, Z., Lee, D. H., Nguyen, J. T., et al (2005) Real-time quantification of microRNAs by stem–loop RT–PCR. *Nucleic Acids Res* **33**, e179.
59. Alwine, J.C., Kemp, D.J., and Stark, G.R. (1977) Method for detection of specific RNAs in agarose gels by transfer to diazobenzyloxymethyl-paper and hybridization with DNA probes. *Proc Natl Acad Sci USA* **74**, 5350–4.
60. Vagin, V.V., Sigova, A, Li, C., Seitz, H., Gvozdev, V., and Zamore, P.D. (2006) A distinct small RNA pathway silences selfish genetic elements in the germline. *Science* **313**, 320–24.
61. Monya, B. (2010) MicroRNA profiling: separating signal from noise. *Nat Methods* 7, 687–92.
62. Linsen, S. E., de Wit, E., Janssens, G., Heater, S., Chapman, L., Parkin, R. K., et al (2009) Limitations and possibilities of small RNA digital gene expression profiling. *Nat Methods* **6**, 474–6.
63. Bolstad, B. M. (2004) Low Level Analysis of High-density Oligonucleotide Array Data: Background, Normalization and Summarization. Dissertation. University of California, Berkeley
64. Casella, G. and Berger, R. L. (2001). Statistical Inference (2nd Ed.). Duxbury Press.
65. Cleveland, W. S. (1979). Robust Locally Weighted Regression and Smoothing Scatterplots. *J Am Stat Assoc* **74**, 829–36.
66. Cleveland, W. S. and Devlin, S. J. (1988). Locally-Weighted Regression: An Approach to Regression Analysis by Local Fitting. *J Am Stat Assoc* **83**, 596–610.
67. Bolstad, B. M., Irizarry, R. A, Astrand, M., and Speed, T. P. (2003) A comparison of normalization methods for high density oligonucleotide array data based on variance and bias *Bioinformatics* **19**, 185–93.
68. Dudoit, S., Yang, Y. H., Callow, M. J., and Speed, T. P. (2002) Statistical methods for identifying genes with differential expression in replicated cDNA microarray experiments. *Stat Sinica* **12**, 111–39.
69. Yang, Y. H., Dudoit, S. D., Luu, P, and Speed, T. P. (2001) Normalization for cDNA Microarray Data. *SPIE BioE* **6**.
70. Yang, Y.H., Dudoit, S., Luu, P., Lin, D.M., Peng, V., Ngai. J., et al (2002). Normalization for cDNA microarray data: a robust composite method addressing single and multiple slide systematic variation. *Nucleic Acids Res* **30**, e15.
71. Chen, C., Ridzon, D. A., Broomer, A. J., Zhou, Z., Lee, D. H., Nguyen, J. T., et al (2005) Real-time quantification of microRNAs by stem–loop RT–PCR, *Nucleic Acids Res* **33**, e179.
72. Saeed, A. I., Sharov, V., White, J., Li, J., Liang, W., Bhagabati, N., Braisted, J., et al (2003) TM4: a free, open-source system for microarray data management and analysis. *Biotechniques* **34**, 374–8.
73. Eisen, M. B., Spellman, P. T., Brown, P. O., and Botstein, D. (1998) Cluster analysis and display of genome-wide expression patterns. *Proc Natl Acad Sci USA* **95**, 14863–8.

Chapter 12

MicroRNA Expression Analysis Using the Illumina MicroRNA-Seq Platform

Shujun Luo

Abstract

Direct sequencing of RNA molecules using next-generation sequencing (NGS) technology has revolutionized the analysis of transcriptome with its massively parallel throughput and low cost. Here, we describe Illumina's microRNA-Seq, a method for sequencing microRNA using the Illumina Genome Analyzer system. The sequence data generated from this method enables direct identifying and profiling of microRNAs in any given organism. It also sheds light in understanding the biogenesis and modification of microRNA.

Key words: MicroRNA, RNA, Next-generation sequencing, Illumina

1. Introduction

MicroRNA is a unique class of noncoding RNA in the transcriptome, with an average size of ~22 nt, and regulates gene expression post-transcriptionally (1). Majority of microRNAs matures through three stages, primary miRNA transcript (pri-miRNA), precursor miRNA (pre-miRNA), and matured miRNA. RNA polymerase II transcribes pri-miRNA from specific genomic DNA locations that fold into a hairpin structure. Drosha, an RNase III enzyme, digests the pri-miRNA hairpin into pre-miRNA, and dicer, another RNase III enzyme, further digests the pre-miRNA into matured microRNA. The biogenesis process of microRNA provides defined groups to the ends of the microRNA fragment, with monophosphate on the 5′ end and hydroxyl on the 3′ end (1).

By utilizing the special end groups of microRNA molecules, the Illumina miRNA-Seq sample prep ligates two adaptors sequentially to the 3′ and 5′ ends of microRNA in the context of 1 μg of

Jian-Bing Fan (ed.), *Next-Generation MicroRNA Expression Profiling Technology: Methods and Protocols*, Methods in Molecular Biology, vol. 822, DOI 10.1007/978-1-61779-427-8_12, © Springer Science+Business Media, LLC 2012

total RNA, subsequently reverse transcribes, and then PCR amplifies the microRNA through the ligated common link sequences. The sequences of the resulting microRNA library are deciphered with Illumina next-generation sequencing (NGS) (2).

Single molecules of a microRNA library DNA are hybridized to primers grafted onto the glass surface of a flow cell. Isothermal bridge PCR is carried out to amplify the DNA in situ. The colonized DNA further serves as the template for sequencing by synthesis using fluorescent reversible terminator deoxyribonucleotides.

Millions of microRNA sequences can be generated in a single Illumina Genome Analyzer run. The sequences can be used for profiling expression level of microRNA, discovering novel microRNA, and studying mechanism of microRNA biogenesis and modification (3).

2. Material

2.1. Total RNA

We recommend the use of a EtOH precipitation-based total RNA purification method for miRNA-Seq to ensure maximal yield of microRNA (see Notes 1–3).

2.2. Illumina v1.5 Small RNA Sample Prep Kit

1. Ultrapure water.
2. 10× v1.5 sRNA 3′ adapter.
3. SRA 5′ adapter.
4. T4 RNA ligase.
5. RNase OUT.
6. 25 mM dNTP mix.
7. SRA RT primer.
8. Primer GX1.
9. Primer GX2.
10. Phusion polymerase (Finnzymes).
11. 5× Phusion HF Buffer (Finnzymes).
12. 25 bp ladder.
13. 10× gel elution buffer.
14. Glycogen.
15. Resuspension buffer.

2.3. Additional Material

1. Clean 0.2, 0.5, and 2.0-mL nuclease-free microcentrifuge tubes.
2. T4 RNA Ligase 2, truncated, with 10× T4 RNL2 truncated reaction buffer (NEB).

3. 100 mM $MgCl_2$ (a 100 mM solution can be prepared from 1 M $MgCl_2$ (USB) or any molecular-grade substitute).
4. 10 mM ATP (Epicentre).
5. SuperScript II Reverse Transcriptase with 100 mM DTT and 5× First Strand Buffer (Invitrogen).
6. 6× DNA loading dye.
7. 5× Novex TBE buffer.
8. 6% Novex TBE PAGE gel, 1.0 mm, 10 well.
9. Clean scalpels.
10. 21-gauge needles.
11. Ultrapure ethidium bromide, 10 mg/mL.
12. 3 M NaOAc, pH 5.2.
13. -20°C 100% ethanol.
14. Room-temperature 70% ethanol.

3. Methods

3.1. Dilute Oligos and $MgCl_2$ for MicroRNA-Seq Protocol

1. Dilute the 10× v1.5 sRNA 3′ adapter by mixing 1 μL of adapter with 9 μL of nuclease-free water.
2. Dilute the SRA RT primer by mixing 1 μL of primer with 4 μL of nuclease-free water.
3. Dilute the 1 M $MgCl_2$ solution by mixing 100 μL with 900 μL of nuclease-free water.

3.2. Ligate v1.5 Small RNA 3′ Adapter and 5′ Adapter to MicroRNAs

1. Set up the 3′ adaptor ligation reaction in a nuclease-free 200 μL PCR tube using the following: 5.2 μL 1 μg total RNA and 1 μL diluted 3′ RNA adaptor.
2. Incubate the tube in a thermal cycler at 70°C for 2 min, and then transfer immediately to ice.
3. Add the following reagents to the tube and mix them well: 1 μL 10× ligation buffer; 0.8 μL 100 mM $MgCl_2$; 1.5 μL T4 RNA ligase, truncated; 0.5 μL RNaseOut (40 U/μL).
4. Incubate the tube in a thermal cycler at 22°C for 1 h.
5. With 5 min remaining, prepare 5′ adaptor for ligation by heating it at 70°C for 2 min and then transferring it to ice.
6. Add the following reagents to the 3′ ligation mixture from Subheading 3.2 step 4 and mix well: 1 μL 10 mM ATP; 1 μL SRA 5′ adaptor; 1 μL T4 RNA ligase.
7. Incubate the tube in a thermal cycler at 20°C for 1 h and keep it at 4°C until the next day if necessary.

3.3. Reverse Transcribe and PCR Amplify Adaptor-Ligated MicroRNAs

1. Dilute 25 mM dNTP to 12.5 mM by mixing 1 μL of 25 mM dNTP with 1 μL of nuclease-free water.
2. Set up a reverse transcription reaction in a nuclease-free 200 μL PCR tube: 4 μL 3′ and 5′ adaptor-ligated RNA; 1 μL diluted SRA RT primer.
3. Incubate the tube in a thermal cycler at 70°C for 2 min and then transfer immediately to ice.
4. Add the following reagents to the tube and mix well: 2.0 μL 5× first-strand buffer; 0.5 μL 12.5 mM dNTP; 1 μL 100 mM DTT; 0.5 μL R NaseOut (40 U/μL); 1.0 μL SuperScript II RT (200 U/μL).
5. Incubate the tube in a thermal cycler at 48°C for 3 min, and then 44°C for 1 h.
6. Add the following reagents to the tube and mix well: 10 μL 5× cloned Phu buffer; 0.5 μL primer GX1; 0.5 μL primer GX2; 0.5 μL 25 mM dNTP mix; 0.5 μL phusion polymerase; 28 μL H_2O.
7. Amplify the material with following PCR cycling conditions:

   1 cycle of 30 s at 98°C; followed by 12 cycles of 10 s at 98°C; 30 s at 60°C; 15 s at 72°C; followed by 10 min at 72°C.

3.4. Gel Purify MicroRNA Library

1. Determine the volume of 1× TBE buffer needed. Dilute the 5× TBE buffer to 1× for use in electrophoresis.
2. Assemble the gel electrophoresis apparatus as per the manufacturer's instructions.
3. Mix 1 μL of 25 bp ladder with 1 μL of 6× DNA loading dye.
4. Mix 50 μL of amplified cDNA construct with 10 μL of 6× DNA loading dye.
5. Load 2 μL of mixed 25 bp ladder and loading dye in one well on the 6% PAGE gel.
6. Load two wells with 25 μL each of mixed amplified cDNA construct and loading dye on the 6% PAGE gel.
7. Run the gel for 30–35 min at 200 V or until the front dye exits the gel.
8. Remove the gel from the apparatus.
9. Dilute the 10× gel elution buffer into a fresh tube by mixing 1 volume of 10× gel elution buffer with 9 volumes of nuclease-free water.
10. Puncture the bottom of a sterile, nuclease-free, 0.5-mL microcentrifuge tube four to five times with a 21-gauge needle.
11. Place the 0.5-mL microcentrifuge tube into a sterile, round-bottom, nuclease-free, 2-mL microcentrifuge tube.
12. Pry apart the cassette and stain the gel with ethidium bromide in a clean container for 2–3 min.

13. View the gel on a Dark Reader transilluminator or a UV transilluminator.
14. Using a clean scalpel, cut out the bands corresponding to approximately the adapter-ligated constructs derived from the 22 and 30 nt small RNA fragments. The band containing the 22 nt RNA fragment with both adapters is a total of 93 nt in length. The band containing the 30 nt RNA fragment with both adapters is 100 nt in length.
15. Place the band of interest into the 0.5-mL microcentrifuge tube from step 1.
16. Centrifuge the stacked tubes at 14,000 rpm in a microcentrifuge for 2 min at room temperature to move the gel through the holes into the 2-mL tube.
17. Add 100 μL of 1× gel elution buffer to the gel debris in the 2-mL tube.
18. Elute the DNA by rotating the tube gently at room temperature for 2 h.
19. Transfer the eluate and the gel debris to the top of a Spin-X filter.
20. Centrifuge the filter for 2 min at 14,000 rpm.
21. Add 1 μL of glycogen, 10 μL of 3 M NaOAc, and 325 μL of -20°C 100% ethanol.
22. Immediately centrifuge to 14,000 rpm for 20 min in a benchtop microcentrifuge.
23. Remove and discard the supernatant, leaving the pellet intact.
24. Wash the pellet with 500 μL of room-temperature 70% ethanol.
25. Remove and discard the supernatant, leaving the pellet intact.
26. Dry the pellet using the speed vac.
27. Resuspend the pellet in 10 μL of resuspension buffer.
28. Load 1 μL of DNA onto the Agilent DNA 1000 chip to check the insert size and concentration of the library.

3.5. Sequencing the MicroRNA Library with the Illumina Genome Analyzer System

DNA fragments from the microRNA library are denatured into single strands, and annealed to complementary oligonucleotides that are pre-grafted on the flow cell surface. The single DNA molecule is amplified in situ by solid face isothermal bridge amplification to form a cluster, which is ~1 μm in diameter. Millions of clusters are formed simultaneously on the flow cell surface. The DNA in each cluster is linearized by cleavage within one adaptor sequence and denatured, generating single-stranded template for sequencing. The sequencing primer is hybridized to the template, and fluorescence-labeled reversible terminator deoxyribonucleotide is incorporated by modified DNA polymerase. Images of the fluorescence of each cluster are analyzed to generate the sequence.

4. Notes

1. It can be difficult to QC microRNA in total RNA due to its small size and low abundance. We recommend using precipitation-based methods over column-based methods for total RNA purification for maximal recovery of microRNA. A positive control that is known to contain microRNA should be included for first-time users.
2. Low amounts of microRNA in starting material can result in poor yield. Different species or tissues might have different amounts of microRNA in total RNA.
3. Since the total RNA is used directly for enzymatic reactions, we recommend resuspending the total RNA in nuclease-free water or Tris buffer to avoid inhibition of downstream reactions.

References

1. Bartel, D. P. (2004) MicroRNAs: genomics, biogenesis, mechanism, and function. *Cell* **116**, 281–97.
2. Lau, N. C., Lim, L. P., Weinstein, E. G. and Bartel, D. P. (2001) An abundant class of tiny RNAs with probable regulatory roles in *Caenorhabditis elegans*. *Science* **294**, 858–62.
3. Chiang, H. R., Schoenfeld, L. W., Ruby, J. G., Auyeung, V. C., Spies, N., Baek, D., et al (2010) Mammalian microRNAs: experimental evaluation of novel and previously annotated genes. *Genes Dev* **24**, 992–1009.

Chapter 13

Next-Generation Sequencing of miRNAs with Roche 454 GS-FLX Technology: Steps for a Successful Application

Ana Raquel Soares, Patrícia M. Pereira, and Manuel A.S. Santos

Abstract

MicroRNAs (miRNAs) are a class of small RNAs (sRNAs) of approximately 22 nucleotides in length that control eukaryotic gene expression at the translational level. They regulate a wide variety of biological processes, namely developmental timing, cell differentiation, cell proliferation, the immune response, and infection. Their identification is essential to understand eukaryotic biology. Their small size, low abundance, and high instability complicated early identification, however new generation genome sequencing approaches, such as the Roche 454 Pyrosequencer, allow for both miRNA identification and for generating miRNA profiles in a given sample. This technique avoids cloning steps in bacteria and is a fast and bias-minimized tool to discover novel miRNAs and other sRNAs on a genome-wide scale. Prior to sequencing, cDNA libraries are built for each sample using total RNA as starter material. Each cDNA library can be tagged with specific identifier sequences that allow sequencing different samples in the same chip run. Here, we describe the protocols for the construction of sRNA cDNA libraries for 454 sequencing, and we include tips for overcoming problems often encountered during cDNA library preparation.

Key words: microRNA, 454 sequencing, Small RNA isolation, cDNA libraries, PCR amplification

1. Introduction

Since the discovery of the first miRNA in 1993 in *C. elegans* (1), thousands of mature miRNAs have been uncovered in several species, suggesting that they appeared early in eukaryotic evolution and play fundamental roles in gene expression control. The first approaches to miRNA discovery used cloning and sequencing of individual RNAs through traditional molecular methods. However, the quantification of miRNA expression has been technically challenging and rather expensive due to their small size, low abundance, low stability, and contamination with other cellular

Jian-Bing Fan (ed.), *Next-Generation MicroRNA Expression Profiling Technology: Methods and Protocols*, Methods in Molecular Biology, vol. 822, DOI 10.1007/978-1-61779-427-8_13, © Springer Science+Business Media, LLC 2012

RNAs and RNA fragments. Besides, the identification of rare or tissue-specific miRNAs was not achieved with traditional methodologies. To overcome these limitations, next-generation DNA sequencing, namely the Roche 454 GS-FLX platform developed strategies to allow miRNA identification. These sequencing methodologies have been successfully applied to both miRNA identification and quantification (2–6). The enormous sequencing power of these technologies has overcome most of the technical hurdles associated with miRNA identification and dramatically increased the number of miRNAs deposited in public databases, such as miRBase (7). These new methodologies are also promoting large-scale initiatives to identify most eukaryotic miRNAs, to understand their evolution and to identify target genes and gene networks regulated by them.

The identification and quantification of miRNAs in zebrafish (ZF) using next-generation DNA sequencing was carried out in our laboratory (6), using the 454 Genome Sequencer FLX from Roche. It enabled us to sequence thousands of individual miRNA molecules, and to identify and quantify them, without traditional cloning in bacteria.

Double-stranded cDNA library preparation for sequencing is accomplished after ligation of specific adapters to the sRNA molecules that contain specific priming sites for sequencing. After preparation, double-stranded cDNA libraries are ready to be clonally amplified via emulsion PCR (emPCR) and sequenced. One of the most interesting features of this methodology is the very high single-read accuracy (99.5%), which enables the differentiation of similar miRNA sequences (8). Each sRNA molecule within a mixture is sequenced individually because of the binding to DNA capture beads. The specific adapters appended to the sRNAs are recognized by the DNA capture beads. These are then clonally amplified by emPCR. For this, beads are placed in a water-in-oil emulsion, forming hundreds of thousands of PCR microreactors, which allows for the entire collection of fragments and beads to be amplified in parallel. When sequenced on the Genome Sequencer FLX, each clonally amplified fragment generates its own unique sequence (8). Besides, the 250 bp reads obtained in the sequencing run are long enough to completely read through the adapters added to the cDNAs during the library preparation in addition to the sRNAs. This provides an ideal quality control for each read.

Using this technology, we were able to successfully sequence sRNAs from different ZF embryonic stages and from adult tissues in a single sequencing run. Novel sRNA molecules were discovered and the total number of zebrafish miRNAs was increased from 192 to 217. By sequencing sRNAs from different embryonic stages and adult tissues, we were also able to generate miRNA expression profiles for the different developmental time points and tissues.

Preparation of cDNA libraries for high-throughput sequencing can be challenging due to the small size of miRNAs, lack of a poly(A) tail and low miRNA concentrations in cells and tissues. To overcome these challenges, we optimized protocols for the construction of sRNA cDNA libraries for 454 high-throughput sequencing. All steps, including RNA isolation, sRNA fractionation, adapter ligations, the reverse transcriptase reaction, and PCR amplification are described in detail below. An overview of emPCR, sequencing and data extraction is also described.

2. Materials

RNases can be introduced into RNA solutions at any of the multiple steps of RNA preparation and should be avoided through careful handling of all equipment used. All materials and solutions should be handled under RNase-free conditions. For routine work, the work bench should be cleaned with RNaseZap prior to use. All solutions should be prepared with RNase-free water. Disposable, sterile gloves, plastic wear, and pipettes should be reserved for RNA work.

2.1. RNA Isolation Reagents

1. TRIzol® reagent from Invitrogen. Store at 4°C.
2. Chloroform. Store at room temperature.
3. Isopropyl alcohol. Store at room temperature.
4. Ethanol: 75% solution in RNase-free water. Store at room temperature.

2.2. RNA Fractionation

2.2.1. Denaturing Polyacrylamide Gel Reagents

1. Polyacrilamide (PAA) 40% stock.
2. Urea.
3. TBE (Tris/Borate/EDTA) 10×: add 108 g Tris base, 55 g boric acid, and 9.3 g EDTA to 800 mL RNase-free water in an RNase-free 1 L flask. Allow dissolution of all components. Adjust the volume to 1 L with additional RNase-free water. Adjust pH to 8.3. Store at room temperature.
4. Ammonium persulfate (APS): 10% solution in RNase-free water. Store at 4°C.
5. Tetramethylethylenediamine (TEMED).

2.2.2. Denaturing PAA Running Buffer

1. TBE 1×. Dilute TBE 10× for a final concentration of TBE 1×. Store at room temperature.
2. Loading buffer: bromophenol blue and xylene cyanol.
3. Size markers: 22 nt cy3-labeled oligo.

4. Electrophoresis apparatus: glass plates (20 mm wide and 18 mm long), spacers, and gel comb (1.5-mm thick), vertical electrophoresis tank.
5. SybrGreen: dilute 1 μL SybrGreen solution in 100 μL TBE 1× to stain the gel.
6. Edge Biosystems DTR Gel filtration columns.
7. Absolute ethanol.
8. Glycogen.

2.3. 3′ Adapter Ligation Reagents

1. 10× T4 RNA ligase buffer without ATP.
2. DMSO.
3. 3′ Adapter 50 μM: AMP-5′p-5′p/CTGTAGGCACCATCA-ATdi-deoxyC-3′ (Integrated DNA Technologies).
4. T4 RNA ligase (Ambion).

2.4. 5′ Adapter Ligation Reagents

1. 10× T4 RNA ligase buffer.
2. DMSO (Dimethyl sulfoxide).
3. RNA Nelson's linker 100 μM - 5′-ATCGTrArGrGrCrAr CrCrUrGrArArA-3′ (IBA-Go).
4. T4 RNA ligase (Ambion).

2.5. Phenol Extraction Reagents

1. Phenol:Chloroform:Isoamyl alcohol (24:24:1). Store at 4°C.
2. Chloroform. Store at room temperature.
3. Absolute ethanol. Store at room temperature.
4. Sodium acetate 3 M, pH 5.2. Dilute 12,3 g sodium acetate in 50 mL RNase-free water and adjust the pH to 5.2. Store at room temperature.

2.6. cDNA Synthesis Reagents

1. dNTPs 10 mM. Store at −20°C.
2. cDNA primer: specific 3′ primer:

5′-GCCTTGCCAGCCCGCTCAG*ATTGATGGTG CCTACAG*-3′.

This primer includes the sequence of the specific 454 adapter and the complementary 3′ linker (*italics*).

3. cDNA synthesis mix: 10× RT buffer, 25 mM $MgCl_2$, 0.1 mM DTT, RNaseOUT.
4. RNase H (Invitrogen).

2.7. PCR Amplification Reagents

All components are stored at −20°C.

1. Taq Polymerase 2U (Fast Start High Fidelity Roche enzyme).
2. dNTPs 10 mM.
3. $MgCl_2$ 100 mM.

4. 10× Taq Polymerase reaction buffer.
5. Forward primer 10 μM:

   5′-GCCTCCCTCGCGCCATCAG**ATC***ATCGTAGGC-ACCTGAAA*-3′ (specific 454 adapter A+optional TAG (**bold**)+5′ linker sequence (*italics*)).

   Reverse primer 10 μM:

   5′-GCCTTGCCAGCCCGCTCAG*ATTGATGGTGCC-TACAG*-3′ (specific 454 adapter B+complementary 3′ linker sequence (*italics*)).
6. Probe elution buffer (Ambion). Store at 4°C.

3. Methods

RNA is initially isolated from the tissue of interest and then fractionated by (PAA) gel electrophoresis (Fig. 1). The RNA fraction in the size range of 15–30 nt corresponding to miRNAs is then extracted

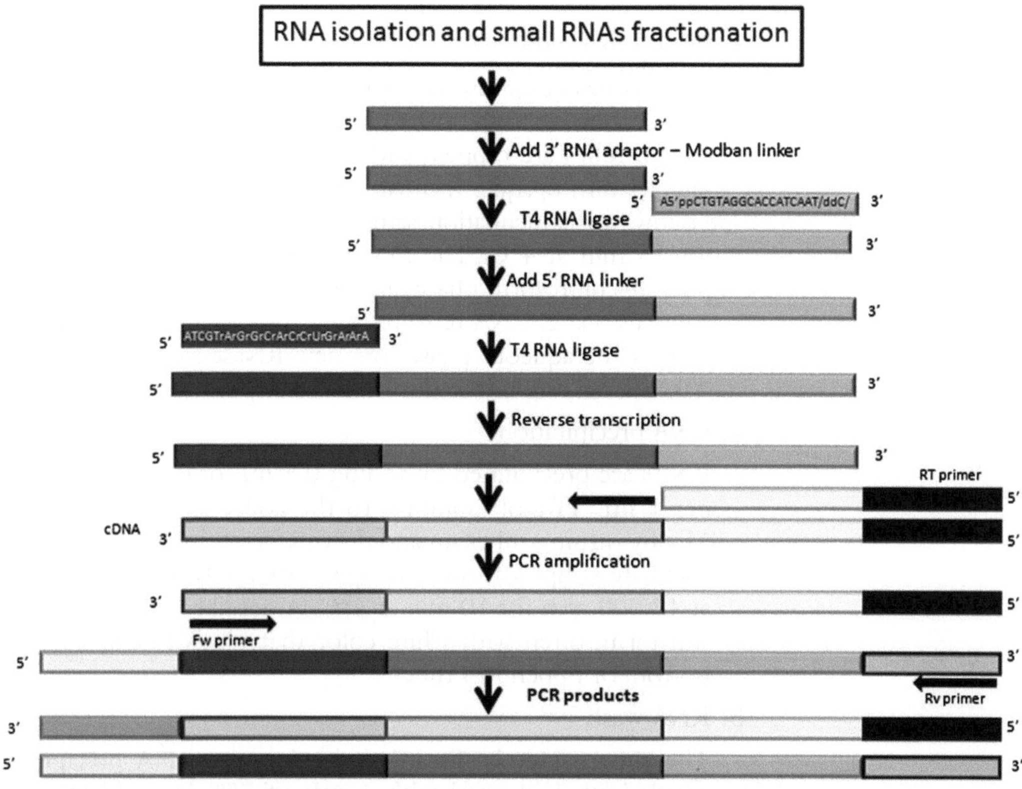

Fig. 1. 454 cDNA library preparation workflow. RNA is isolated using TRIzol® and fractionated on 10% denaturing PAA gels. Small RNAs are purified from these gels and then ligated to a 3′ adapter and to a 5′ linker. cDNA is prepared by reverse transcription and amplified using minimal rounds of PCR (22 cycles), to avoid library bias. PCR products are subjected to clonal amplification by emulsion PCR and then pyrosequenced using a 454 genome sequencer.

and eluted from the gel. Recovered sRNAs pools are then ligated to a 3′ adapter followed by a 5′ adapter ligation, using T4 RNA ligase. These adapters introduce primer-binding sites for reverse transcription and PCR amplification. After PCR amplification, the quantity and quality of DNA are assessed and the libraries can be sequenced (Fig. 1).

3.1. Total RNA Isolation

Samples are harvested and immediately frozen in liquid nitrogen in 2 mL RNase-free screw cap tubes suitable for the tissue lyser. RNA isolation is carried out using TRIzol®, a monophasic solution of phenol and guanidine isothiocyanate.

1. Homogenization

   Tissue samples are disrupted to a fine powder using a mortar and pestle under liquid nitrogen. When working with cell samples, the above grinding step is not required.

2. RNA is isolated by adding 1 mL of TRIzol® per 100 mg of tissue and two zirconium oxide spherical beads. The Precellys 24 rotator (Bertin) or equivalent should be used and three cycles of 10 s at 2300×*g* should be carried out to allow complete tissue dissociation. Always keep samples on ice between homogenization cycles.

3. Phase separation

   After tissue disruption, samples are incubated for 5 min at room temperature, then 0.2 mL chloroform per 1 mL of TRIzol® is added, tubes are capped, with vigorous shaking by hand until samples turn opaque. Incubate at room temperature for 3 min. Following this incubation, samples are centrifuged at 12,000 rpm for 15 min at 4°C. The mixture separates into a lower red, phenol–chloroform phase, an interphase and a colorless upper aqueous phase. RNA remains exclusively in the aqueous phase. Collect the aqueous phase to a new RNase-free 1.5 mL tube. Take special care not to disturb the interphase or lower phase.

4. RNA precipitation

   RNAs are precipitated by adding 0.5 mL of isopropyl alcohol per 1 mL TRIzol® solution to the collected aqueous phase. Mix by turning tubes up and down several times (see Note 1). Incubate samples at room temperature for 10 min. Centrifuge at 12,000 rpm for 10 min at 4°C. A gel-like RNA precipitate will form, often with white color that binds to the side and bottom of Eppendorf tubes.

5. RNA wash

   The supernatant is discarded, leaving the RNA precipitate, which is then washed with 1 mL of 75% ice cold ethanol. Samples should be mixed by vortexing until the pellet detaches from the bottom of the tube and then centrifuged at 7,500 rpm for 10 min at 4°C.

6. Redissolving RNA

   The supernatant is carefully removed to avoid disrupting the pellet. Briefly, the RNA pellet is dried using a speed-vacuum dryer for 1–2 min, until it becomes transparent. It is important not to let the pellet dry completely as over drying decreases RNA solubility. The dried pellet is then resuspended in 30–40 μL RNase-free water, quantified using a NanoDrop spectrophotometer (see Note 2) and stored at −80°C.

7. RNA integrity

   Due to the omnipresence of RNases and RNA instability, it is essential to control RNA integrity before RNA-dependent applications. To assess RNA integrity, ~2 μg of each sample can be analyzed on 1.5% agarose gels or on the Bioanalyzer (Agilent), using the RNA nano or pico kit. The Bioanalyzer allows for easy and fast analysis, using 1 μL of RNA sample only. It gives the RNA spectrum and the RNA integrity number (RIN), which should be higher than 7.

3.2. Small RNA Fractionation

Total RNA is fractionated on 10% denaturing PAA gels. sRNAs are eluted from gels by excising the bands corresponding to 15–30 nt. The RNA size is determined by a co-migrating cy3-labeled RNA size marker in the same gel. Visualization of this marker allows for the identification of gel pieces that contain the RNA of desired length (Fig. 2).

1. Glass (20 × 18 mm) and spacers (1.5 mm) should be cleaned with RNaseZap to avoid RNase contamination and rinsed with RNase-free water. Use a vertical electrophoresis system.

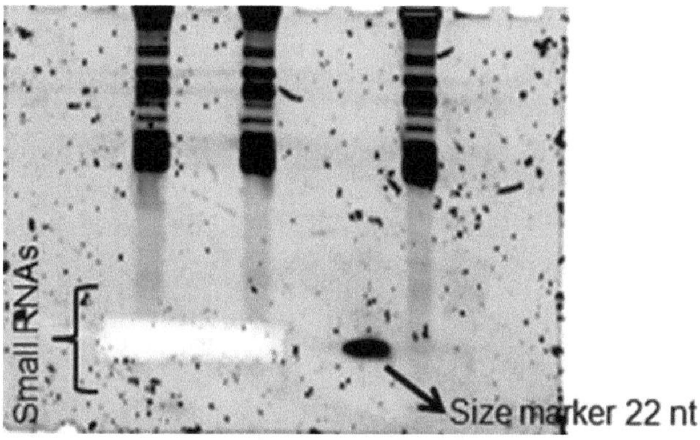

Fig. 2. sRNA extraction from acrylamide gels. Total RNA is fractionated on 10% denaturing PAA gels. Since the sRNA fraction is not easily detected even with SyBr Green staining, a 22 nt cy3-labeled oligo is co-migrated with the RNA samples. A band is excised from the gel for each RNA sample 5 mm above and below the size marker. Alternatively, two labeled markers of the desired sizes can be used to increase accuracy of band identification and excision.

2. Preparation of a 10% denaturing PAA gel (7 M urea)

   Weigh 21 g urea and transfer to an RNase-free flask. Add 5 mL TBE 10×, 12.5 mL PAA, and RNase-free water to a total volume of 50 mL. Mix until the components are dissolved. Transfer 5 mL of this solution into a new RNase-free tube to prepare the gel base.

3. Mix 60 μL APS (10%) and 6 μL of TEMED for polymerization of the gel base. Lay the glass plates, already assembled with the spacers, on a test tube rack. With a pipette, pour the above PAA solution into the bottom of the glass plates (~1 cm gel base). Allow the gel base to polymerize for 15 min (see Note 3).

4. When the base is polymerized, lay the glass plates at a 10° angle, add 450 μL APS 10%, and 40 μL of TEMED to the remaining 45 mL gel solution, mix by swirling. Fill the space between the glass plates to the top. Immediately insert the appropriate comb (1.5-mm thick) and allow the gel to polymerize for 1 h at room temperature (see Note 4).

5. Remove the comb from the gel and rinse with RNase-free water. Prerun the gel at 200 V for 15–30 min. The gel is now ready for electrophoresis and loading.

6. Sample loading

   Concentrate RNA samples to ensure each gel well is loaded with 50–100 μg of total RNA. For each sample, add 10 μL RNA sample (containing between 50 and 100 μg RNA) plus 10 μL of 2× loading dye. Mix, centrifuge briefly and load 20 μL of sample in each well (see Note 5). In unused wells, load size markers (5 μg is sufficient) by adding 10 μL sample plus 10 μL 2× loading dye.

7. Run the gel for 1 h 30 min at 275 V, and then stain it with SyBr Green (see Note 6).

3.3. Small RNA Extraction from the Gel

1. Cut RNA bands in the size range of 15 and 30 nt (2 mm above and below the size marker) with a clean, RNase-free razor blade (Fig. 2). Put each gel slice in a clean 1.5 mL tube and crush it with a pipette tip or using a homogenizer pestle. Add 200 μL RNase-free water and keep crushing the gel slice until a fine slurry is obtained. Incubate the mixture at 70°C for 10 min.

2. Meanwhile, prepare a DTR column for each slice. Centrifuge the column for 2 min at 3000 rpm. Transfer the column to a new 1.5 mL clean tube. Vortex the mixture (slurry) and transfer the entire volume to the DTR column.

3. Centrifuge at 3,000 rpm for 3 min. Discard the DTR column and keep the eluate. Add 5 volumes of absolute ethanol and

1 μL glycogen (20 μg) to the eluate. Mix by inversion and allow for precipitation at −80°C overnight (see Note 7). Spin tubes at 12,000 rpm for 10 min. Pour out the supernatant and dry the RNA. Proceed to 3′ adapter ligation.

3.4. 3′ Adapter Ligation

After recovering the sRNA fraction from gel slices, the sRNAs can be ligated with 3′ and 5′ linkers in two separate reactions. In the first reaction, the 3′ ligation is carried out using the entire sample from above. For this:

1. Resuspend the sRNA precipitate in 13 μL RNase-free water.
2. Assemble the 3′ adapter ligation in 20 μL reactions containing 13 μL of the resuspended sRNA fraction, 2 μL 10× T4 RNA ligase buffer without ATP (see Note 8), 3 μL DMSO, 1 μL 3′ adapter 50 μM (AMP-5′p-5′p/CTGTAGGCACCATCAATdi-deoxyC-3′) and 1 μL T4 RNA ligase. Incubate the reaction at room temperature for 2 h.
3. Add 20 μL of 2× loading dye, heat at 65°C for 5 min to stop the reaction and load the total volume (40 μL) on a 10% denaturing PAA (prepared as described in Subheading 3.2).
4. Run the samples alongside a 40-nt oligo (5 μg oligo in 20 μL + 20 μL loading dye). Run gel as described previously at 275 V for 1 h 30 min.
5. Excise the RNA bands by cutting gel slices 2 mm above and below the 40-nt band with a clean, RNase-free razor blade after staining with SyBrGreen, as described previously (see Note 9).
6. The 3′ linked sRNA fraction is then recovered as described previously for the sRNA fraction described in Subheading 3.3, using DTR columns.

3.5. 5′ Adapter Ligation

1. Dissolve the pellet obtained in the previous step in 13 μL RNase-free water.
2. The 5′ adapter ligation is assembled in 20 μL reactions containing 13 μL of 3′ligated sRNA fraction, 2 μL 10× T4 RNA ligase buffer (with ATP), 2 μL DMSO, 1 μL 100 μM RNA Nelson's linker (5′ATCGTrArGrGrCrArCrCrUrGrArArA3′), and 2 μL T4 RNA ligase (see Note 10). Incubate the reaction at 37°C for 1 h.
3. It is not necessary to gel purify this reaction, rather extract it with phenol. Add 80 μL RNAse-free water to the 3′/5′ linked RNA to a final volume of 100 μL, then add 100 μL of phenol:chloroform:isoamyl alcohol 25:24:1 and mix by vortexing. Centrifuge at 12,000 rpm for 15 min at 4°C. Keep the aqueous phase which contains the RNA.

4. Perform a second extraction by adding 1 volume (100 μL) of chloroform. Mix by vortexing, centrifuge at 12,000 rpm for 15 min at 4°C, and keep the aqueous phase.
5. Precipitate the RNA by adding 5 volumes absolute ethanol and 0.1 volume of sodium acetate (NaOAc) pH 5.2 to the above aqueous phase, mix and incubate overnight at −80°C.
6. Centrifuge at 12,000 rpm for 15 min at 4°C and discard the supernatant. Wash the pellet with 1 mL 75% ice cold ethanol. Mix the sample by vortexing until the pellet is detached from the bottom of the tube. Centrifuge at 7,500 rpm for 10 min at 4°C.
7. Redissolving RNA.

Transfer all the supernatant carefully to avoid disrupting the pellet. Dry the RNA pellet briefly, using a speed-vacuum dryer (1–2 min) until it becomes transparent. As before, it is important not to let the pellet dry completely as over drying decreases RNA solubility. Resuspend the dried pellet in 7 μL RNase-free water and use this fraction to synthesize cDNA.

3.6. cDNA Synthesis

The 5′ and 3′ ligated RNAs contain both RNA and DNA domains. The RNA domain is converted to DNA by reverse transcription. The protocol described below has been optimized for SuperScript™ III reverse transcriptase (Invitrogen).

1. cDNA reactions (10 μL) are assembled in 1.5 mL Eppendorf tubes. Add the 7 μL of ligated RNA from the previous step, 1 μL dNTPs 10 mM, 1 μL specific 3′ cDNA primer (5′ GCCTTGCCAGCCCGCTCAG*ATTGATGGTGCCTACAG* 3′) (specific 454 adapter B + complementary 3′ linker sequence (*italics*)), 50 μM and 1 μL RNase-free water. Incubate for 5 min at 65°C and then on ice for 1 min.
2. Add 10 μL of cDNA synthesis mix to the above samples. This mix contains 10× RT buffer, 4 μL $MgCl_2$ 25 mM, 2 μL DTT 0.1 mM, and 1 μL RNaseOUT. Incubate samples for 1 h at 50°C plus 5 min at 85°C and 1 min on ice. Centrifuge briefly to ensure that the reaction components are at the bottom of the tube. Add 1 μL RNase H and incubate for 20 min at 37°C. This reaction can then be stored at −20°C until used.

3.7. PCR Amplification

In this step, the above cDNAs are amplified and specific adapters (A + B) are appended to the cDNA copy of the sRNA. These adapters are essential for pyrosequencing, since they are recognized downstream by the DNA capture beads and allow clonal amplification by emPCR of the sequences. Specific fusion primers of the A and B adapters and the sRNA cloning adapters from the 3′ and 5′ linkers are used in a round of PCR amplification. The synthesized double-stranded library of sRNA molecules is then ready for amplification by emPCR and sequenced. To avoid library bias, a minimal number of PCR cycles are performed as follows:

A 25 μL PCR reaction is assembled in 0.5 mL sterile tubes with 0.25 μL Fast Start High Fidelity Roche enzyme, 0.5 μL dNTPs 10 mM, 0.5 μL $MgCl_2$ 100 mM, 2.5 μL 10× Fast Start enzyme buffer, 1.25 μL forward primer 10 μM 5′ GCCTCCCTCGCGCCATCAG**ATC***ATCGTAGGCACCTGAAA* 3′ (specific 454 adapter A + optional TAG (**bold**) + 5′ linker sequence (*italics*)) (see Note 11), 1.25 μL reverse primer 10 μM 5′ GCCTTGCCAGCCCGCTCAG*ATTGATGGTGCCTACAG* 3′ (specific 454 adapter B + complementary 3′ linker sequence (*italics*)) and 5 μL of the previously prepared cDNA.

1. The tubes are inserted in a thermocycler and the samples are subjected to the following amplification cycle: 1 cycle *denaturation* at 96°C, 1 min; 22 cycles *denaturation* at 96°C, 10 s, *annealing* at 50°C, 1 min, *elongation* at 72°C, 20 s; 1 cycle *final elongation* at 72°C, 3 min; *hold* at 10 min.
2. The PCR products are analyzed on 4% agarose gels (Fig. 3). The corresponding bands (~100 nt) are excised with clean, sterile razor blades, and the DNA is eluted from the gel with a Qiaquick kit from Qiagen (see Note 12).

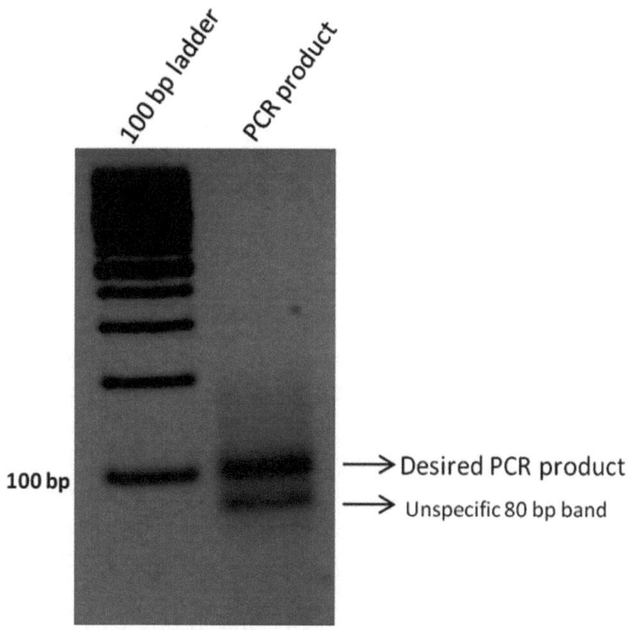

Fig. 3. Amplification of cDNA libraries by PCR. After amplification of cDNA, PCR products are analyzed on 4% agarose gels. The desired products should be ~100 bp in size. Sometimes a nonspecific band of ~80 nt is also observed. This corresponds to adapter ligation products lacking sRNA inserts. If this band appears, gel purification of amplified DNA must be carried out to avoid sequencing DNA fragments without miRNA inserts. The 100-nt size bands are excised from gels with a clean razor blade and DNA is eluted with a Qiaquick kit (Qiagen).

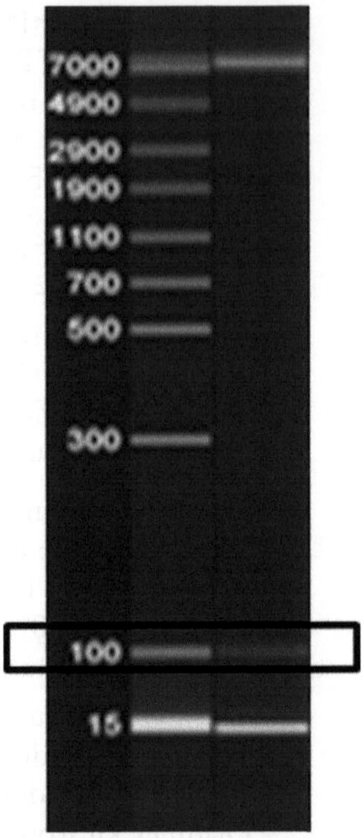

Fig. 4. Bioanalyzer analysis of PCR products prior to 454 sequencing. After clean up of the PCR products, each DNA sample should be analyzed using the Bioanalyzer to determine DNA quantity and quality. Only 1 band of ~100 nt should be observed in this step.

3. Clean PCR products are checked for integrity and concentration using the Bioanalyzer RNA 6000 NanoChip from Agilent Technologies (Fig. 4) prior to sample quantification for deep sequencing. Samples are now ready to be clonally amplified via emPCR and sequenced.

3.8. emPCR and Sequencing

After amplification of the cDNA libraries, samples are subjected to emPCR. In this step, each sRNA molecule within a mixture is sequenced individually because of binding to the DNA capture beads. The DNA capture beads recognize the specific adapters appended to the sRNAs during library amplification, which are then clonally amplified by emPCR. Beads are placed in a water-in-oil emulsion, containing hundreds of thousands of PCR microreactors. Each microreactor contains all the reagents necessary for the amplification, which allows for the entire collection of fragments and beads to be amplified in parallel, resulting in bead-immobilized clonally amplified DNA fragments. This avoids the introduction of competing or contaminating sequences. After the round of emPCR, the samples are ready for sequencing. The Roche 454 GS-FLX platform offers multiple sequencing formats, so one can choose according to the experimental objectives. We used mini-chips that allow sequencing for up to 100,000 reads.

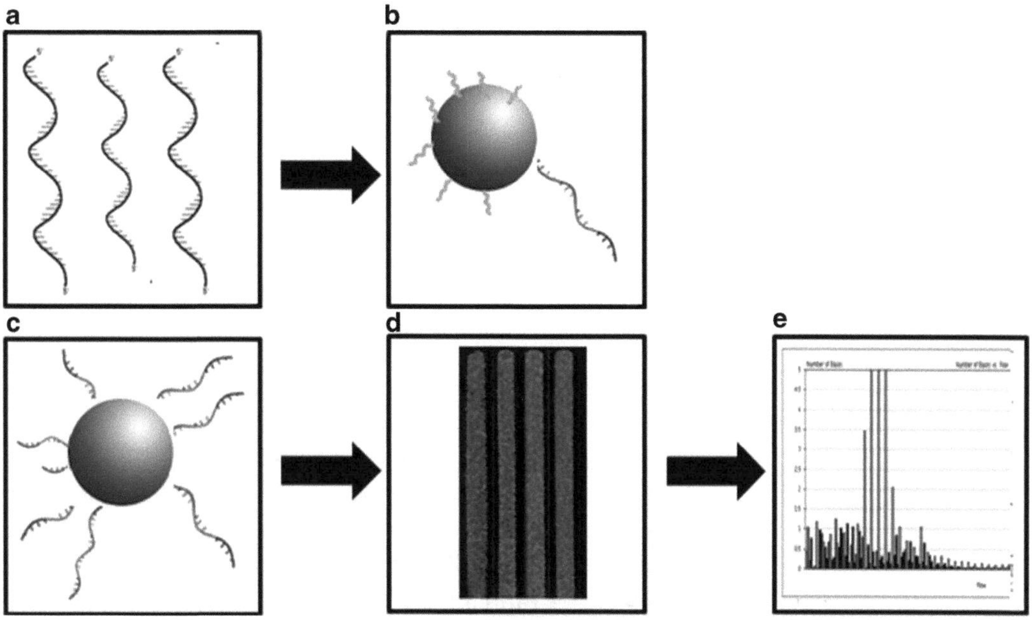

Fig. 5. Emulsion PCR and high-throughput sequencing. After cDNA library preparation, samples are ready for emPCR. Specific adapters that were previously appended to each sRNA (**a**) are recognized by DNA beads (**b**). Beads are placed in a water-in-oil emulsion, containing hundreds of thousands of PCR microreactors. Each microreactor contains the reagents necessary for DNA amplification, which allows for the entire collection of fragments and beads to be amplified in parallel (**c**). After emPCR, samples are added to the DNA bead incubation mix containing DNA polymerase and are layered onto a sequencing chip (PicoTiterPlate™). Each chip well contains a single bead. Once placed in the instrument, sequencing reagents flow across the chip wells which allow the sequencing of hundreds of thousands of beads with millions of copies of DNA. (**d**) Image of the sequencing device after sequencing. Each fragment generates its own flowgram, corresponding to its sequence (**e**). Adapted from Thomas Jarvie and Timothy Harkins, *Small RNA analysis using the Genome Sequencer™ FLX System*, Nature Methods.

Library beads are added to the DNA bead Incubation Mix that contains DNA polymerase and are layered onto the sequencing device (PicoTiterPlate™). This device consists of a set of wells where bead deposition occurs. In each well only one bead is deposited. Once placed in the instrument, sequencing reagents flow across the plate wells, allowing each of the hundreds of thousands of beads with millions of copies of DNA to be sequenced in parallel. Each clonally amplified fragment generates its own unique sequence (8). The 250-bp reads obtained in the sequencing run are long enough to completely read through the adapters added to the cDNAs during library preparation. This provides an ideal quality control of each read. A scheme of the emPCR and sequencing steps are shown in Fig. 5.

3.9. Data Extraction

After sequencing, base calling and quality trimming, sequence reads are carried out using the Genome Sequencer FLX software. Raw images are processed to remove background noise and to normalize data. TAGs and adapter sequences that were used for library construction are then identified and trimmed. After trimming, the resulting reads should be >15 nt. These are then retrieved for downstream analysis and identification of miRNAs using bioinformatic approaches.

4. Notes

1. It is important to mix the aqueous phase containing the RNA with isopropyl alcohol prior to centrifugation. If the isopropyl alcohol is not properly mixed with RNA, no precipitation or very low yield of RNA will be obtained.
2. If the sample concentration is too high (NanoDrop spectra out of range), dilute it in a high volume of RNase-free water. The ideal concentration should be between 1.5 and 3 μg/μL. The 260/230 and 260/280 ratios should be higher than 1.8.
3. The APS (10%) and TEMED that are added to the 5 mL base gel allow for rapid polymerization, sealing the bottom of the plates, and avoiding acrylamide leaks when pouring the main gel.
4. Polymerized PAA gels can be stored overnight at 4°C wrapped in cling film to avoid evaporation and gel drying. Gels can be run over the next few days.
5. There is no need to denature the samples before loading because gels contain 7 M urea as well as the loading buffer. This disrupts RNA secondary structure without preheating.
6. The low concentration of miRNAs makes it difficult to observe clear bands after staining. For this reason, size markers must be run alongside RNA samples. SyBr Green staining is optional since size markers are labeled with cy3. These markers are observable to the naked eye as pink bands. However, to ensure that the correct bands are excised, it is good practice to stain gels and locate bands. This also allows checking RNA integrity. Staining with SyBr Green is preferable to staining with ethidium bromide because it is not carcinogenic and does not cross-link during UV irradiation.
7. The use of DTR columns is crucial for cDNA library preparation, as it is difficult to obtain high amounts of sRNAs from gel extractions. We have tested other methods, but this was the one that yielded the highest amount of sRNAs and successfully removed urea and other salts with minimal loss of RNA.
8. The 3′ adapter ligation is carried out in the absence of ATP to avoid circularization of the RNA fragments. This 3′ adapter contains pre-adenylated deoxyoligonucleotides, blocked at the 3′ end, to avoid their circularization. The pre-adenylation of

the adapter eliminates the need to use ATP during ligation. This minimizes the problem of 5′ adenylation of the pool of sRNAs that could lead to circularization.

9. Free 3′ linker will compete with 3′ linked sRNAs in the 5′ ligation step and must be removed. Successfully ligated RNAs are between 33 and 48 nt long, 15–30 nt fragments constitute the sRNA fraction and the additional 18 nt fragment constitutes the 3′ linker. Free adapter is easily observable after gel staining as an 18 nt band. To ensure that gel slices are excised in the correct position, a 40-nt oligo should be run alongside the ligated RNA samples. Often the 3′ ligated material is not concentrated enough to be visible after gel staining.
10. In our experience, T4 RNA ligase from Ambion produced the best results under our experimental conditions. It is possible that some sRNAs ligate differentially, and for this reason sequencing data should be validated by northern blot or qPCR analysis.
11. The 454 sequencing technology allows for sequencing different samples in a single run. In order to distinguish sRNAs from different samples, the latter must be tagged, using a 5′ fusion primer (forward primer of the PCR amplification) containing the specific 454 adapter A plus choice TAG (a unique combination of 3 nt for each sample) plus the 5′ linker sequence. In our case, sample 1 corresponding to RNA extracted from brain had an ATC tag and sample 2 derived from heart had a TTC tag. Therefore, tagged purified PCR products can be mixed in equal amounts prior to 454 sequencing. These TAGs are traced downstream during bioinformatic analysis of the sequencing data.
12. It is essential to run the final PCR products on gels because an extra ~80 nt band is often detected. This band corresponds to adapter ligation products lacking the sRNA inserts. If this band appears, gel purification must be carried out to avoid nonspecific sequencing products. As an alternative to agarose gels, PCR fragments can be fractionated on 10% denaturing PAA gels. The correct size bands (~100 nt) can be excised from gels using a clean razor blade. DNA can be eluted from the gel with Probe Elution buffer (Ambion), following the manufacturer's instructions. Denaturing PAA gels have a better resolution than agarose gels and separate better similar sized DNA bands.

Acknowledgments

ARS and PMP are supported by the Portuguese Foundation for Science and Technology (FCT). We are also thankful to the Centre for Neurosciences of the University of Coimbra for providing a

PhD fellowship to ARS and to Biocant for making available the Genome Sequencer FLX. This project was funded by FCT/FEDER project SAU-MMO/55476/2004.

References

1. Lee, R. C., Feinbaum, R. L., and Ambros, V. (1993) The C. elegans heterochronic gene lin-4 encodes small RNAs with antisense complementarity to lin-14. *Cell* **75**, 843–54.
2. Berezikov, E., Thuemmler, F., van Laake, L. W., Kondova, I., Bontrop, R., Cuppen, E., et al. (2006) Diversity of microRNAs in human and chimpanzee brain. *Nat Genet* **38**, 1375–7.
3. Ciaudo, C., Servant, N., Cognat, V., Sarazin, A., Kieffer, E., Viville, S., et al. (2009) Highly dynamic and sex-specific expression of microRNAs during early ES cell differentiation. *PLoS Genet* 5, e1000620.
4. Klevebring, D., Street, N. R., Fahlgren, N., Kasschau, K. D., Carrington, J. C., Lundeberg, J., et al. (2009) Genome-wide profiling of populus small RNAs. *BMC Genomics* **10**, 620.
5. Ruby, J. G., Jan, C., Player, C., Axtell, M. J., Lee, W., Nusbaum, C., et al. (2006) Large-scale sequencing reveals 21U-RNAs and additional microRNAs and endogenous siRNAs in C. elegans. *Cell* **127**, 1193–1207.
6. Soares, A. R., Pereira, P. M., Santos, B., Egas, C., Gomes, A. C., Arrais, J., et al. (2009) Parallel DNA pyrosequencing unveils new zebrafish microRNAs. *BMC Genomics* **10**, 195.
7. Griffiths-Jones, S., Saini, H. K., van, D. S., and Enright, A. J. (2008) miRBase: tools for microRNA genomics. *Nucleic Acids Res* **36**, D154–8.
8. Droege, M. and Hill, B. (2008) The Genome Sequencer FLX System-longer reads, more applications, straight forward bioinformatics and more complete data sets. *J Biotechnol* **136**, 3–10.

Chapter 14

Methods for Small RNA Preparation for Digital Gene Expression Profiling by Next-Generation Sequencing

Sam E.V. Linsen and Edwin Cuppen

Abstract

Digital gene expression (DGE) profiling techniques are playing an eminent role in the detection, localization, and differential expression quantification of many small RNA species, including microRNAs (1–3). Procedures in small RNA library preparation techniques typically include adapter ligation by RNA ligase, followed by reverse transcription and amplification by PCR. This chapter describes three protocols that were successfully applied to generate small RNA sequencing SOLiD™ libraries. The Ambion SREK™-adopted protocol can be readily used for multiplexing samples; the modban-based protocol is cost-efficient, but biased toward certain microRNAs; the poly(A)-based protocol is less biased, but less precise because of the A-tail that is introduced. In summary, each of these protocols has its advantages and disadvantages with respect to the ease of including barcodes, costs, and outcome.

Key words: microRNAs, DGE Profiling, Sequencing by ligation

1. Introduction

Among various high-throughput sequencing platforms available, the SOLiD™ platform delivers deep coverage and accuracy, which is essential for small RNA sequencing. On the SOLiD™ platform, sequencing is established by "sequencing by ligation": a set of four fluorescently labeled di-base DNA probes competes for ligation to the sequencing primer with the interrogated DNA fragment as a template (4). Specificity of the di-base probe is achieved by interrogating every first and second base in each ligation reaction. This method allows for strong noise-reduction (5) and ensures reliable identification of nearly-identical microRNA family members as well as editing events (6, 7).

This chapter describes three different approaches to construct small RNA libraries from total RNA. After isolation of the small

Jian-Bing Fan (ed.), *Next-Generation MicroRNA Expression Profiling Technology: Methods and Protocols*, Methods in Molecular Biology, vol. 822, DOI 10.1007/978-1-61779-427-8_14, © Springer Science+Business Media, LLC 2012

RNA of interest, essentially each library is prepared in three steps, i.e., adapter ligation, reverse transcription, and enrichment PCR, before it is submitted to the sequencing platform-dependent sample preparation. All protocols assume that the small RNA species of interest contain 5′ P and 3′ OH groups, like microRNAs. If modifications cannot be excluded, a dephosphorylation step, followed by a phosphorylation step, may facilitate cloning (e.g., when a P group is located at the 3′ terminus). Total RNA isolation is not discussed here: any standard or commercial RNA isolation protocol suffices as long as it accounts for the selection of small RNAs (see Note 1 for handling samples and RNA). Importantly, the small RNAs should be retained during precipitation and we therefore recommend applying the precipitation strategy described in the methods.

The reason for the presence of three protocols is that each has its advantages and disadvantages with respect to the possibility to include barcodes, cost, and outcome. Importantly, it has been demonstrated that library preparation methods introduce severe capture biases which affect the final read distribution irrespective of the applied sequencing platform (8), which is most likely due to characteristics of enzymes used at various steps in the procedures. Currently, it is unclear which library preparation method reflects the physiological distribution best (or, whether a method exists that does fully reflect the actual molecule count). Every step, every enzyme, and every method exposes preferences toward certain molecules, and therefore we strongly recommend following exactly identical procedures for each library within one experiment where samples are compared. It should also be noted, that the observed biases disqualify sequencing as a method to obtain absolute, or ranked, expression levels within a sample. In addition, libraries constructed by different protocols, or established by different types of enzymes, should not be compared. Comparing datasets that have been generated by different library methods is noninformative. Notably, this observation is valid for any small RNA profiling technique, including microarrays and high-throughput real-time PCR. In spite of this limitation, digital gene expression (DGE) profiling is very well suited to deduce differential expression of both known and unknown, small RNAs. It has been shown that differential expression levels obtained on different sequencing platforms, or with different techniques, actually correlate very well (8–10).

2. Materials

2.1. Gel Electrophoresis and Staining

1. Mini-protean gel system (Bio-Rad) or similar gel systems for running preparative denaturing polyacrylamide gels.
2. Gel solutions. The following recipes can be followed to make 80 mL gel mix. To prepare the 15% denaturing PAA mixture,

add 33.6 g ureum to 8 mL 10× TBE, 30 mL 40% 19:1 acrylamide/bisacrylamide mixture and 14 mL milliQ H_2O. The 10% denaturing PAA mixture contains of 33.6 g ureum to 8 mL 10× TBE, 20 mL 40% 19:1 PAA, and 24 mL milliQ H_2O. The 6% native PAA gel mixture contains 10 mL 50% glycerol, 8 mL TBE, 12 mL 40% 19:1 PAA, and 50 mL milliQ H_2O. Use 5 mL for each gel if you make use of the Bio-Rad system with 0.75 mm spacers. To each lot of 5 mL, add 50 μL 10% APS (in H_2O) and 5 μL TEMED. Run the gels in 1× TBE. Store the gel mixture at 4°C.

3. SybrGold/TBE (see Note 2) and a Safe Imager (Invitrogen) for visualizing RNA and DNA without introducing damage.
4. DNA ladder: the ultra low range (ULR) (Fermentas), with bands between 15 and 300 nt. Comes with 6× DNA loading buffer.
5. miSPIKES and piSPIKES (5 μM each): these can clearly indicate the position of the small RNA species on gel.
6. 2× RNA loading buffer: can be obtained commercially or prepared manually: 90% formamide, 10% 10× TBE, 0.5 g bromophenol blue in 15 mL.
7. Clean razor blades for gel excision.
8. Lonza gel system, 2.2% Lonza gels, either single or double tiers.

2.2. PAA Gel Excision and Elution

1. SpinX columns from Costar: these are used to separate gel fragments from the eluate (see Note 3).
2. Elution buffer: 0.3 M NaCl. For RNA: in H_2O. For DNA, buffered in 10 mM Tris pH 8 (see Note 4).
3. Components of precipitation mix: EtOH, 5 M ammonium acetate, GlycoBlue from Ambion: blue-labeled glycogen that act as a carrier for small RNAs and DNAs.

2.3. Consumables

2.3.1. Miscellaneous Consumables

1. Nuclease-free H_2O.
2. 1.5 mL and 200 μL PCR tubes. Keep them RNAse free by only touching them with new gloves.
3. Latex gloves.

2.3.2. Consumables for SREK™

1. SREK™ adapter mix, ligation buffer, ligation enzyme (Ambion).

2.3.3. Consumables for Modban and poly(A)

1. T4 RNA ligase (Ambion).
2. 10× T4 RNA ligase with ATP (Ambion).
3. Phusion Taq polymerase (Finnzymes).

2.3.4. Consumables for Modban only

1. 10× T4 RNA ligase buffer w/o ATP (500 mM Tris pH 7.8, 100 mM $MgCl_2$, 100 mM DTT).

2.3.5. Consumables for poly(A) only

1. Poly(A) polymerase (*E. coli* derived, from Ambion). It comes with buffer and $MnCl_2$.
2. Periodic Acid (Sigma). Used to terminate the poly(A) tail as it cleaves diols.
3. Sephadex G25 spin columns (Roche) to get rid of non-incorporated ATPs after polyadenylation.

2.4. Probes and Primers

2.4.1. PCR Primers

1. Forward lib primer 5′ CCA CTA CGC CTC CGC TTT CCT CTC TATG 3′.
2. Reverse lib primer: 5′ CTG CCC CGG GTT CCT CAT TCT 3′.

2.4.2. SREK™ 20 Barcodes

1. Barcoded RT Primers

   SOLiD™ RT Primer BC1 5′-CTGCCCCGGGTTCCTCATTCTCTGTGTAAGAGGCTGCTGTACGGCCAAGGCG-3′

   SOLiD™ RT Primer BC2 5′-CTGCCCCGGGTTCCTCATTCTCTAGGGAGTGGTCTGCTGTACGGCCAAGGCG-3′

   SOLiD™ RT Primer BC3 5′-CTG CCC CGG GTT CCT CAT TCT CTA TAG GTT ATA CTG CTG TAC GGC CAA GGC G-3′

   SOLiD™ RT Primer BC4 5′-CTG CCC CGG GTT CCT CATTCT CTG GAT GCG GTC CTG CTG TAC GGC CAA GGC G-3′

   SOLiD™ RT Primer BC5 5′-CTG CCC CGG GTT CCT CAT TCT CTG TGG TGT AAG CTG CTG TAC GGC CAA GGC G-3′

   SOLiD™ RT Primer BC6 5′-CTG CCC CGG GTT CCT CAT TCT CTG CGA GGG ACA CTG CTG TAC GGC CAA GGC G-3′

   SOLiD™ RT Primer BC7 5′-CTG CCC CGG GTT CCT CATTCT CTG GGT TAT GCC CTG CTG TAC GGC CAA GGC G-3′

   SOLiD™ RT Primer BC8 5′-CTG CCC CGG GTT CCT CAT TCT CTG AGC GAG GAT CTG CTG TAC GGC CAA GGC G-3′

   SOLiD™ RT Primer BC9 5′-CTG CCC CGG GTT CCT CA T TCT CTA GGT TGC GAC CTG CTG TAC GGC CAA GGC G-3′

   SOLiD™ RT Primer BC10 5′-CTG CCC CGG GTT CCT CAT TCT CTG CGG TAA GCT CTG CTG TAC GGC CAA GGC G-3′

SOLiD™ RT Primer BC11 5′-CTG CCC CGG GTT CCT CAT TCT CTG TGC GAC ACG CTG CTG TAC GGC CAA GGC G-3′
SOLiD™ RT Primer BC12 5′-CTG CCC CGG GTT CCT CAT TCT CTA AGA GGA AAA CTG CTG TAC GGC CAA GGC G-3′
SOLiD™ RT Primer BC13 5′-CTG CCC CGG GTT CCT CAT TCT CTG CGG TAA GGC CTG CTG TAC GGC CAA GGC G-3′
SOLiD™ RT Primer BC14 5′-CTG CCC CGG GTT CCT CAT TCT CTG TGC GGC AGA CTG CTG TAC GGC CAA GGC G-3′
SOLiD™ RT Primer BC15 5′-CTG CCC CGG GTT CCT CAT TCT CTG AGT TGA ATG CTG CTG TAC GGC CAA GGC G-3′
SOLiD™ RT Primer BC16 5′-CTG CCC CGG GTT CCT CAT TCT CTG GGA GAC GTT CTG CTG TAC GGC CAA GGC G-3′
SOLiD™ RT Primer BC17 5′-CTG CCC CGG GTT CCT CAT TCT CTG AGG TGG GAG CTG CTG TAC GGC CAA GGC G-3′
SOLiD™ RT Primer BC18 5′-CTG CCC CGG GTT CCT CAT TCT CTG GTG GGA CAC CTG CTG TAC GGC CAA GGC G-3′
SOLiD™ RT Primer BC19 5′-CTG CCC CGG GTT CCT C AT TCT CTG TTA TAC TAA CTG CTG TAC GGC CAA GGC G-3′
SOLiD™ RT Primer BC20 5′-CTG CCC CGG GTT CCT CAT TCT CTG CGA GAT AAT CTG CTG TAC GGC CAA GGC G-3′

2.4.3. Modban and Poly(A)

1. 5′ linker: 5′ CCA CTA CGC CTC CGC TTT CCT CTC TrArU rGrGrG rCrArG rUrCrG rGrUrG rArU 3′

   rX indicates ribonucleotides.

2.4.4. Modban

1. Modban (Cloning linker 1 provided by IDT): AMP-5′p-5′p/CTGTAGGCACCATCAATdi-deoxyC-3′.
2. Modban RT primer: 5′ CTG CCC CGG GTT CCT CAT TCT CTG ATT GAT GGT GCC TAC AG 3′.

2.4.5. Poly(A)

1. Poly(A) RT primer: 5′ CTG CCC CGG GTT CCT CAT TCT CTT TTT TTT TTT TTT TTT TTT TTT TTV 3′.

2.5. Devices

1. Thermocycler with heated lid.
2. Benchtop centrifuge, maximal centrifugation force >12,000 × *g*.
3. Optional: Bioanalyzer, RNA pico chips, DNA chips (Agilent).

3. Methods

The protocols described in this chapter are directed to capture microRNAs; for other small RNA molecule libraries, these protocols have to be modified according to the respective molecule size. The microRNA isolation procedure is identical for each of the three protocols. Total RNA can be obtained by TRIzol extraction or by using commercial kits, such as the miRVAna kit (Ambion). For these preparatory steps, we recommend to always follow the same protocol for each sample within one experiment to circumvent method-specific biases, although at present such biases have not been studied. In the case of TRIzol or phenol/chloroform extractions, the precipitation mix should be adapted for retaining small fragments: the presence of a carrier, such as glycogen, a high alcohol concentration, and sufficient incubation time at −80°C, will retain small fragments. For every precipitation step, we recommend to apply the precipitation protocol as described in this section. If desired, one can choose to set aside a small aliquot of the sample after each elution. These can later be analyzed on the BioAnalyzer (Agilent) to verify the presence of the correctly sized product and the quality of the sample.

3.1. Precipitation of RNA or DNA

1. Add 10% 5 M ammonium acetate, 300% absolute EtOH, and 0.5 μL GlycoBlue (Ambion) to the sample.
2. Incubate at −80°C for 30 min, or overnight. If the sample is frozen when taken out of −80°C, allow the sample to thaw on ice and mix the sample by inverting the tube about five times.
3. Spin the sample at >12,000 × *g* for 25 min at 4°C.
4. Remove the supernatant using a pipette, add 70% EtOH and spin at >12,000 × *g* for 5 min at 4°C.
5. Remove the supernatant and spin the pellet briefly in a bench top centrifuge to collect droplets of remaining fluid in the bottom of the tube.
6. Remove the fluid using a P2 or P20 pipette and air-dry the pellet for 2 min. Pellets that dry longer may be difficult to redissolve.

3.2. Small RNA Collection

3.2.1. Fractionation

MicroRNAs are isolated from total RNA by excision from a 15% denaturing PAA gel. It is recommended to start with 5 μg of total RNA, although we also successfully used amounts as low as 100 ng. This procedure consists of a fractionation (steps 1–3), gel fragmentation (steps 4 and 5) and elution (steps 6 and 7) step.

1. Add 1 volume 2× RNA loading buffer to the sample (after which it is incubated at 65°C for 5 min and stored on ice), mi-spike/pi-spike mix (2 μL) and ladder (2 μL) (see Notes 5–6).

2. Load the samples onto the gel (see Note 7) and run the gel at 200 V for 1 h. The distance between two samples should at least span one empty well, to avoid contamination among samples.
3. Stain the gel in 1× TBE/SybrGold (Invitrogen) for >2 min and expose the gel on a clean glass slide on a Safe Imager (Invitrogen). Verify the microRNA fraction by checking the ladder together with the spike-in controls and excise the 17–30 nt bands (see Note 8).
4. Put the gel bands into a 1.5-mL Eppendorf tube, the bottom of which has been perforated twice with a 19-gauge needle.
5. Place this tube tightly into a 1.5-mL Eppendorf tube and spin this combination down at >12,000 × *g* for 2 min, in order to fragment the gel and collect these fragments in the lower tube. Make sure that the whole fragment is spun through the holes.
6. Add 300 μL of 0.3 M NaCl to the tube and elute the RNA by rotating the tube gently at room temperature for 4 h (or overnight at 4°C).
7. Transfer the eluate and the gel debris onto the top of a SpinX filter (see Note 9) and spin at full speed for 2 min.
8. Precipitate as indicated (Subheading 3.1) and dissolve the RNA in 3 μL nuclease-free H_2O.

3.3. Introducing the Universal Sequencing Adapters

3.3.1. Option 1: A Modified SREK™ Protocol

This protocol utilizes components of the Small RNA Expression kit (SREK™, Ambion), which hybridizes a mix of partially double-stranded 5′ and 3′ adapters in a directional orientation to both termini of a microRNA simultaneously, prior to ligation. Although relatively expensive, one advantage of this method is the possibility to introduce barcodes into the libraries, without custom modification of the adapters and potential loss of dedicated sequence length. The modified protocol that is described here ensures an advantageous dynamic range as compared with the modban method (see below), biases toward certain microRNAs are mild. Our slight modifications to the manufacturer's protocol have lead to an improved read quality from the first to the last nucleotide. Two adapter mixes can be applied; adapter mix A positions the sequence primer on the forward strand, while adapter mix B places it on the reverse-complementary strand, altering the orientation of the sequence reads. As part of the bias in cloning efficiencies is also determined by the base sequence of the adapters that are used, it is not recommended to compare libraries constructed with adapter mix A with adapter mix B in the same experiment. Although not dramatically different, we found that the read distribution of identical samples does not correlate as well as those from two replicate libraries that have been constructed with one adapter mix.

1. Mix 1.5 μL of the purified small RNAs, 1 μL adapter mix A or B and 1.5 μL hybridization solution together and incubate this mixture at 65°C for 5 min.
2. Put the mixture on ice and then carefully add 5 μL ligation buffer (this buffer is very viscous) and 1 μL ligation enzyme.
3. Pipette the mixture up and down five times and incubate it for 4 h at RT or overnight at 16°C.
4. Combine 5 μL of the mixture with 5 μL 2× RNA loading buffer, place it at 65°C for 5 min and put it on ice.
5. Load the ligated samples and ladder into wells of a 10% denaturing PAA gel (see Note 7). Run the gel at 200 V for 40 min.
6. Stain the gel with ~10 mL TBE containing SybrGold. Expose the gel under the safe imager and excise the 60–90 nt fragment (see Note 8).
7. Fragment, elute, and precipitate as described in Subheading 3.2. Dissolve in 10 μL H_2O.

3.3.2. Option 2: The Modban Protocol

This protocol is adapted from Lau et al. (11), who ligated adapters to both termini of the small RNA. After isolation of the small RNAs of interest, the 3′ adapter, i.e., the commercially available modban linker, is introduced first. The modban linker itself serves as a phosphate donor during ligation and is modified at the 3′ terminus, which ensures that the 5′ end of the modban adapter can only link to the 3′ termini of the microRNAs; in the absence of free ATP, no other product can be formed and small RNA molecules cannot concatenate or circularize. After removal of the non-incorporated adapters, the 5′ adapter is subsequently introduced. While this method is cheaper than the SREK™-based method, it requires one more gel excision and is therefore more laborious. Also, barcodes are not automatically introduced (see Note 10). The dynamic range is smaller than the other methods discussed as the modban method is strongly biased toward members of the *let-7* family: in most cases, >50% of all resulting reads derive from these microRNAs, resulting in reduced informative capacity for other microRNAs.

1. To 1.5 μL of the purified microRNAs, add 5 μL 2× buffer without ATP, 0.6 μL 10 μM modban linker, 1 μL T4 10 U/μL T4 RNA ligase (Ambion), and 0.4 μL RNAseOut (Invitrogen).
2. Incubate this mixture at 20°C for 6 h in a thermal cycler; the mixture can subsequently be kept at 4°C overnight.
3. Add 10 μL 2× gel loading buffer to the mixture, mix, and heat this sample at 65°C for 5 min to resolve secondary RNA structures.

Put the samples on ice prior to loading them onto a 15% TBE-Urea gel.

4. Load the sample on gel (see Note 7): one of the lanes should contain 2 μL of the ladder, which is dissolved in 2 μL 2× RNA loading buffer, but has not been heated to 65°C. Run the gel at 200 V for 1 h.
5. Stain the gel in SybrSafe/TBE for at least 2 min.
6. Dissect, under the Safe Imager, the gel band corresponding to 30–50 nt fragment sizes (see Note 8).
7. Elute and precipitate the ligated RNA fragments as described in Subheading 3.2. Dissolve the nucleotide pellet in 5.7 μL H_2O.
8. In the next steps, the 5′ adapter is ligated. Add the sample (5.7 μL) to 1.3 μL of the 5′ linker (5 μM), 1 μL of 10× RNA ligation buffer with ATP (Ambion), 1 μL T4 RNA ligase (10 U/μL, Ambion), and 1 μL RNAseOut (Invitrogen).
9. Incubate the mixture at 20°C for 6 h in a thermal cycler. The reaction mixture can be held overnight at 4°C.
10. Add 10 μL 2× RNA loading buffer to the sample, heat it to 65°C for 5 min, and store it on ice.
11. Load the sample onto a 10% PAA gel (see Note 7). Load 2 μL DNA ladder combined with 2 μL RNA loading buffer that has not been heated in one of the wells. Run the gel at 200 V for 1 h.
12. Stain the ladder in TBE/SybrSafe for at least 2 min and transfer the gel onto a safe imager.
13. Dissect the band that corresponds to 70–90 nt products (see Note 8).
14. Purify the product as described in Subheading 3.2 and dissolve the nucleotide pellet in 10 μL H_2O.

3.3.3. Option 3: Cloning Based on Polyadenylation of the 3′ Termini

Introducing a 3′ poly(A) tail is a cheap and efficient way to introduce an adapter to the 3′ terminus. Compared to the modban protocol, the preferences for certain microRNAs are less profound. A disadvantage is the reduced reliability to define the 3′ terminus of the microRNA because of the poly(A) tail and the lack of a straightforward barcoding approach (see Note 10). The polyadenylation step is followed by a 5′ adapter ligation. The 3′ adenosine is terminated by periodic acid (12), to prevent circularization or concatenation ligation of polyadenylated microRNAs in the subsequent 5′ adapter ligation step.

1. At first, the 3′ termini of the small RNAs are polyadenylated. Add 1.5 μL sample to 3.5 μL H_2O, 2 μL 5× PAP buffer (Ambion), 1 μL 10 mM ATP, 1 μL 25 mM $MnCl_2$, and 1 μL

E. coli-cloned poly(A) polymerase (Ambion). Mix well and incubate this mixture at 37°C for 30 min.

2. Heat inactivate the enzyme for 2 min at 70°C, add 40 μL H_2O and remove non-incorporated adenosines by spinning the solution over a G25 column (GE).
3. Expose the RNA solution to periodic acid (40 min @ 0°C 10 mM $NaIO_4$ in a total volume of 100 μL) in order to terminate the 3′ nucleotide group (12).
4. Precipitate the sample as described in Subheading 3.1 and dissolve the RNA in total of 5.7 μL of DEPC-treated water.
5. The adapter ligation to the 5′ terminus is identical to the 5′ adapter ligation in the modban protocol (Subheading 3.3.2, steps 8–14). Since the poly(A) tail can reach up to hundreds of adenosines, collect the complete fraction containing fragments larger than 50 nt. Purify the product as described in Subheading 3.3.2; however, add twice the volume of elution buffer and divide the eluate into two tubes prior to precipitation. Dissolve the pellet in 10 μL H_2O.

3.4. Preparing cDNA

1. Add to 5 μL of the RNA-adapter product (from Subheadings 3.3.1 to 3.3.3), 2 μL RT buffer (Promega), 1.5 μL 10 mM each dNTPs, and 0.5 μL 10 μM primer. Use barcoded primer if adapter mix A is used or FWD lib primer if adapter mix B is used, or the appropriate RT primer for either the modban or the poly(A) protocol.
2. Incubate this mixture at 72°C for 5 min and put it on ice.
3. Add 1 μL (200 U) RT MMLV-enzyme (Promega) to the mixture and perform the reverse transcription at 37°C for 30 min.

3.5. PCR Amplification

1. To amplify the cDNA products, a PCR is performed (see Note 11). The PCR is performed in 100 μL, containing 2 μL RT product, 77.8 μL H_2O, 10 μL 10× PCR buffer, 1.2 μL proofreading Taq (from the SREK™ kit or Phusion Taq (Finnzymes)), 8 μL 2.5 mM each dNTPs and 1 μL 22 μM each FWD and REV primer mix, or FWD and barcoded primer in the case of SREK™-adapter mix B libraries.
2. Start the PCR program with 95°C for 5 min (98°C in case of Phusion Taq), then cycle 95°C (98°C for Phusion Taq) for 30 s, 62°C for 30 s, and 72°C for 30 s, and finish at 72°C for 7 min. Cycle 15 times at first, put the PCR product on ice and check 3 μL of the PCR product by running it with DNA loading buffer for 4 min at 284 V on a 2.2% Lonza gel (or similar system). The expected product size is around 110 bp (be aware of byproducts of about 90 bp in size); if this product is visible, one can proceed with the next step (steps 4).

3. If this product is not visible, at most five extra PCR cycles can be performed (set the number of cycles between 1 and 5 of the original PCR program and restart) and the verification procedure can be repeated. Only proceed if a product with the desired length is visible.
4. Precipitate the sample as described (Subheading 3.1) and dissolve in 10 μL.
5. Prepare a 6% native PAA gel (see Note 7). Add DNA loading buffer to 5 μL the samples (see Note 12), as well as to the ladder (1 μL per gel). Load the samples and run the gel at 200 V for 40 min.
6. Stain the gel with 1× TBE and SybrGold for at least 2 min.
7. Expose the gel under the Safe Imager and excise, in the case of microRNA libraries, the 110–120 bp fragment (SREK™) or the 100–110 bp band (modban and poly(A)) (see Note 8).
8. Fragment the gel as described in Subheading 3.2. Elute the library for 2 h at RT in 100 μL 0.3 M NaCl, buffered with Tris (pH 8).
9. Precipitate the library as described in Subheading 3.1. Dissolve the pellet in 20 μL 10 mM Tris (pH 8).

4. Notes

1. Working with RNA requires cautious handling in a clean environment. Not only should the exposure to RNAses be avoided, also temperature, electrolytes, ice-crystals and pH affect the stability of these fragile molecules. Degradation of larger RNA species, such as rRNA and tRNA can contaminate the small RNA sample, which will result in many noninformative sequence reads. Furthermore, such contaminants put additional challenges on subsequent bioinformatic analysis, a part of DGE profiling experiment for which the required effort is routinely underestimated. Contamination by degraded RNA may induce cross-hybridization during mapping (13), false-positive predictions, and loss of statistical power. In order to prevent these issues, one should make sure that the input material is of supreme quality, preferably obtained from fresh tissues that have been snap-frozen in liquid nitrogen until further handling. Thawed and refrozen tissues will likely introduce many unwanted small RNAs in the small RNA pool and should, if possible, be avoided. Always store RNA or gel fragments containing the RNA at −80°C. Gel elution can be paused at −80°C and continued at a later stage.

2. Mix SybrGold into TBE first. Then, add the gel to this mixture. If SybrGold is added when the gel is already present in the TBE, it will not distribute evenly through the gel.
3. Be alert when closing spinX tubes. After adding the precipitation mix to the spinX tube, shake the tube carefully while closing it tightly between thumb and index finger, to prevent precipitation mix from spilling out.
4. Since RNA degradation occurs faster at higher pH, the 0.3 M NaCl elution buffer should only be buffered with Tris pH 8 for DNA.
5. RNA loading buffer can be loaded in excess. If volumes are small (e.g., less than 5 μL), it is recommended to add loading buffer up to a total volume of 10 μL. It loads better.
6. If the DNA ladder is to be loaded onto a denaturing gel, it can also be loaded in RNA loading buffer. In that case, do not heat the mixture prior to loading.
7. Prior to loading, rinse the wells intensively with TBE to remove gel slurry. To verify that each well is empty, 1 μL of loading buffer may be loaded.
8. Excise the gel on a clean surface to avoid contamination. Glass slides or cellophane plastic may be used.
9. Gel debris is hard to pipette. Place a 1-mL tip into the bottom of the tube and aspirate gently. Some eluate may be lost in this procedure. It is possible to add elution buffer onto the gel debris twice to avoid loss. In this case, modify precipitation mix amounts accordingly.
10. During RT-PCR of an SREK™-prepared library, a SOLiD™-specific barcode is introduced that is read from the internal adapter sequence present on the SREK™ 3′ adapter. These barcodes are decoded in the SOLiD™ sequencing run, either before or after the library sequencing. In case of the modban or poly(A) protocols, barcodes could be introduced by inserting short sequences at the 3′ terminus of the 5′ linker, which would then be reflected by the first letters in the read.
11. The PCR amplification increases the cDNA to a quantifiable or detectable amount. However, PCR also introduces biases. Only minute amounts of cDNA, about 400 pg, are required for the SOLiD™ emulsion PCR. If quantification on such amounts of cDNA could be accurately performed, e.g., using quantitative PCR, it would be preferable to skip the PCR amplification entirely. Note, this is not possible when SREK™-adapter mix B is used, since no barcodes and reverse primers have been introduced up to this point.
12. Add excess DNA loading buffer: a 1× dosage may leave the sample floating in the well, instead of forcing it down to the bottom of the well.

Acknowledgements

The authors would like to thank Ewart de Bruijn and Nico Lansu for critically reading and contributing to this manuscript.

References

1. Berezikov, E., Cuppen, E., and Plasterk, R. H. (2006) Approaches to microRNA discovery *Nat Genet* **38** Suppl, S2–7.
2. Ruby, J. G., Jan, C., Player, C., Axtell, M. J., Lee, W., Nusbaum, C., et al. (2006) Large-scale sequencing reveals 21U-RNAs and additional microRNAs and endogenous siRNAs in C. elegans *Cell* **127**, 1193–207.
3. Tarasov, V., Jung, P., Verdoodt, B., Lodygin, D., Epanchintsev, A., Menssen, A., et al. (2007) Differential regulation of microRNAs by p53 revealed by massively parallel sequencing: miR-34a is a p53 target that induces apoptosis and G1-arrest *Cell Cycle* **6**, 1586–93.
4. http://solid.appliedbiosystems.com
5. Shendure, J. and Ji, H. (2008) Next-generation DNA sequencing *Nat Biotechnol* **26**, 1135–45.
6. Kawahara, Y., Zinshteyn, B., Sethupathy, P., Iizasa, H., Hatzigeorgiou, A. G., and Nishikura, K. (2007) Redirection of silencing targets by adenosine-to-inosine editing of miRNAs *Science* **315**, 1137–40.
7. Linsen, S. E., de Wit, E., de Bruijn, E., and Cuppen, E. (2010) Small RNA expression and strain specificity in the rat *BMC Genomics* **11**, 249.
8. Linsen, S. E., de Wit, E., Janssens, G., Heater, S., Chapman, L., Parkin, R. K., et al. (2009) Limitations and possibilities of small RNA digital gene expression profiling *Nat Methods* **6**, 474–6.
9. Chen, Y., Gelfond, J. A., McManus, L. M., and Shireman, P. K. (2009) Reproducibility of quantitative RT-PCR array in miRNA expression profiling and comparison with microarray analysis *BMC Genomics* **10**, 407.
10. Willenbrock, H., Salomon, J., Søkilde, R., Barken, K. B., Hansen, T. N., Nielsen, F. C., et al. (2009) Quantitative miRNA expression analysis: comparing microarrays with next-generation sequencing *RNA* **15**, 2028–34.
11. Lau, N. C., Lim, L. P., Weinstein, E. G., and Bartel, D. P. (2001) An abundant class of tiny RNAs with probable regulatory roles in Caenorhabditis elegans *Science* **294**, 858–62.
12. Kurata, S., Ohtsuki, T., Suzuki, T., and Watanabe, K. (2003) Quick two-step RNA ligation employing periodate oxidation *Nucleic Acids Res* **31**, e145.
13. de Hoon, M. J., Taft, R. J., Hashimoto, T. Kanamori-Katayama, M., Kawaji, H., Kawano, M., et al. (2010) Cross-mapping and the identification of editing sites in mature microRNAs in high-throughput sequencing libraries *Genome Res* **20**, 257–64.

Chapter 15

Profiling of Short RNAs Using Helicos Single-Molecule Sequencing

Philipp Kapranov, Fatih Ozsolak, and Patrice M. Milos

Abstract

The importance of short (<200 nt) RNAs in cell biogenesis has been well documented. These short RNAs include crucial classes of molecules such as transfer RNAs, small nuclear RNA, microRNAs, and many others (reviewed in Storz et al., Annu Rev Biochem 74:199–217, 2005; Ghildiyal and Zamore, Nat Rev Genet 10:94–108, 2009). Furthermore, the realm of functional RNAs that fall within this size range is growing to include less well-characterized RNAs such as short RNAs found at the promoters and 3′ termini of genes (Affymetrix ENCODE Transcriptome Project et al., Nature 457:1028–1032, 2009; Davis and Ares, Proc Natl Acad Sci USA 103:3262–3267, 2006; Kapranov et al., Science 316:1484–1488, 2007; Taft et al., Nat Genet 41:572–578, 2009; Kapranov et al., Nature 466:642–646, 2010), short RNAs involved in paramutation (Rassoulzadegan et al., Nature 441:469–474, 2006), and others (reviewed in Kawaji and Hayashizaki, PLoS Genet 4:e22, 2008). Discovery and accurate quantification of these RNA molecules, less than 200 bases in size, is thus an important and also challenging aspect of understanding the full repertoire of cellular and extracellular RNAs. Here, we describe the strategies and procedures we developed to profile short RNA species using single-molecule sequencing (SMS) and the advantages SMS offers.

Key words: Single-molecule sequencing, Short RNAs, Promoter-associate short RNAs, Polyadenylated short RNAs

1. Introduction

In a cell, a final functional product of a precursor RNA species is often a shorter RNA derived from it. The most obvious examples of this event include splicing and 3′ end RNA processing to generate mature long mRNAs. However, a large number of other classes of functional RNAs that are not protein coding are made from longer precursors. Many of such RNAs are in the realm of less than 200 nt, even though this range is somewhat arbitrary and simply

Jian-Bing Fan (ed.), *Next-Generation MicroRNA Expression Profiling Technology: Methods and Protocols*,
Methods in Molecular Biology, vol. 822, DOI 10.1007/978-1-61779-427-8_15, © Springer Science+Business Media, LLC 2012

based on methods for fractionation. Most commonly biochemical column-based methods are used to fractionate RNA into species that are shorter or longer than 200 nt. Such RNAs are represented by tRNAs, small nuclear (sn)RNAs, small nucleolar (sno) RNAs, micro (mi)RNAs, and others (1–9). The latter miRNA class is probably one of the extreme examples in which very long (kbs) precursor RNA is cleaved by two enzymatic steps into a final functional product of 21–23 nt (2). Therefore, a full understanding of the cellular repertoire and functional cellular products must include the study of complex, short RNA (sRNA) population that cannot be achieved by profiling only long RNAs (3–9). In addition, extracellular sRNAs appear to be a promising class of biomarker molecules (10, 11). The following chapter describes the methods to enable successful profiling of this important class of short RNAs.

2. Materials

2.1. RNA Isolation

1. Purification of sRNA from total RNA or cultured cells could be achieved using these kits:

   *mir*Vana™ miRNA Isolation Kit (Ambion).

   miRNeasy Mini Kit (Qiagen).

   RNA/DNA kit (Qiagen) – suitable for preparation of large quantities of sRNAs from cultured cells.
2. Alternatively, sRNA of a desired fraction could be purified using TBE-Urea polyacrylamide gel electrophoresis and overnight elution (12).

2.2. cDNA Synthesis

2.2.1. General Method

1. *Escherichia coli* PolyA polymerase (Ambion).
2. 100 mM CTP (Roche).
3. ThermoScript reverse transcriptase (15 U/μL Invitrogen) – also, (see Note 5).
4. Phenol/chloroform/isoamyl alcohol (Ambion).
5. 5 M Ammonium acetate (Ambion).
6. cDNA synthesis primer – custom made from Integrated DNA Technologies sequence: TCG CGA GCG GCC GCG GGG GGG GGG GGrG rGrG. Important – last three bases are ribonucleotides.
7. RNAse A (Ambion).

2.2.2. Method for Profiling 3′ polyA sRNAs

1. SuperScript III reverse transcriptase (Invitrogen).
2. USER enzyme (New England Biolabs).
3. dTU-V cDNA synthesis primer – custom made from Integrated DNA Technologies sequence: TTTTUTTUTUTTTUTTTT UTTTUTTV.

4. RNAse H (Invitrogen).
5. RNAse 1f (New England Laboratories).

2.2.3. Reagents Common to Both Methods

1. 100 and 70% ethanol (Sigma).
2. RNAse inhibitors: ANTI-RNAse (Ambion) or RNAseOUT (Invitrogen).
3. 10 mM dNTPs (Invitrogen).
4. AMPure® beads (Agencourt).
5. Magnetic stand for 1.5-mL tubes.
6. PCR machine. The protocol below has been tested on the BioRad Tetrad 2 Thermal Cycler.

2.3. Sequencing of cDNA

1. 20 U/μL Terminal Transferase (New England Biolabs).
2. dATP (Helicos BioSciences).
3. 1 mM Biotin-ddATP (Perkin Elmer).
4. 10 mg/mL Bovine serum albumin (BSA) (New England Biolabs).
5. Quant-tT™ OliGreen® ssDNA Reagent (Invitrogen).
6. NanoDrop 3300 (Thermo Scientific).
7. HeliScope™ Single Molecule Sequencer (Helicos BioSciences Corporation).
8. Helicos® Flow Cells (Helicos BioSciences Corporation).

2.4. Data Analysis

1. A suite of unix-based Helicos processing tools that could be freely downloaded from here: http://open.helicosbio.com/mwiki/index.php/Releases.
2. Computer hardware. In general, it is suggested to have at least 5 GB per CPU core for alignments to the human genome using the Helicos aligner indexDPgenomic (13).
   (a) A cluster, desirable if processing of multiple channels is required. An example of specification would be a combination of dual and quad-core 3 GHz CPUs (i.e., Dell 1950) with memory ranging from 8 to 32 GB.
   (b) A standalone unix system if processing of only a few channels is required with a similar specs to the ones listed above.

3. Methods

Profiling of short RNAs has several unique challenges that the researcher should consider. First, the sRNA fraction of the desired length range must be isolated, otherwise, the signal could be derived from a long RNA that overlaps a short RNA. Second, most (not all) of the sRNAs of interest lack an easy molecular handle such as the 3′ polyA tail of the mRNAs that is needed for conversion into cDNAs. Third, they are often too short for efficient conversion into cDNA using random hexamers. Fourth, certain sRNAs have modifications at their 5′ (and 3′) ends that interfere with subsequent molecular manipulation and thus can go undetected by certain methods. Fifth, some classes of sRNAs have strong secondary structures (see Note 4) that preclude them from being detected by enzymatic methods that are typically conducted under mild (nondenaturing) conditions. Sixth, some sRNAs like miRNAs have lengths that are too short for efficient mapping to very complex genomes, thus making discovery work challenging. Seventh, most of the methods used rely on ligation and PCR amplification that can skew both the composition of the population and the quantification of the RNA species. Eighth, sRNA fraction could be dominated be relatively few, very highly abundant RNA classes, such as rRNA, snRNA, and snoRNAs which will require significant depth for complete characterization of the sRNA population. This can be avoided by further selections such as selection of a specific size range enriched in miRNAs (~18–25 bases) or selection of those with 3′ polyA tails (Subheading 3.3).

Below, we provide two general methods for the detection of sRNAs using single molecule sequencing: one can be used for detecting any sRNA that has a 3′ OH and the second for detecting 3′ polyA sRNAs. To circumvent some of the challenges presented above, we start with an isolated sRNA fraction followed by (1) the addition of 3′ polyC tail using polyA polymerase and cDNA synthesis using a polyG-containing oligonucleotide or (2) cDNA synthesis using a polyU oligonucleotide without tailing of RNA. The resulting cDNA could then be purified and sequenced directly after polyA tailing and 3′ blocking. The diagram of the first method is shown in Fig. 1.

Both methods do not require ligation or amplification and are also ambiguous as to the status of the 5′ end of an sRNA – thus sRNAs with any modification at their 5′ end can be detected. We also show the results one could expect when profiling a population of sRNAs less than 200 nt in the human HeLaS3 cells and pros and cons of each method.

3.1. Isolation of sRNA Fraction

1. The sRNA fraction could be isolated using a variety of methods (see Notes 1 and 2). If a general <200 nt fraction is required, commercially available kits like mirVana (Ambion); miRNeasy, or RNA/DNA from Qiagen could be used. We have used these kits following the manufacturer's guidelines with satisfactory outcomes.
2. If it is desired to isolate sRNAs in a specific size range, a TBE-Urea denaturing polyacrylamide gel-electrophoresis could be considered as an alternative purification method. We have followed the protocols described in (12).

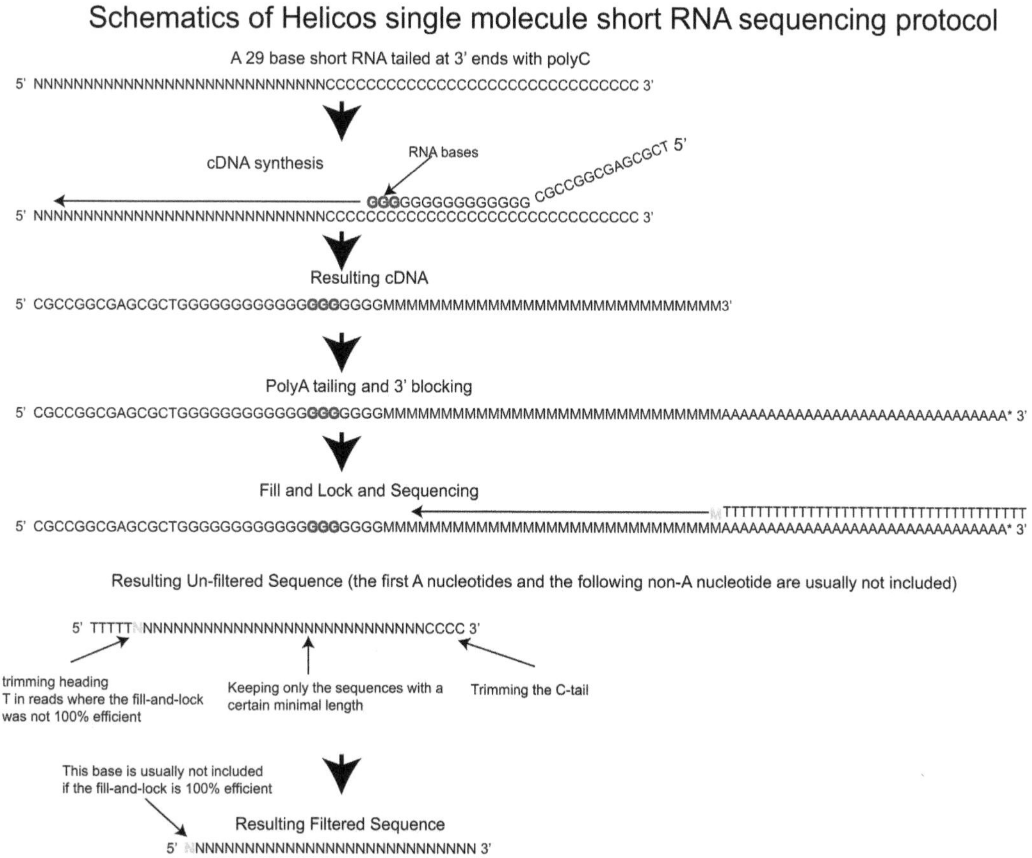

Fig. 1. A diagram depicting the Helicos small RNA method for the preparation of cDNA for sequencing to the downstream processing of raw sequence reads.

3.2. General Method for Profiling of sRNAs

3.2.1. Tailing of RNA with 3′ polyC

1. Bring RNA to 30 μL in water in a PCR tube: 10 ng to 5 μg of short RNAs can be used with this protocol.
2. Incubate for 2 min at 85°C in a PCR machine and then put on ice for at least 2 min.
3. Add the following reagents: 10 μL of 5× *E. coli* PolyA polymerase buffer; 5 μL of 25 mM $MnCl_2$; 1 μL of 100 mM CTP; 1 μL of Anti-RNAse or RNAseOUT, and 3 μL of 2 U/μL *E. coli* PolyA polymerase. Mix well (do not vortex) and incubate for 3 h at 37°C.
4. Add 40 μL of water and 10 μL of 5 M ammonium acetate.
5. Extract twice with phenol/chloroform/isoamyl by vortexing vigorously for 30 s.
6. Precipitate by adding 3× volumes of 100% EtOH for a minimum of 30 min at −80°C.
7. Centrifuge at 4°C for 30 min at top speed. Wash once with 70% EtOH. Vacuum-dry and resuspend in 30.5 μL water.

3.2.2. cDNA Synthesis

1. Add 1 μL of 100 μM cDNA synthesis primer: TCG CGA GCG GCC GCG GGG GGG GGG GGrG rGrG (*the last three bases – RNA*) to the 30.5 μL of RNA from the step above and mix. The presence of three RNA bases at the 3′ end is to ensure that the oligo is not tailed by terminal transferase, as TdT does not utilize RNA as a substrate.
2. Incubate for 2 min at 70°C in a PCR machine: fast ramp to 4°C and incubate for 5 min at 4°C.
3. While the samples are at 4°C, add the following: 10 μL of 5× ThermoScript cDNA Synthesis buffer; 5 μL of 0.1 M DTT; 2.5 μL of 10 mM dNTPs, and 1 μL of ThermoScript reverse transcriptase. If you need to scale up the protocol to 3–5 μg of RNA, you could use more ThermoScript at the cDNA synthesis step: ~1 μL/1 μg of RNA.
4. Slow Ramp (30 min) to 60°C followed by 1.5-h incubation. The following conditions were used on the BioRad cycler: 0.1°C/s ramp from 4 to 42°C, incubate for 15 min followed by 0.1°C/s ramp from 42 to 50°C, incubate for 15 min and followed by 0.1°C/s ramp from 4 to 60°C and incubation at 60°C for 90 min.
5. Fast ramp to 75°C followed by incubation for 15 min to inactivate the reverse transcriptase.

3.2.3. Purification of cDNA

1. Treat the cDNA synthesis reaction with 1 μL of RNAse A for 30 min at 37°C.
2. Mix the AMPure beads suspension really well to ensure that the magnetic beads are resuspended. Add 150 μL of the suspended AMPure beads and incubate for 30 min with mixing at room temperature.
3. Capture the beads using the magnetic stand for 5 min.
4. Carefully remove and discard the supernatant.
5. Wash twice with 200 μL of 70% EtOH.
6. Dry the pellet for 30–45 min at room temperature or 30 min at 37°C.
7. Elute cDNA twice with 20 μL of water (pipet up and down at least 20 times) (see Note 3).

3.3. Profiling of sRNAs with polyA Tails

3.3.1. cDNA Synthesis

1. Combine 1 μL of 50 μM dTU-V primer with (1–5 μg) of small RNA and 1 μL of 10 mM dNTPs and water if required in a total volume of 10 μL. Mix and incubate at 65°C for 5 min in a thermocycler, followed by rapid cooling on a prechilled aluminum block kept in an ice and water slurry (~0°C). Let the samples sit on the cold aluminum block for 1 min before proceeding with the next step.

2. Add to the above reaction 2 μL of 10× SuperScript III reaction buffer, 4 μL of 25 mM $MgCl_2$, 2 μL of 0.1 M DTT, 1 μL of SuperScript III, and 1 μL of RNAseOut while keeping the samples on the cold aluminum block. Mix the solution well after adding the reagents. Total volume is now 20 μL.
3. Incubate the 20-μL cDNA synthesis mix with the following temperature and time conditions: fast ramp to 50°C, incubate for 50 min followed by fast ramp to 85°C, incubate for 5 min and keep the samples at 4°C.
4. To remove the dTU-V primer sequences, add 1 μL USER enzyme to the reaction, mix and incubate at 37°C for 15 min. Total volume should now be 21 μL.
5. To digest away the RNA, add 1 μL of RNAse H and 1 μL of RNAse If, mix and incubate for 20 min at 37°C. The total volume is now 23 μL.

3.3.2. Purification of cDNA

1. While USER/RNAse treatment is in progress, warm up AMPure beads (180 μL per sample) to room temperature.
2. After the USER/RNAse treatment step, add 180 μL of warmed-up AMPure beads to the 23 μL cDNA reaction. Incubate with shaking at room temperature for 40 min to 1 h to bind cDNA to the beads.
3. Following the binding step, capture the beads using the magnetic stand for 5 min. Remove the supernatant with a pipettor. Wash the beads twice with 500 μL of 70% EtOH (no need to resuspend the beads in ethanol during the ethanol washes).
4. After the 70% ethanol washes, dry the pellet for 45 min at RT or 30 min at 37°C in a clean warm room or oven. At the end of the drying step, ensure that there is no liquid is visible in the tube (the pellet often assumes a cracked appearance).
5. To elute the cDNA from the beads, add 20 μL of nuclease-free water, and pipet up and down at least 10–20 times. Keep the sample at room temperature or 37°C for 5 min.
6. Place the tube on the magnet for 5 min and collect the eluate (~18 μL).
7. Repeat the elution step with 20 μL nuclease-free water and incubate at room temperature or 37°C for 5 min. Place the tube on the magnet for 5 min, collect the eluate again (~19 μL), and combine with the first eluate (step 6 above). Total volume of the cDNA is now ~37 μL.

3.4. Preparation of cDNA for Sequencing

The cDNA has to be tailed at the 3′ end with polyA residues using terminal transferase (TdT) and blocked at the 3′ end so that it can bind to the oligo-dT present at the surface of a flow cell (14). This step is common for the general protocol and for the polyA protocol.

Amounts of cDNA are quantified using either the regular NanoDrop if the expected concentration is at least 5–10 ng/μL or Quant-tT™ OliGreen® ssDNA kit and NanoDrop 3300 (Thermo Scientific). Two major differences are dictated by the amounts of cDNA.

3.4.1. Tailing of Small Amounts of cDNA

If only small amounts (<10 ng) of cDNA are available or expected, use the following:

1. Prepare cDNA to be tailed (<10 ng) in 10.8 μL of water. If cDNA is in larger volume – dry it down using speed-vac. Add 2 μL of 10× TdT buffer and 2 μL of 2.5 mM $CoCl_2$. Incubate at 95°C for 5 min in a thermocycler for denaturation, followed by rapid cooling on a prechilled aluminum block kept in an ice and water slurry (~0°C).
2. Add the following: 4 μL of 50 μM dATP; 0.2 μL of BSA, and 1 μL of TdT diluted 1:4 (final concentration 5 U/μL) in 1× TdT buffer. Final volume should be 20 μL.
3. Mix well and incubate in a PCR machined at 37°C for 60 min followed by 70°C for 10 min and 4°C forever (or ice).
4. Heat to 95°C for 5 min, transfer on ice for a minimum of 2 min.
5. Add the following: 1 μL of 10× TdT buffer; 1 μL of 2.5 mM $CoCl_2$; 0.5 μL of 200 μM Biotin-ddATP; 6.5 μL of water, and 1 μL of TdT diluted 1:4 (final concentration 5 U/μL) in 1× TdT buffer. Final volume should be 30 μL.
6. Mix well and incubate in a PCR machined at 37°C for 30 min followed by 70°C for 20 min and 4°C forever (or ice). Use directly for sequencing or plate assay to measure the concentration of tailed material (Subheading 3.5).

3.4.2. Tailing of Regular Amounts of cDNA

If larger (≥50 ng) of cDNA are available, use the following:

1. Prepare cDNA to be tailed (50–200 ng or 2–3 pmol, see Note 3) in 33.8 μL of water. Incubate at 95°C for 5 min in a thermocycler for denaturation, followed by rapid cooling on a prechilled aluminum block kept in an ice and water slurry (~0°C).
2. Add the following: 5 μL of 10× TdT buffer; 5 μL of 2.5 mM $CoCl_2$; 5 μL of 50 μM dATP, and 1.2 μL of TdT. Final volume should be 50 μL.
3. Mix well and incubate in a PCR machined at 42°C for 60 min followed by 70°C for 10 min and 4°C forever (or ice).
4. Add 0.6 μL of 1 mM biotin-ddATP to the tailing reaction from above.
5. Heat to 95°C for 5 min, transfer on ice for a minimum of 2 min.
6. Add 1.2 μL of TdT.

7. Mix well and incubate in a PCR machined at 37°C for 30 min followed by 70°C for 10 min and 4°C forever (or ice). Use directly for sequencing or plate assay to measure the concentration of tailed material (Subheading 3.5).

3.5. Estimation of Concentration of polyA Tailed cDNA

Helicos scientists have developed the OptiHyb™ Assay to determine the concentration of polyA-tailed templates and to allow the loading of the samples at optimal loading densities for single molecule sequencing on the Helicos® Genetic Analysis System. The protocols and reagents for this assay can be obtained from Helicos BioSciences Corporation (http://www.helicosbio.com).

3.6. Mapping of Reads to the Genome or Sequences of Known sRNAs

The sequence information obtained from the HeliScope Sequencer could be used for two general purposes: counting abundances of known sRNAs (digital gene expression) and discovery of new sRNAs. We have found that with an error rate of 3–5% that at present accompanies single-molecule sequencing, a read has to be at least 25 bases so that it can be reliably aligned to a genome as complex as human even though shorter reads with this error rate could be aligned to less complex genomes. Thus, discovery of novel miRNAs via mapping of *individual* reads to the genome may not be possible with a current state of the technology. However, one can envision that algorithms that perform clustering of reads prior to alignments could alleviate this problem. Also, the decrease in the error rate of SMS that will inevitably happen as the technology matures will likely alleviate this problem as well. However, currently digital gene expression of known sRNAs by aligning reads only to the sequences of the sRNA whose expression is desired to be quantified rather than to the whole genome may be more appropriate because accurate alignments to a smaller reference set would be possible with a smaller length of reads – 18 or 20 bases. Below we describe different steps for the two analytical approaches. The various unix-based programs and pipelines for the analysis of the HeliScope data including the ones referenced below could be downloaded freely here: http://open.helicosbio.com/mwiki/index.php/Releases. Their descriptions could be found here: http://open.helicosbio.com/helisphere_user_guide/.

3.6.1. Filtering of the Reads

1. Output from a single-molecule sequencing experiment using the Helicos platform is represented by a distribution of the read lengths, typically ranging from 6 to 70 bases (14, 15). The reads have to be filtered to remove (Fig. 1) (1) very short reads; (2) to trim the 5′ polyT sequence that corresponds to the polyA-tail added by TdT to the 3′ end of cDNA; (3) to trim the sequence that corresponds to trailing polyC-tail added by the polyA-polymerase to the 3′ end of a nascent RNA prior to the cDNA synthesis in the general method or the sequence that corresponds to 3′ polyA if the method for profiling of

sRNAs with 3′ polyA tails was used; (4) artifactual reads. The filtering is done by a program called "filterSMS" using different parameters.

2. Depending on the goals of the experiments and the genomes that the sequences are to be aligned to later, one can change to filter the sequences based on different minimal lengths. The standard default minimal length is 25 bases, however, one may choose to use less stringent minimal lengths such as 20 bases, for example, if one is interested in detecting miRNAs and/or if the alignments are to be done against smaller references such as sequences of known sRNAs or *E.coli* genome.
3. Filtering of preceding and/or trailing homopolymeric tails (e.g., polyT or polyC) could be done with different parameters, such as specifying the minimal homopolymeric content of either tail. The default parameter is 75%.
4. Removing artifactual reads is done by filtering out reads that have repetition of a base-addition order sequence (CTAG); removing reads with high AT content and reads having high fraction of repeats of certain dinucleotides.
5. Usage of any of these parameters is optional, for example, filtering only by read length could be employed if the entire raw sequence is desirable as an output.

3.6.2. Sequence Alignment

1. The unique property of single molecule sequencing is that the error profile is dominated by indels rather the substitutions as in other sequencing platforms (14). Thus, an aligner that is tolerant to these types of errors should be employed. We recommend using the indexDPgenomic aligner (13) developed at Helicos and freely available for download (http://open.helicosbio.com/mwiki/index.php/Releases). Other aligners such as Mosaik (http://bioinformatics.bc.edu/marthlab/Mosaik), BWA (16), and Shrimp (17) could be used as well.
2. One of the basic parameters used by the indexDPgenomic is the normalized score computed as in the example below:

   Tag sequence CCTCCGTGTTGTTCCAGCC-CAGTGCTC GCAGG

   Ref sequence C-TCCGTGTTGTTCCAGCCACAGTGCTCG CAGG

   Length of alignment block: 33

   Length of tag sequence: 32

   Number of matches: 31

   Number of errors: 2

   Score: $(31 \times 5) - (2 \times 4) = 155 - 8 = 147$

Normalized score = 147/32 = 4.59375

3. Output of the indexDPgenomic is a binary alignment file that could be further filtered using a filterAlign program to select only, for example, for reads that could only be aligned to the genome once (uniquely aligning reads), reads that align with a certain minimal normalized score, etc. The binary alignment files could also be converted into the various text formats, including the standard BED or SAM/BAM formats using the following programs: printAlignmentFile; align2txt or align2sam.
4. Filtering of the reads, sequence alignment, and filtering of the alignment files could be combined by running the basic pipeline. Various specifications for the filterSMS, indexDPgenomic, and filterAlign could be incorporated into a configuration file that could then be invoked by the basic pipeline.
5. Expected results based on mapping of the reads obtained using both the general protocol and the protocol for profiling of 3′ polyA sRNA can be seen in Table 1. Note that the 3′ poly protocol is heavily enriched in novel sRNAs, specifically in PASRs and TASRs.

3.6.3. Digital Gene Expression

1. This analysis requires the same tools as above. Just like in the example of the basic pipeline above the DGE pipeline, combines the various processing steps and, as one of it outputs generates a text *.count.txt file that contains counts for each reference.
2. There are three different types of counts generated for each reference in the "Min," "Frac," or "RMC" columns. The difference is in how they use the nonuniquely mapping reads. The "Min" column lists only counts based only on the reads that align uniquely to each reference. The "Frac" and "RMC," on the other hand, use the nonunique reads as well. The "Frac" evenly splits the nonunique reads among the different references they align to, while "RMC" used Baysean-based approach to assign nonunique reads to different mapping positions (15).

4. Notes

1. The RNA should be free from genomic DNA contamination and, therefore, DNAse I treatment is recommended. The RNA should be pure before the DNAse I digestion so that the enzymatic step is not inhibited. Thus, an ethanol precipitation step before the DNAse I treatment is recommended. Otherwise, incomplete DNAse I digestion could create oligonucleotides that are themselves a better substrate for sequencing than the genomic DNA.

Table 1
Results of the mapping of HeLaS3 sRNAs to the human genome with a minimal length of a trimmed read = 25 bases and minimal normalized score = 4.5

	General method			Method to profile 3′ polyA sRNAs		
Category	Number of reads: three channels	Fraction of uniquely mapping reads (%)	Fraction of uniquely mapping novel reads (%)	Number of reads: two channels	Fraction of uniquely mapping reads (%)	Fraction of uniquely mapping novel reads (%)
Total filtered reads	54,183,628			32,719,870		
All mapping reads	17,938,108			10,103,423		
Uniquely mapping	10,353,110	100		7,131,693	100	
Complete ribosomal repeat unit	5,974,541	57.7		4,593,222	64.4	
Mitochondrial	380,982	3.7		101,060	1.4	
chrY	8,855	0.1		1,345	0	
Repeats[a]	2,600,485	25.1		1,031,027	14.5	
Sno-miRNAs	997,801	9.6		49,408	0.7	
Selected group of small, non-coding RNAs (RNAse P RNA, U12, etc.)	105,204	1		57,845	0.8	
RNA genes	100,178	1		22,500	0.3	
Predicted sno	26,610	0.3		243	0	
Novel	158,454	1.5		1,275,043	17.9	
PASRs[b]	30,762	0.3	19.4	583,907	8.2	45.8
TASRs[c]	16,152	0.2	10.2	322,220	4.5	25.3

[a] As annotated by the RepeatMasker on the UCSC Browser track
[b] PASRs and TASRs were defined as regions ±500 bp from the respectively 5′ and 3′ end of a transcript annotated by UCSC Genes track
[c] The reads were first overlapped with PASR regions and then those that do not overlap the PASR regions were overlapped with the TASR regions

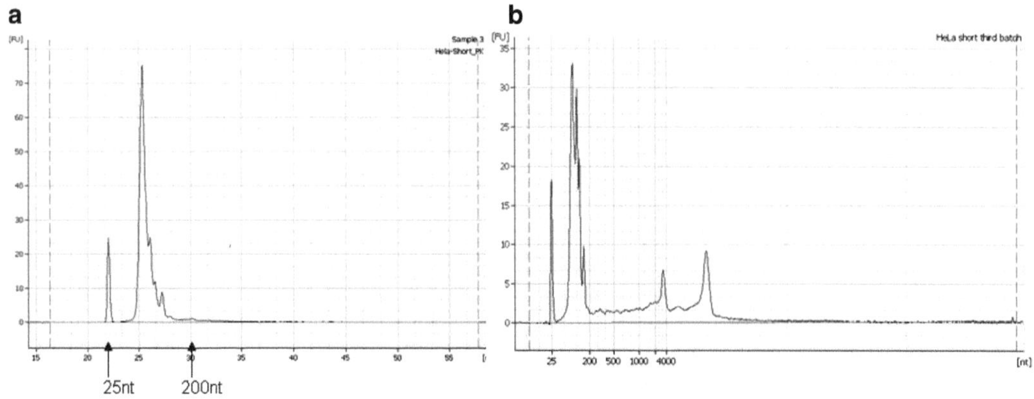

Fig. 2. BioAnalyzer profiles of (**a**) pure small RNA fraction and (**b**) the fraction contaminated with long RNAs.

2. Isolation of sRNA fraction is an important step. Care should be taken to remove the long RNA fraction otherwise the signal from the sRNA fraction could be contaminated by the reads coming from the overlapping long RNA. The sRNA fraction should be checked on the BioAnalyzer to ensure absence of the long RNA fraction. See Fig. 2, for example, of pure [part (a)] small RNA fraction and the one contaminated with long RNAs [part (b)].
3. It is hard to estimate the yield of cDNA after purification since the oligonucleotide used to prime cDNA synthesis will co-purify with the small RNA cDNAs. We typically tail ~200 ng of cDNA assuming that ~50% of that is the oligonucleotide. Due to the presence of three ribonucleotides at its 3′ terminus, the oligonucleotide will not be tailed by TdT.
4. We typically do not detect the pre-miRNAs species (the products of the Drosha cleavage) most likely due to their very stable secondary structure.
5. We have not tested different reverse transcriptases and the ones listed in this report may or may not be optimal for the detection of certain RNA species.

Acknowledgment

We wish to thank Sharon Bleakney for the help with manuscript editing.

References

1. Storz, G., Altuvia, S., and Wassarman, K. M. (2005) An abundance of RNA regulators. *Annu Rev Biochem* **74**, 199–217.
2. Ghildiyal, M., and Zamore, P. D. (2009) Small silencing RNAs: an expanding universe. *Nat Rev Genet* **10**, 94–108.
3. Affymetrix, ENCODE, Transcriptome, Project;, Cold, Spring, et al. (2009) Post-transcriptional processing generates a diversity of 5′-modified long and short RNAs. *Nature* **457**, 1028–32.
4. Davis, C. A., and Ares, M., Jr. (2006) Accumulation of unstable promoter-associated transcripts upon loss of the nuclear exosome subunit Rrp6p in Saccharomyces cerevisiae. *Proc Natl Acad Sci USA* **103**, 3262–7.
5. Kapranov, P., Cheng, J., Dike, S., Nix, D. A., Duttagupta, R., Willingham, A. T., et al. (2007) RNA maps reveal new RNA classes and a possible function for pervasive. transcription *Science* **316**, 1484–8.
6. Taft, R. J., Glazov, E. A., Cloonan, N., Simons, C., Stephen, S., Faulkner, G. J., et al. (2009) Tiny RNAs associated with transcription start sites in animals. *Nat Genet* **41**, 572–8.
7. Kapranov, P., Ozsolak, F., Kim, S. W., Foissac, F., Lipson, D., Hart, C., et al. (2010) Novel class of human RNAs associated with gene termini suggests an uncharacterized RNA copying mechanism. *Nature* **466**, 642–6.
8. Rassoulzadegan, M., Grandjean, V., Gounon, P., Vincent, S., Gillot, I., and Cuzin, F. (2006) RNA-mediated non-mendelian inheritance of an epigenetic change in the mouse. *Nature* **441**, 469–74.
9. Kawaji, H., and Hayashizaki, Y. (2008) Exploration of small RNAs. *PLoS Genet* **4**, e22.
10. Mitchell, P. S., Parkin, R. K., Kroh, E. M., Fritz, B. R., Wyman, S. K., Pogosova-Agadjanyan, E. L., et al. (2008) Circulating microRNAs as stable blood-based markers for cancer detection. *Proc Natl Acad Sci USA* **105**, 10513–8.
11. Chen, X., Ba, Y., Ma, L., Cai, X., Yin, Y., Wang, K., et al. (2008) Characterization of microRNAs in serum: a novel class of biomarkers for diagnosis of cancer and other diseases. *Cell Res* **18**, 997–1006.
12. Sambrook, J., and Russell, D. W. (2001) Molecular cloning: a laboratory manual, Cold Spring Harbor Laboratory Press, Cold Spring Harbor, N.Y.
13. Giladi, E., Healy, J., Myers, G., Hart, C., Kapranov, P., Lipson, D., et al. (2010) Error tolerant indexing and alignment of short reads with covering template families. *J. Comput. Biol.* **17**, 1397–411.
14. Harris, T. D., Buzby, P. R., Babcock, H., Beer, E., Bowers, J., Braslavsky, I., et al. (2008) Single-molecule DNA sequencing of a viral genome. *Science* **320**, 106–9.
15. Lipson, D., Raz, T., Kieu, A., Jones, D. R., Giladi, E., Thayer, E., et al. (2009) Quantification of the yeast transcriptome by single-molecule sequencing. *Nat Biotechnol* **27**, 652–8.
16. Li, H., and Durbin, R. (2009) Fast and accurate short read alignment with Burrows-Wheeler transform. *Bioinformatics* **25**, 1754–60.
17. Rumble, S. M., Lacroute, P., Dalca, A. V., Fiume, M., Sidow, A., and Brudno, M. (2009) SHRiMP: accurate mapping of short color-space reads. *PLoS Comput Biol* **5**, e1000386.

Chapter 16

deepBase: Annotation and Discovery of MicroRNAs and Other Noncoding RNAs from Deep-Sequencing Data

Jian-Hua Yang and Liang-Hu Qu

Abstract

Recent advances in high-throughput deep-sequencing technology have produced large numbers of short and long RNA sequences and enabled the detection and profiling of known and novel microRNAs (miRNAs) and other noncoding RNAs (ncRNAs) at unprecedented sensitivity and depth. In this chapter, we describe the use of deepBase, a database that we have developed to integrate all public deep-sequencing data and to facilitate the comprehensive annotation and discovery of miRNAs and other ncRNAs from these data. deepBase provides an integrative, interactive, and versatile web graphical interface to evaluate miRBase-annotated miRNA genes and other known ncRNAs, explores the expression patterns of miRNAs and other ncRNAs, and discovers novel miRNAs and other ncRNAs from deep-sequencing data. deepBase also provides a deepView genome browser to comparatively analyze these data at multiple levels. deepBase is available at http://deepbase.sysu.edu.cn/.

Key words: Deep sequencing, MicroRNA, ncRNA, Small RNA, Expression pattern, Next-generation sequencing

1. Introduction

Deep-sequencing technologies have been applied to investigate various small noncoding RNA (ncRNA) transcriptomes and have reshaped the overall RNomics research landscape (1–5). The application of such technologies has identified various new and different ncRNA classes, such as microRNAs (miRNAs), small interfering RNAs (siRNAs), and Piwi-interacting RNAs (piRNAs) (4–7). The increasing interest in miRNAs and other ncRNAs and the increasing

Jian-Bing Fan (ed.), *Next-Generation MicroRNA Expression Profiling Technology: Methods and Protocols*,
Methods in Molecular Biology, vol. 822, DOI 10.1007/978-1-61779-427-8_16, © Springer Science+Business Media, LLC 2012

amount of deep-sequencing data have generated a strong demand among researchers for an integrated database that would facilitate the annotation and analysis of these massive data sets.

The deepBase database (8) is the most comprehensive integration and analysis of public deep-sequencing data and provides the comprehensive annotation and identification of miRNAs and other ncRNAs from deep-sequencing data. The deepBase database has the following four aims:

1. Mapping all published deep-sequencing data for various genomes and providing various graphical interfaces and detailed information to explore these data. Currently, deepBase contains more than 370 million sequence reads from 237 small RNA libraries of seven organisms: human, mouse, chicken, *Ciona intestinalis*, *Drosophila melanogaster*, *Caenorhabditis elegans*, and *Arabidopsis thaliana*.
2. Annotating and classifying all mapped deep-sequencing data, providing experimental evidence for miRBase-annotated (9) miRNAs and other known ncRNAs, and allowing users to revise known miRNAs and discover novel small RNAs derived from these known miRNAs and ncRNAs. Currently, deepBase classifies mapped small RNAs into four catalogues according to their genomic annotation: ncRNA-associated small RNAs (nasRNAs), promoter-associated small RNAs (pasRNAs), exon-associated small RNAs (easRNAs), and repeat-associated small RNAs (rasRNAs) (4–7).
3. Constructing miRNAs and other ncRNAs expression patterns from deep-sequencing data and facilitating novel discoveries in miRNA and other ncRNA gene regulation.
4. Predicting novel miRNAs and other ncRNAs from deep-sequencing data and providing web-based tools to discover novel miRNAs and other ncRNAs from deep-sequencing data.

To facilitate future updates and development, the deepBase database structure and software architecture is flexibly designed to incorporate new deep-sequencing data, miRNAs, and other ncRNAs (the database structure is briefly described in Note 1). The Automatic Mapping, Annotating and Mining Tools (AutoMAMT) in deepBase are run on our high-performance computer servers. The deepBase will be regularly updated and extended to include more species and deep-sequencing data using AutoMAMT tools. All data generated from the deepBase database can be freely downloaded and retrieved (see Notes 2 and 3).

2. Materials

2.1. Hardware

Unix, Windows, or Macintosh workstation with an Internet connection.

2.2 Software

1. Web browser: An up-to-date Internet browser, such as Internet Explorer (http://www.microsoft.com/windows/internet-explorer/worldwide-sites.aspx); Firefox (http://www.mozilla.org/firefox); Safari (http://www.apple.com/safari); or Google Chrome (http://www.google.com/chrome).
2. snoSeekerNGS software: snoSeekerNGS is an improved version of snoSeeker (see Note 4), used for discovering snoRNAs from pooled deep-sequencing data (8, 10). SnoSeekerNGS is freely available from the following URL: http://deepbase.sysu.edu.cn/SnoSeekerNGS.php. Choose the appropriate platform for a binary distribution or a graphical user interface (GUI) distribution. For the binary distribution, unpack it and copy the binary to the desired directory. For the GUI distribution, double-click it and follow the on-screen installation procedure.

3. Methods

The methods presented in this chapter describe how to use the deepBase web interface to explore known miRNAs and ncRNAs, their expression pattern, and long RNA clusters and how to discover miRNAs and snoRNAs and run a comparative analysis of these data using the deepView genome browser.

3.1. Exploring Known MicroRNAs and Other ncRNAs from Deep-Sequencing Data

The analysis of small RNAs associated with miRNA and ncRNAs precursors has identified many novel nasRNAs, such as miRNA-offset RNAs (moRs) (11), tRNA-derived RNA fragments (tRFs) (12), and miRNA-like snoRNAs (13). Moreover, recent studies have shown that many miRBase-annotated miRNA genes failed to generate sequenced RNAs with miRNA-like features (14, 15). To explore these nasRNAs and refine known miRNA genes, we developed web interfaces to view these nasRNAs.

The following steps describe how to use the deepBase nasRNA website to evaluate known miRNAs (this workflow is also applicable for other ncRNAs to identify nasRNAs) from high-throughput deep-sequencing data:

1. Click "nasRNA->Browse" to open the nasRNA page.
2. Select the ncRNA type, clade, genome, and database of interest. Click the "Browse" button. For example, choose the "miRNA->mammal->human->hg19" to see a list of all known miRNAs

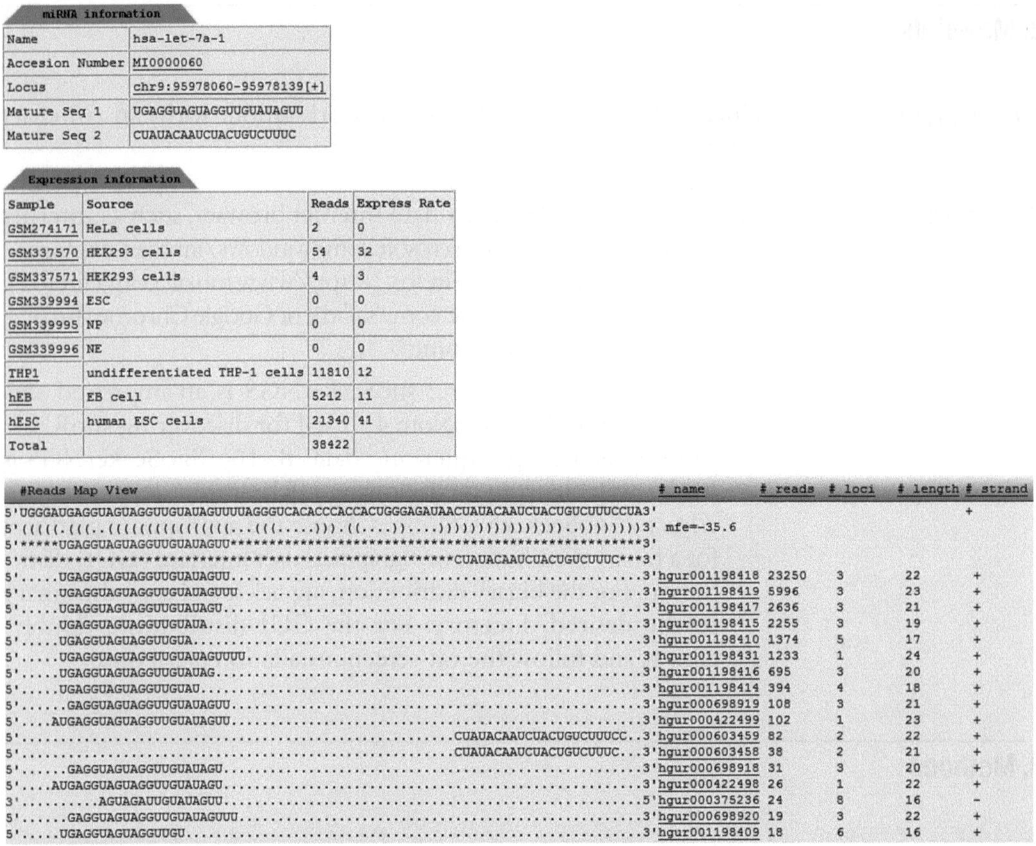

miRNA information

Name	hsa-let-7a-1
Accesion Number	MI0000060
Locus	chr9:95978060-95978139[+]
Mature Seq 1	UGAGGUAGUAGGUUGUAUAGUU
Mature Seq 2	CUAUACAAUCUACUGUCUUUC

Expression information

Sample	Source	Reads	Express Rate
GSM274171	HeLa cells	2	0
GSM337570	HEK293 cells	54	32
GSM337571	HEK293 cells	4	3
GSM339994	ESC	0	0
GSM339995	NP	0	0
GSM339996	NE	0	0
THP1	undifferentiated THP-1 cells	11810	12
hEB	EB cell	5212	11
hESC	human ESC cells	21340	41
Total		38422	

```
#Reads Map View                                                                      # name           # reads  # loci  # length  # strand
5'UGGGAUGAGGUAGUAGGUUGUAUAGUUUUAGGGUCACACCCACCACUGGGAGAUAACUAUACAAUCUACUGUCUUUCCUA3'                                                +
5'(((((.(((..((((((((((((((((...(((.....))).((....))....)))))))))))))))..))))))))3' mfe=-35.6
5'*****UGAGGUAGUAGGUUGUAUAGUU*****************************************************3'
5'********************************************************CUAUACAAUCUACUGUCUUUC***3'
5'.....UGAGGUAGUAGGUUGUAUAGUU.....................................................3'hgur001198418 23250  3   22  +
5'.....UGAGGUAGUAGGUUGUAUAGUUU....................................................3'hgur001198419 5996   3   23  +
5'.....UGAGGUAGUAGGUUGUAUAGU......................................................3'hgur001198417 2636   3   21  +
5'.....UGAGGUAGUAGGUUGUAUA........................................................3'hgur001198415 2255   3   19  +
5'.....UGAGGUAGUAGGUUGUA..........................................................3'hgur001198410 1374   5   17  +
5'.....UGAGGUAGUAGGUUGUAUAGUUUU...................................................3'hgur001198431 1233   1   24  +
5'.....UGAGGUAGUAGGUUGUAUAG.......................................................3'hgur001198416 695    3   20  +
5'.....UGAGGUAGUAGGUUGUAU.........................................................3'hgur001198414 394    4   18  +
5'......GAGGUAGUAGGUUGUAUAGUU.....................................................3'hgur000698919 108    3   21  +
5'....AUGAGGUAGUAGGUUGUAUAGUU.....................................................3'hgur000422499 102    1   23  +
5'........................................................CUAUACAAUCUACUGUCUUUCC..3'hgur000603459 82     2   22  +
5'........................................................CUAUACAAUCUACUGUCUUUC...3'hgur000603458 38     2   21  +
5'......GAGGUAGUAGGUUGUAUAGU......................................................3'hgur000698918 31     3   20  +
5'....AUGAGGUAGUAGGUUGUAUAGU......................................................3'hgur000422498 26     1   22  +
3'.........AGUAGAUUGUAUAGUU.......................................................5'hgur000375236 24     8   16  -
5'......GAGGUAGUAGGUUGUAUAGUUU....................................................3'hgur000698920 19     3   22  +
5'.....UGAGGUAGUAGGUUGU...........................................................3'hgur001198409 18     6   16  +
```

Fig. 1. A sample screenshot of miRNA loci, showing a description of the miRNA general information, the miRNA relative expression rate, and deep-sequencing reads mapped to microRNA (miRNA) precursor.

from miRBase (release v13) in the human genome. The miRNA name, the genomic coordinates in the human genome (UCSC hg19), and the number of deep-sequencing reads are indicated in a table. The user can click on the title of the table to sort miRNAs according to various features, such as the number of reads, miRNA names, and the genomic coordinates.

3. Click on the miRNA name within the table to launch a detailed page providing further information on that miRNA.
4. The detailed information for a miRNA includes a description of the miRNA general information, the miRNA relative expression profiles, and deep-sequencing data mapped to the miRNA precursor. See Fig. 1 for a sample screenshot.
   (a) Clicking the "Accession Number" will link to miRBase to get more information. Click the genome coordinates to view the genomic context of the miRNA gene locus in our deepView genome browser.

(b) In the "expression information" section, the expression rate for miRNA is provided to test for a differential expression pattern among different tissues and cell lines. More details for expression patterns are described in Subheading 3.2. Clicking the sample accession number will link to the NCBI GEO database (16) for further information.

(c) In the "Reads Map View" section, all mapped reads are listed according to sequential read number. Users can view miRNA 5′ and 3′ heterogeneity, isomiR (17), read number, read locus number, and read length and determine whether the annotated miRNAs are bonafide genes. Clicking the read name will display read information, including read length, read location, and read number from various libraries.

(d) The "references" section enables the retrieval of the primary articles yielded the deep-sequencing data. Click the article title link to visit NCBI PUBMED website.

3.2. MiRNA and ncRNA Expression Patterns in Deep-Sequencing Data

Before the effect of ncRNAs on gene regulation can be globally studied, a robust method for profiling the expression level of each ncRNA in a sample is required. We integrated various deep-sequencing data from different sequenced samples and constructed relative miRNA and other ncRNA expression profiles. The steps described below explain how to use the expression profiles:

1. Click "Heatmap" to open the expression pattern page.
2. As described in Subheading 3.1, step 2, select the ncRNA type, clade, genome, and database of interest. And then click the "Browse" button. For example, choose the "microRNA->mammal->human->hg19" to see the miRNA expression pattern. The expression pattern was visualized as a heat map: vertical and horizontal axes represent miRNAs and tissues/cell lines, and colors represent miRNA expression levels (Fig. 2, see Note 5). Clicking the tissue will link to the NCBI GEO database for more information on this small RNA library. Click the miRNA to launch a detailed page providing further information on that miRNA (Subheading 3.1). Users can change the number of items to view all or partial miRNA expression patterns at the top of the page.
3. Move the mouse cursor onto the Heatmap and a tooltip box will be displayed. Users can view the ncRNA name, tissue, GEO accession number, normalized expression level, and sequenced read numbers.

3.3. Discovering miRNAs from Deep-Sequencing Data

The emergence of high-throughput next-generation deep-sequencing technologies has led to the identification of hundreds of miRNAs in a multitude of species in many species in a multitude of species. However, it remains a daunting challenge to identify miRNAs

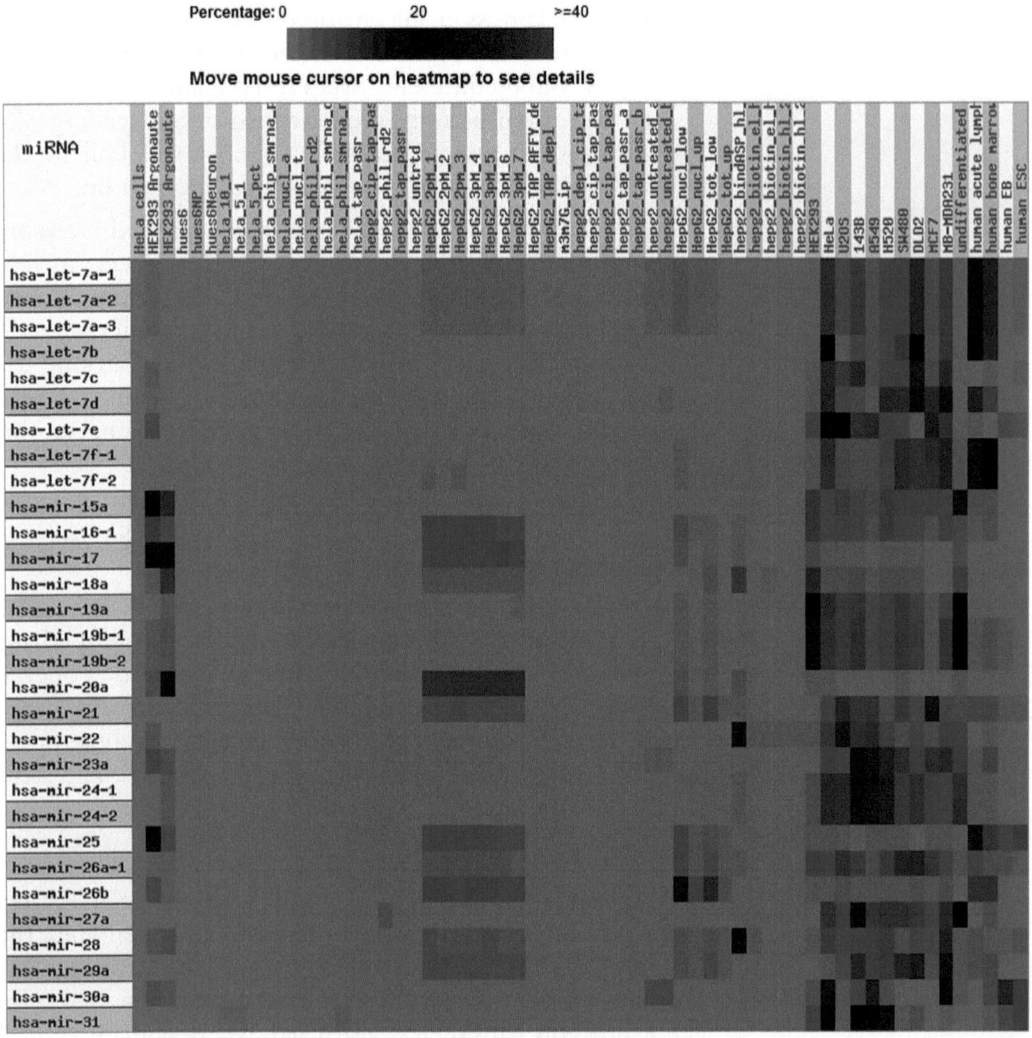

Fig. 2. miRNA expression pattern generated from deep-sequencing data. Vertical and horizontal axes represent miRNAs and tissues/cell lines, and colors represent miRNA expression levels.

from the millions of short reads generated by high-throughput sequencing technologies. In this section, we summarize the features and workflow used to identify miRNA from deep-sequencing data and summarize how to browse miRNA candidates predicted by an improved miRDeep (see Note 6) using our deepBase web interface.

3.3.1. Workflow for Identifying miRNAs from Deep-Sequencing Data

The general workflow summary from a series of recent publications (15, 18–20) is as follows:

1. Deep-sequencing data processing. Remove 5′ and 3′ adapters or barcodes from raw deep-sequencing data. The same reads were collapsed into unique reads.

2. Mapping processed reads to reference genome. The user can select various ultrafast aligners, such as Bowtie (21) and BWA (22). To avoid repeat-associated sequences, reads with multiple matching positions in the genome were discarded.
3. Filter small RNA significantly overlapping annotated genomic elements, such as known ncRNAs, genomic repeats, and coding sequences.
4. Extract potential miRNA precursors from the genome. Mapped reads, extended by an additional 20–50 nt in both the 5′ and 3′ directions, were extracted as potential precursors. Fold the potential precursors using RNAfold (23) or Mfold (24), and discard the potential precursors that did not fold into a hairpin.
5. Filter candidates using read distribution features derived from known miRNAs (15, 18–20). (1) The majority of reads overlapping with mature miRNA have the same 5′ ends. (2) A corresponding candidate miRNA* base-pair with the 2-nt 3′-overhang provides particularly compelling evidence for Dicer-like processing from a miRNA hairpin. (3) The length of dominant reads is from 21 to 25 nt. (4) Almost all of the reads within the precursor are processed from the same strand direction. (5) The number of dominant read is often larger than 10 or the read exists in multiple different deep-sequencing experiments.

3.3.2. Browsing miRNAs Predicted by miRDeep

We applied the improved miRDeep (18) to identify miRNAs from pooled small RNA libraries. We provide enhanced resolution and identify more novel miRNAs owing to the integration of the larger number of small RNA libraries of diverse tissues and cell lines. All of the predicted results were stored in a mySQL database and are displayed on our deepBase website. Users can browse these data with the following steps:

1. Select "prediction->miRDeep" and the clade, genome, and database of interest to open the list of predicted miRNAs. Users can sort the candidates by clicking the candidate name, genome coordinates, and miRDeep score.
2. Click a candidate to launch a detail page for this candidate. An example detail page is shown in Fig. 3. The page contains miRDeep score information, the expression information of the candidate in various libraries, and reads mapped to a potential precursor. The miRDeep score includes the total score, nucleus score, frequency score, mfe score, and randfold score (18).
3. Click locus to view the genomic context of the candidate miRNA locus in the deepView genome browser (described in Subheading 3.6). Users can click the predicted mature miRNA, miRNA*, or isomiRs to view read information, including the read length, read number, read location, and read number in various libraries.

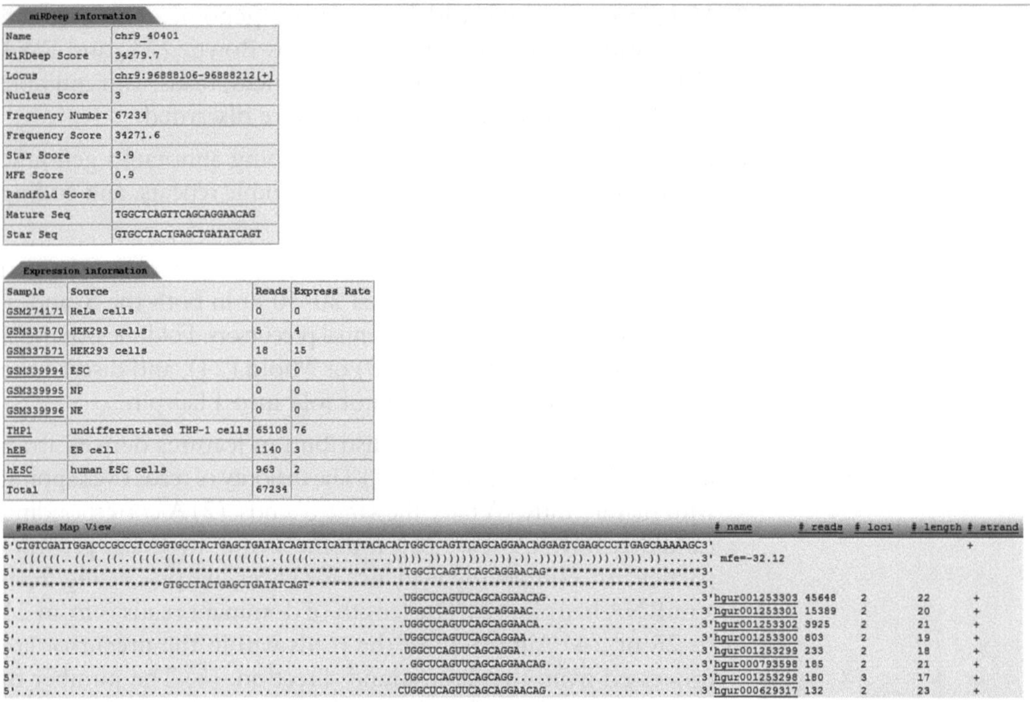

Fig. 3. The page for a miRNA candidate predicted by miRDeep, showing miRDeep score information, expression information of the candidate in various libraries, and reads mapped to the potential precursor.

3.4. Discovering snoRNAs from Deep-Sequencing Data

Traditional snoRNA-search programs are usually based on sequence complementarity to rRNAs or snRNAs and sequence conservation among species and are therefore limited to the detection of guide and conserved snoRNAs and not orphan or species-specific snoRNAs. In this section, we introduce how to identify snoRNA genes from deep-sequencing data using our snoSeekerNGS program (8, 10).

1. Genome preparation. You can download the genome from public genome centers, such as NCBI Ftp or the UCSC Bioinformatics web site, TAIR.
2. Mapping small RNAs (the format of small RNA is described in Note 7) from one or multiple experiments to the genome using an aligner, such as Bowtie (21) and BWA (22). Alignment reads must be saved in bowtie or SAM (25) format.
3. Click snoSeekerNGS to launch the GUI interface. Select snoSeekerNGS->CDseekerNGS or snoSeekerNGS->ACAseekerNGS to predict the C/D and H/ACA snoRNA gene, respectively. For example, choose the "snoSeekerNGS->CDseekerNGS" to launch a dialog. In the dialog, users can click the "Browse" button to load the mapped read file, the genome file, the output file, and the modification file. The user can set the option in the option text box. For example, the option-G 37.5 tells

CDseekerNGS to just output the guide C/D snoRNAs with a score ≥37.5 bits (see Note 8). You can click the about button for more options for CDseekerNGS (Fig. 4a). The same steps are applicable for ACAseekerNGS to predict H/ACA snoRNAs.

4. Click the "OK" button. After a short time, the program will return the results (Fig. 4b).

We have applied snoSeekerNGS to pooled deep-sequencing data and have predicted thousands of snoRNAs. Users can browse these candidates by click "prediction->snoSeeker." Users can also download our local version of snoSeeker with a GUI to predict snoRNAs from deep-sequencing data (Subheading 2.2).

3.5. Exploring Long RNA Clusters from Deep-Sequencing Data

Analysis of genomic RNA clusters can be used to identify novel miRNAs, piRNAs, and other ncRNAs (4–6). In this section, we describe how to browse and filter RNA clusters using our cluster web interface.

1. Click "cluster" to open the cluster page.
2. Select the organism of interest, read number, unique number, Dominant read number, Dominant read length, Dominant read percentage, Locus number of dominant read, and Dominant length read percentage. Click the "Browse cluster" button. For example, choose the "mammal->human->hg19" to see a list of all clusters in the human genome. The cluster name, the genomic coordinates in the human genome (UCSC hg19), unique read number, and the number of deep-sequencing reads are indicated in a table. The user also can click on the title of the table to sort these RNA clusters.
3. Selecting different filter options in the "cluster" page will generate different results. According to the read distribution features of miRNAs described in Subheading 3.3.1, we set the parameter to filter cluster, and we can identify some novel miRNA-like/associated clusters as well as known miRNAs.
4. Click on the cluster name within the table to launch a detailed page providing further information on that cluster.
5. The detailed information for a cluster includes a description of the cluster's general information (Fig. 5), the length distribution of sequenced reads and unique reads mapped to the cluster, the distance distribution between two nearby reads, and the frequency of 5′ terminal nucleotide of reads in the cluster. This detailed information will help users to determine whether the cluster yields functional small RNAs.
6. By clicking the cluster name, the user can view which reads are mapped to the clusters. Click the locus to view the genomic context of the cluster locus in the deepView genome browser (Subheading 3.6).

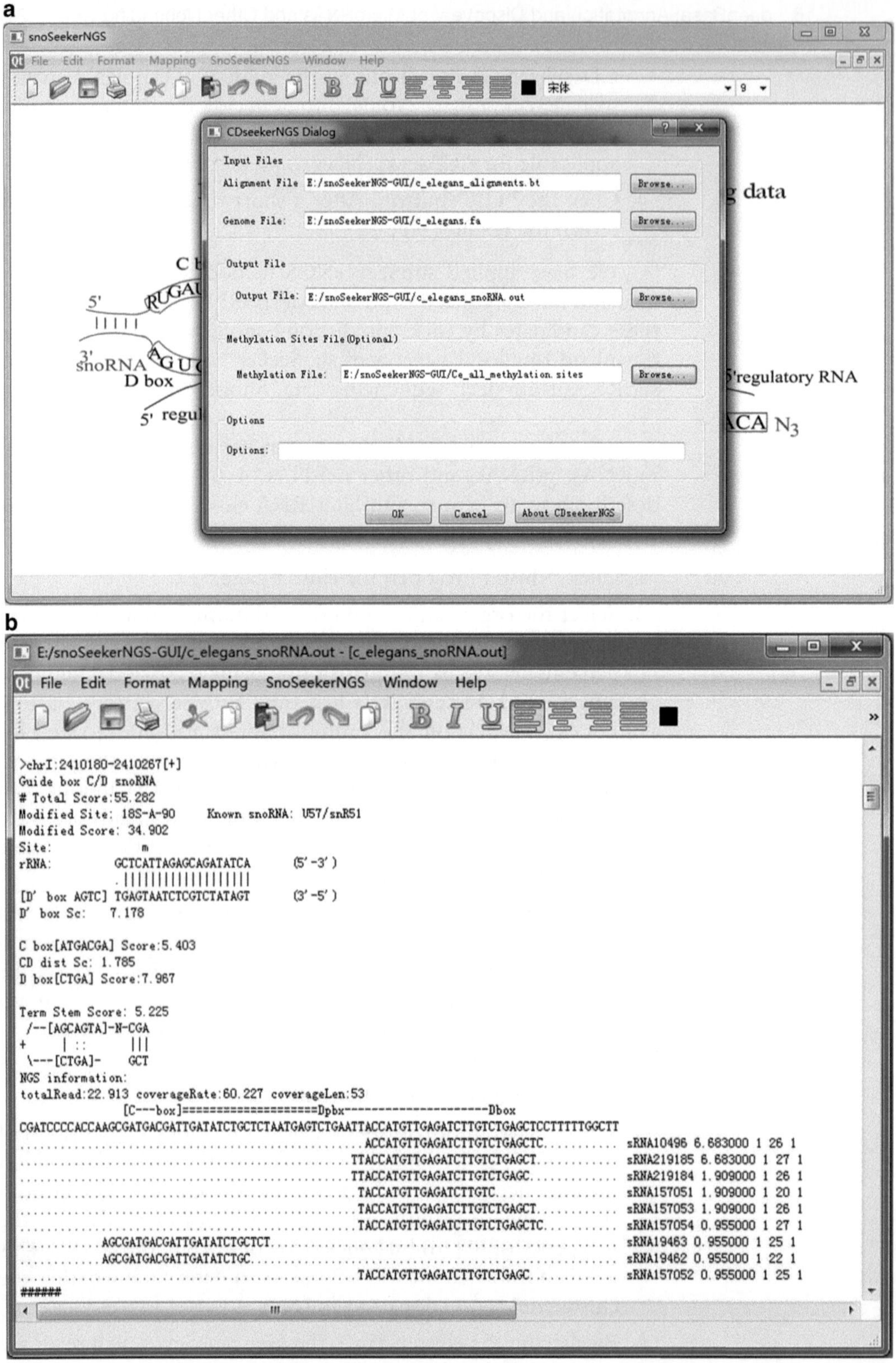

Fig. 4. snoSeekerNGS graphical user interface. (**a**) Input interface for CDseekerNGS. User can load the files by clicking the browser button. (**b**) The output results of the CDseekerNGS program. The information for each candidate includes genomic location, snoRNA type, modified site, duplex between snoRNA and target RNA, box elements, terminal stem, and their scores, followed by alignments of the mapped reads to the candidate snoRNA.

Cluster information	
Name	chr3 fcluster1404 (view reads)
totalReadNum	299419
totalUniqueNum	64
maxReadSeq	UUCAAGUAAUCCAGGAUAGGCU
maxReadNum	223156(74.53)
maxReadLen	22
maxReadLoci	2
highestLenReadNum	Len:22 readNum:223315(74.583)
highestLenUnqiueReadNum	Len:15 readNum:10(15.625)
Locus	chr3:38010884-38010964 (deepView Genome Browser)

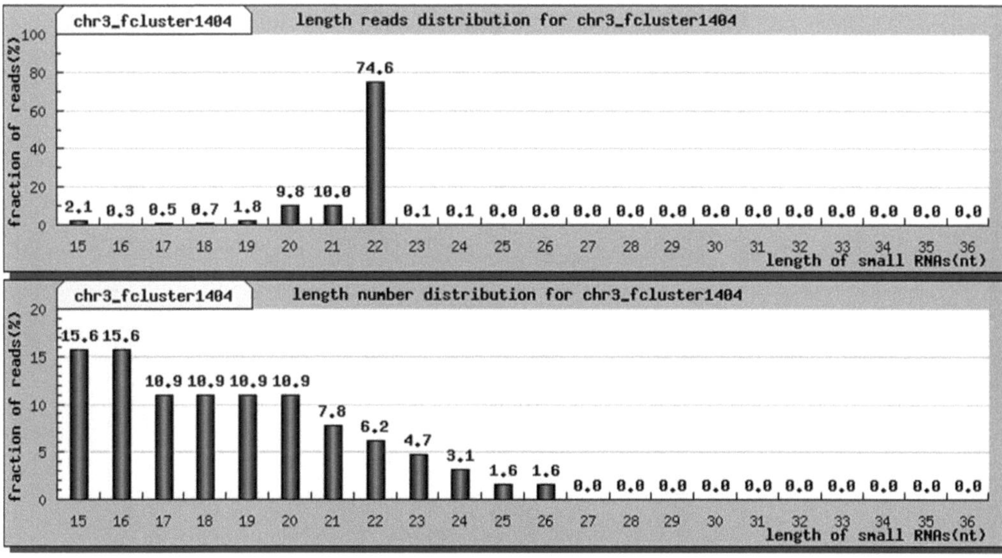

Fig. 5. A sample screenshot of an RNA cluster, showing a description of the cluster general information, the length distribution of sequence reads, and unique reads mapped to the cluster.

3.6. Navigating with the deepView Genome Browser to a Specific Genomic Position

The rapid progress of deep-sequencing has increased the demand for visual tools that offer quick and easy access to deep-sequencing data at various levels. In this section, we introduce a new visualization tool, the deepView genome browser, to facilitate simultaneously comparative analysis of massive data, visual data validation, and hypothesis generation.

1. Click the "Browser" to open the deepView Genome Browser page.
2. Select the genome of interest as described above, and then type a genomic coordinate or protein-gene or ncRNA gene name into the "genomic position" text box or click the "example" button to specify the genome region to display. Click the "submit" button.

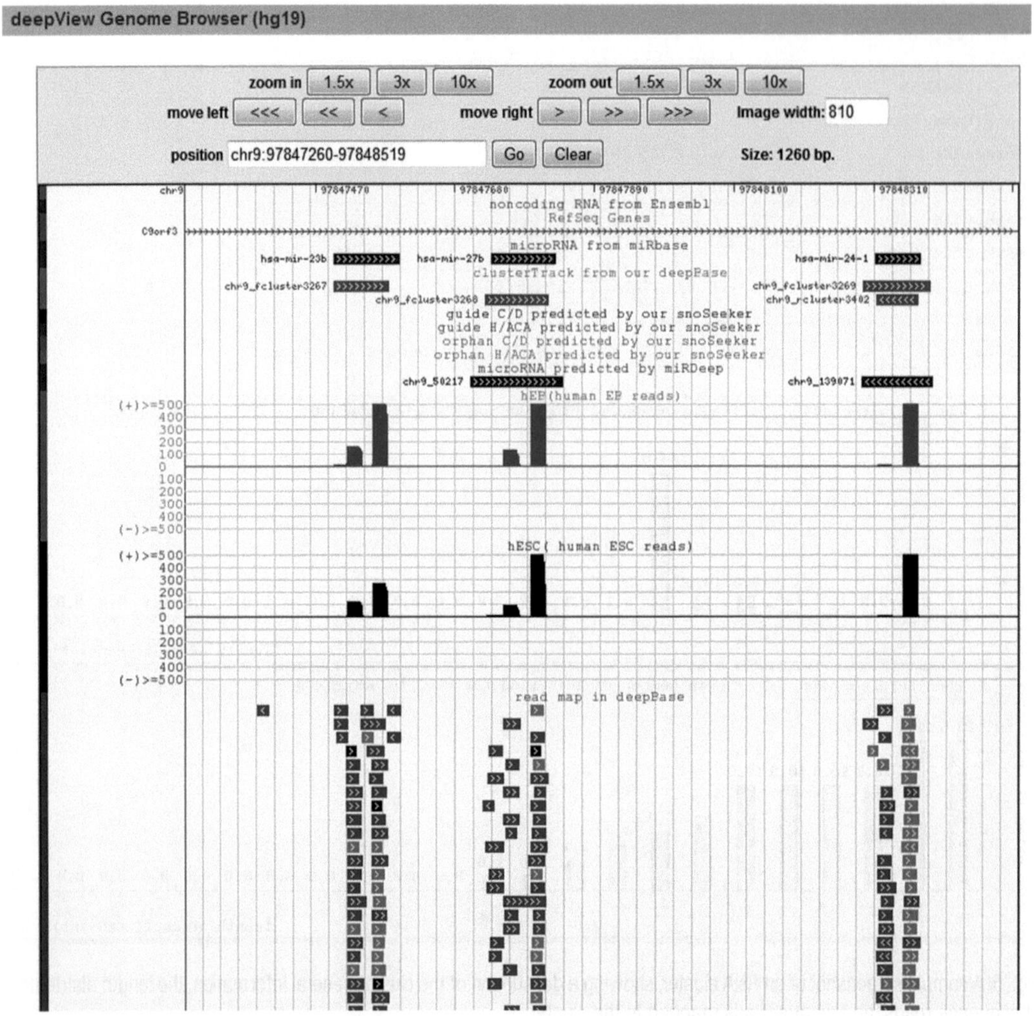

Fig. 6. The deepView Browser page displaying the chr9:97847260–97848519 region in the human genome (UCSC hg19). The navigation buttons are visible at the top of the image. The deepView Browser provides an integrated view of RNA genes from Ensembl or the literature, refSeq Gene, miRNA gene from miRBase v13, RNA clusters, the predicted snoRNAs using snoSeeker, the predicted miRNA genes using miRDeep, strand-specific expression peaks, and reads mapped to the genome.

3. deepView genome browser display page. Figure 6 shows the output of the deepView browser and displays all the track details for the specified chromosome and region. The deepView browser in deepBase provides an integrated view of mapped reads, known and predicted ncRNAs, protein-coding genes, and RNA clusters and their expression peaks (see Note 9). Clicking a track item within the browser launches a detailed page providing further information on that item or links to external resources, such as NCBI, miRBase, UCSC, and TAIR, from which one can obtain more comprehensive information.

4. In the track options section, click a check box or select read number from a pull-down menu, and then click the "refresh tracks" button to change the display mode of the annotation track. Click a track name, and a detailed description page for the track will be displayed. Users can click the "default tracks" button to reset the displayed tracks. Click the "View region at UCSC" to view the genomic context of the locus in the UCSC genome browser.
5. Type a new protein-coding gene, ncRNA gene name or new genomic position, and then click the "Go" button to completely change the deepView genome browser display image.
6. Click the "zoom out" or "zoom in" button to extend or shrink the width of the displayed coordinate range. Click "move left" or "move right" to move the image to the left or right by shifting the genomic coordinates.

4. Notes

1. In deepBase, all of the data are stored in relational tables of MySQL database. A specific naming convention has been used for each mapped unique read. Human mapped reads have been named from hgur000000001 to hgur002388637. Similarly, prefixes of "mm," "gg," "ci," "dm," "ce," and "at" have been used for naming the mouse, chicken, *C. intestinalis*, *D. melanogaster*, *C. elegans*, and *Arabidopsis*, respectively. RNA clusters have been named according to chromosome name, strand orientation, and sequential number (e.g., chr6_fcluster220).

   The database contains deep-sequencing data from the NCBI GEO website or from the supplementary data of the original articles. Small RNAs overlapping with genomic elements were classified into four categories, including nasRNAs, pasRNAs, easRNAs, and rasRNAs in deepBase database.

   All the data stored in the database generally include name, genomic coordination, and strand direction. The data, therefore, can easily be searched by name or genomic coordination and can be displayed in the genome browser.
2. The deepBase data files can be freely downloaded from the "Download" web page. The sequence and annotation data are in UCSC BED and Fasta formats. The RNA cluster file in the BED format includes chromosome, start position, end position, RNA cluster name, sequenced read number, and strand direction. It should be noted that all start coordinates are 0-based in the deepBase database. All RNA clusters can be downloaded in the BED format, compatible with the UCSC database.

Users can upload these files to the UCSC Genome Browser and can display them in the UCSC Genome Browser. Predicted miRNAs and snoRNAs, nsaRNAs, pasRNAs, rasRNAs, and easRNAs can be downloaded in Fasta format, and their genomic coordinates are also displayed in entry name. In addition, genomic element overlapping with small RNAs is also listed in their name.

3. Sequenced reads, miRNA, ncRNAs can be retrieved by entering sequences or name at the "Search" Web page. Other small RNA classes, such as pasRNA, easRNA, and rasRNA can be searched by genomic location or small RNA numbers mapped to them.

   In addition, a BLAST (26) function was provided for performing searches against sets of small RNA sequences. Users can enter query sequences in search box and can then set the various options. The output results are displayed as a table. By clicking the read name view the detailed read information. Click "align" to see the alignment between the query sequence and the object sequence deposited in the database.

4. snoSeeker firstly was developed to highly efficient and specific screen, both guide and orphan snoRNA genes in mammalian genomes (for a more detailed explanation, see ref. 10). An online version of snoSeeker can be accessed at http://genelab.sysu.edu.cn/snoSeeker/. User can download snoSeeker software package at this website.

5. In a heat map, red indicates overexpression in the tissue or cell line, and green indicates underexpression. Except for miRNAs, the Heatmap reflects a rough measure of the total expression of other ncRNAs because most of the reads mapped to the other ncRNA species might be degenerated products (8). The relative contribution of each ncRNA from each sample was calculated as follows: (1) the read number of each ncRNA from each sample was normalized to the total read number of that library (27). (2) The normalized count of a particular ncRNA in a particular sample was divided by the sum of normalized count for that ncRNA across all samples. (3) Those normalized counts were transformed to 100 percentiles, and each bar in heat map represents the normalized level (8).

6. Improved miRDeep. How to improve search speed of miRDeep has been described (8).

7. The format of read without adapters is Fasta format. Each read entry consists of a line beginning with the symbol ">" followed by the read's identifier and one line with the read itself. For example, >sRNA99_630 AAGAGAAGGTGTCTTGCTGGCGGT. Where, "sRNA99" represents a user-definable unique ID for unique reads. The "630" indicates that this read (sRNA99)

has been detected 630 times in the deep-sequencing data set. Both of them were linked by an underline.

8. snoSeekerNGS manual. The snoSeekerNGS distribution contains readme file for snoSeekerNGS that explain all parameters.
9. To study the expression patterns of RNA clusters, genome-wide expression peaks were constructed by plotting the abundance of each genomic position on the genome sequences. The deepView genome browser provides a direct comparison of peak patterns generated from different cell lines and tissues. Different ncRNAs have different expression peaks in the deepView genome browser. For example, miRNA resembled a "two-hill" shape in the browser (Fig. 6). Thus, the two-hill-shaped peak pattern allows the user to fairly easily distinguish true miRNA from other ncRNAs by simple observation of the peak pattern.

Acknowledgments

This research is supported by the National Natural Science Foundation of China (No. 30830066, 30900820); Ministry of Science and Technology of China, National Basic Research Program (No. 2005CB724600, 2011CB811300); the funds from the Ministry of Education of China and Guangdong Province (No. 9451027501002591); and China Postdoctoral Science Foundation (No. 200902348).

References

1. Mardis, E. R. (2008) Next-generation DNA sequencing methods. *Annu Rev Genomics Hum Genet* **9**, 387–402.
2. Mardis, E. R. (2008) The impact of next-generation sequencing technology on genetics. *Trends Genet* **24**, 133–41.
3. Lister, R., Gregory, B. D., and Ecker, J. R. (2009) Next is now: new technologies for sequencing of genomes, transcriptomes, and beyond. *Curr Opin Plant Biol* **12**, 107–18.
4. Okamura, K., and Lai, E. C. (2008) Endogenous small interfering RNAs in animals. *Nat Rev Mol Cell Biol* **9**, 673–8.
5. Ghildiyal, M., and Zamore, P. D. (2009) Small silencing RNAs: an expanding universe. *Nat Rev Genet* **10**, 94–108.
6. Kim, V. N., Han, J., and Siomi, M. C. (2009) Biogenesis of small RNAs in animals. *Nat Rev Mol Cell Biol* **10**, 126–39.
7. Lin, H. (2007) piRNAs in the germ line. *Science* **316**, 397.
8. Yang, J. H., Shao, P., Zhou, H., Chen, Y. Q., and Qu, L. H. (2010) deepBase: a database for deeply annotating and mining deep sequencing data. *Nucleic Acids Res* **38**, D123–30.
9. Griffiths-Jones, S., Saini, H. K., van Dongen, S., and Enright, A. J. (2008) miRBase: tools for microRNA genomics. *Nucleic Acids Res* **36**, D154–8.
10. Yang, J. H., Zhang, X. C., Huang, Z. P., Zhou, H., Huang, M. B., Zhang, S., et al (2006) snoSeeker: an advanced computational package for screening of guide and orphan snoRNA genes in the human genome. *Nucleic Acids Res* **34**, 5112–23.
11. Shi, W., Hendrix, D., Levine, M., and Haley, B. (2009) A distinct class of small RNAs arises from pre-miRNA-proximal regions in a simple chordate. *Nat Struct Mol Biol* **16**, 183–9.

12. Lee, Y. S., Shibata, Y., Malhotra, A., and Dutta, A. (2009) A novel class of small RNAs: tRNA-derived RNA fragments (tRFs). *Genes Dev* **23**, 2639–49.

13. Ender, C., Krek, A., Friedlander, M. R., Beitzinger, M., Weinmann, L., Chen, W., et al (2008) A human snoRNA with microRNA-like functions. *Mol Cell* **32**, 519–28.

14. Chiang, H. R., Schoenfeld, L. W., Ruby, J. G., Auyeung, V. C., Spies, N., Baek, D., et al (2010) Mammalian microRNAs: experimental evaluation of novel and previously annotated genes. *Genes Dev* **24**, 992–1009.

15. Berezikov, E., Liu, N., Flynt, A. S., Hodges, E., Rooks, M., Hannon, G. J., et al (2010) Evolutionary flux of canonical microRNAs and mirtrons in Drosophila. *Nat Genet* **42**, 6–9; author reply 9–10.

16. Barrett, T., Troup, D. B., Wilhite, S. E., Ledoux, P., Rudnev, D., Evangelista, C., et al (2009) NCBI GEO: archive for high-throughput functional genomic data. *Nucleic Acids Res* **37**, D885–90.

17. Morin, R. D., O'Connor, M. D., Griffith, M., Kuchenbauer, F., Delaney, A., Prabhu, A. L., et al (2008) Application of massively parallel sequencing to microRNA profiling and discovery in human embryonic stem cells. *Genome Res* **18**, 610–21.

18. Friedlander, M. R., Chen, W., Adamidi, C., Maaskola, J., Einspanier, R., Knespel, S., et al (2008) Discovering microRNAs from deep sequencing data using miRDeep. *Nat Biotechnol* **26**, 407–15.

19. Hendrix, D., Levine, M., and Shi, W. miRTRAP, a computational method for the systematic identification of miRNAs from high throughput sequencing data. *Genome Biol* **11**, R39.

20. Ruby, J. G., Jan, C., Player, C., Axtell, M. J., Lee, W., Nusbaum, C. (2006) Large-scale sequencing reveals 21U-RNAs and additional microRNAs and endogenous siRNAs in C. elegans. *Cell* **127**, 1193–207.

21. Langmead, B., Trapnell, C., Pop, M., and Salzberg, S. L. (2009) Ultrafast and memory-efficient alignment of short DNA sequences to the human genome. *Genome Biol* **10**, R25.

22. Li, H., and Durbin, R. (2009) Fast and accurate short read alignment with Burrows-Wheeler transform. *Bioinformatics* **25**, 1754–60.

23. Hofacker, I. L. (2003) Vienna RNA secondary structure server. *Nucleic Acids Res* **31**, 3429–31.

24. Zuker, M., and Stiegler, P. (1981) Optimal computer folding of large RNA sequences using thermodynamics and auxiliary information. *Nucleic Acids Res* **9**, 133–48.

25. Li, H., Handsaker, B., Wysoker, A., Fennell, T., Ruan, J., Homer, N., et al (2009) The Sequence Alignment/Map format and SAMtools. *Bioinformatics* **25**, 2078–9.

26. Altschul, S. F., Madden, T. L., Schaffer, A. A., Zhang, J., Zhang, Z., Miller, W., et al (1997) Gapped BLAST and PSI-BLAST: a new generation of protein database search programs. *Nucleic Acids Res* **25**, 3389–402.

27. Ruby, J. G., Stark, A., Johnston, W. K., Kellis, M., Bartel, D. P., and Lai, E. C. (2007) Evolution, biogenesis, expression, and target predictions of a substantially expanded set of Drosophila microRNAs. *Genome Res* **17**, 1850–64.

Chapter 17

PhenomiR: MicroRNAs in Human Diseases and Biological Processes

Andreas Ruepp, Andreas Kowarsch, and Fabian Theis

Abstract

The association of dysregulated microRNAs (miRNAs) and diseases has been shown in a variety of studies. Here, we review a resource denoted as PhenomiR, providing systematic and comprehensive access to such studies. It allows machine-readable access to miRNA and target relations from these studies to study the impact of miRNAs on multifactorial diseases across many samples and biological replicates. We summarize the PhenomiR data structure and its content and show how to access the database and use it in everyday miRNA profile analysis using the R language.

Key words: MicroRNA, Database, Disease, Bioprocess

1. Introduction

MicroRNAs (miRNAs) are endogenous RNAs with a length of around 22 nt. They are post-transcriptional regulators, which can bind to partially complementary sequences on target mRNAs, resulting mostly in translational repression and thereby silencing of the target protein. miRNAs are predicted to regulate the expression of the majority of all mammalian genes (1).

Classically, diseases and genetic disorders are linked to a single genetic polymorphism, resulting, for instance, in dysregulation of a single mRNA. A systematic analysis of genetic disorders has since revealed that many more such diseases are of complex, multifactorial origin, meaning that they are likely associated with the effects of multiple genes and more so related with lifestyle and environmental factors (2). Here, we ask the question if similar multifactorial origins are also reflected on correlation patterns of associated miRNA level.

Jian-Bing Fan (ed.), *Next-Generation MicroRNA Expression Profiling Technology: Methods and Protocols*,
Methods in Molecular Biology, vol. 822, DOI 10.1007/978-1-61779-427-8_17, © Springer Science+Business Media, LLC 2012

We provide a systematic resource denoted as PhenomiR, consisting of literature studies linking dysregulated miRNA profiles with diseases. With a total of 12,189 miRNA expression phenotype relation data points, collected from 628 different experiments, PhenomiR is the most comprehensive resource of its kind. Studies compiled in PhenomiR have, for example, found that miRNA clusters like miR-17-92, miR-106b-93-25, and miR-222-221 have a severe impact on cellular processes and diseases, especially cancer (3, 4). Here, miRNA clusters are defined as miRNAs on close-by genomic locations, which have been shown to be often coexpressed (5). Given a maximum distance of 5 kb, about 34% of human miRNAs appear as miRNA clusters of at least two members. We have previously investigated the influence of differentially regulated miRNA clusters on diseases (6). Using the comprehensive dataset from our PhenomiR database, we asked whether the impact of miRNA clusters on diseases is only restricted to a few examples or if miRNA clusters systematically affect the pathobiology of diseases. We were able to show that dysregulated miRNA clusters are significantly over-represented compared to singular miRNA gene products (not contained in an miRNA cluster). These clusters are tightly associated with many diseases.

In the following, we summarize the content of the PhenomiR database and outline its use both via Web interface and through local installation. Moreover, we briefly review analyses of related miRNA profiles using the R language in Subheading 3. Since for more detailed analyses it may be useful to understand the core PhenomiR data structure and the concept of an entry before examining the following methods, we briefly describe these principles in Note 1.

2. Materials

Methods presented in this chapter deal with both Web-based functionality and installing of the data for manipulation in your local environment. The Web site has further helped pages and resources that should be referred to alongside this guide.

2.1. Web-Based Protocols

By default, PhenomiR is made available via a Web-based interface at the URL http://mips.helmholtz-muenchen.de/phenomir. Its current release is version 2.0 (as of November 2010), with 675 unique miRNAs, mentioned in 362 articles, which discuss 144 diseases and 98 bioprocesses. The database may be browsed via its Java-based Web application front end (see Subheading 3.1). In addition, we provide full data access via download; see the next section.

2.2. For Local Installation

Phenotypic annotations of miRNAs are organized in an MySQL relational database. For local database installation, MySQL table dumps are provided on the Web page: http://mips.helmholtz-muenchen.de/phenomir/download/phenomir.zip.

The data are also portable to other relational databases, such as ORACLE, POSTGRES, and SQLSERVER.

For MYSQL installation, see http://www.mysql.com.

3. Methods

The following methods describe how to use the PhenomiR Web interface to obtain a comprehensive overview about the database (Subheading 3.1), how to use the different search options (Subheading 3.2), and take the user through the information that is available for each PhenomiR entry (Subheading 3.3). Subheading 3.4 describes how to install a local copy of the data in an MySQL relational database.

3.1. Browsing the Content of PhenomiR

Access to the data content of PhenomiR may be performed via two search options (Subheading 3.2) or by inspection of three compiled datasets, including the list of all entries, all diseases, and all miRNAs. The latter option allows obtaining an overview about the complete database content. A user who is interested in diseases for which miRNA expression analyses exist selects the list of diseases (Fig. 1a). The result shows the Online Mendelian Inheritance in Man (OMIM) description of the diseases as well as the number of annotated experiments. The single empty box in this compilation is connected to the list of bioprocess studies. The disease names, e.g. Burkitt lymphoma, are hyperlinked to the list of the respective experiments (entries), where additional information, such as the investigated tissue/cell line, PubMed identifier of the respective article, and the study design, can be found. Members of the entry lists are hyperlinked to the annotated result page of the experiment (Subheading 3.4).

3.2. Searching the PhenomiR Database

The PhenomiR database may be queried online by two different options.

1. The "General search" performs simultaneous queries across several attributes, like "miRNA name", "disease", or "gene name" (Fig. 1b). The "Search type" allows choosing between the option "Strict" using exclusively the exact term and "Fuzzy" which finds all results that contain the search term. Thus, a "Fuzzy" query with the term "lung" reveals a comprehensive collection of lung-related experiments, including miRNA expression

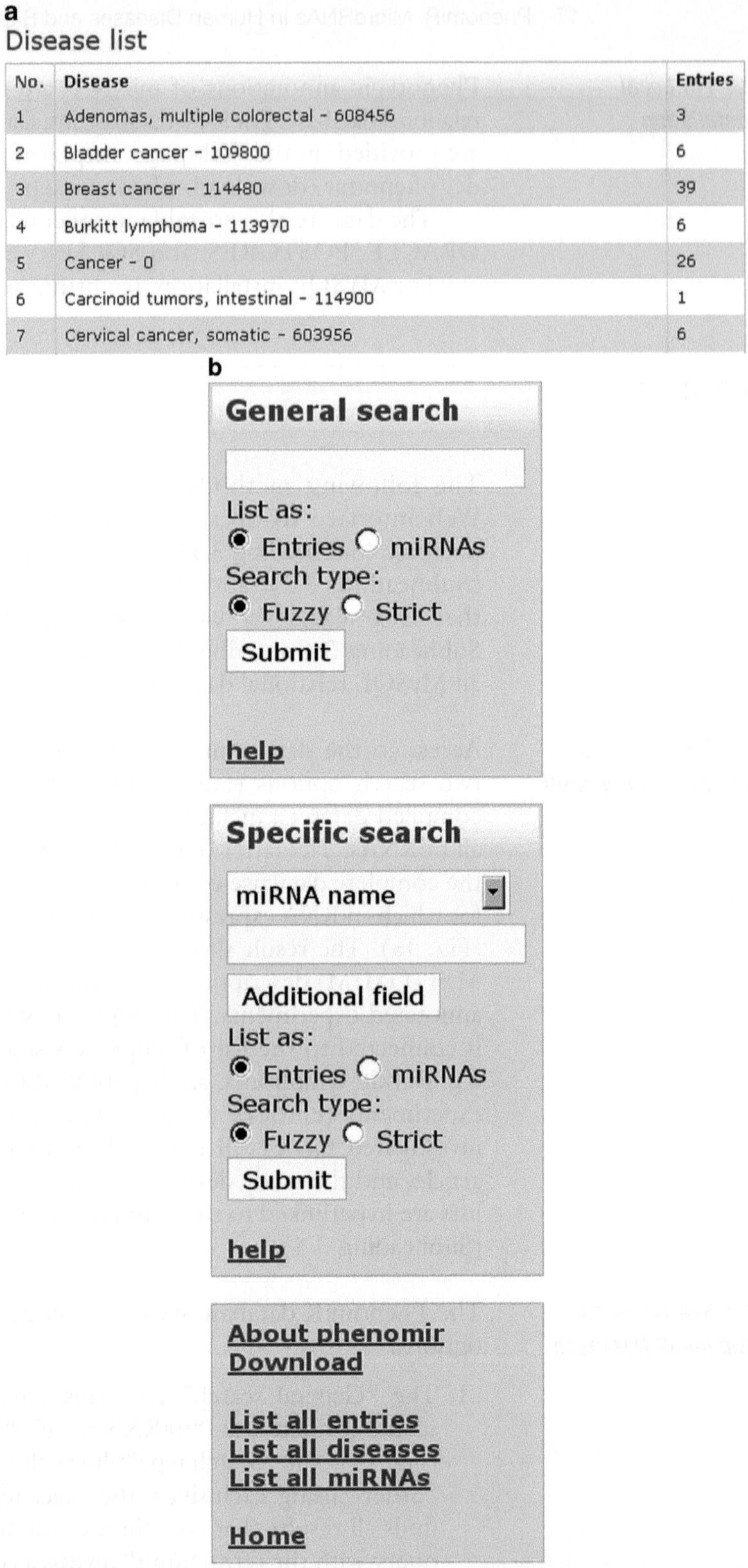

a

Disease list

No.	Disease	Entries
1	Adenomas, multiple colorectal - 608456	3
2	Bladder cancer - 109800	6
3	Breast cancer - 114480	39
4	Burkitt lymphoma - 113970	6
5	Cancer - 0	26
6	Carcinoid tumors, intestinal - 114900	1
7	Cervical cancer, somatic - 603956	6

Fig. 1. Database search and search results in PhenomiR. User might enter the database via predefined lists (**a**), such as the "Disease list". The search options for the user-defined "General search" and the "Specific search" (**b**) are presented on the home page. The entry pages.

c

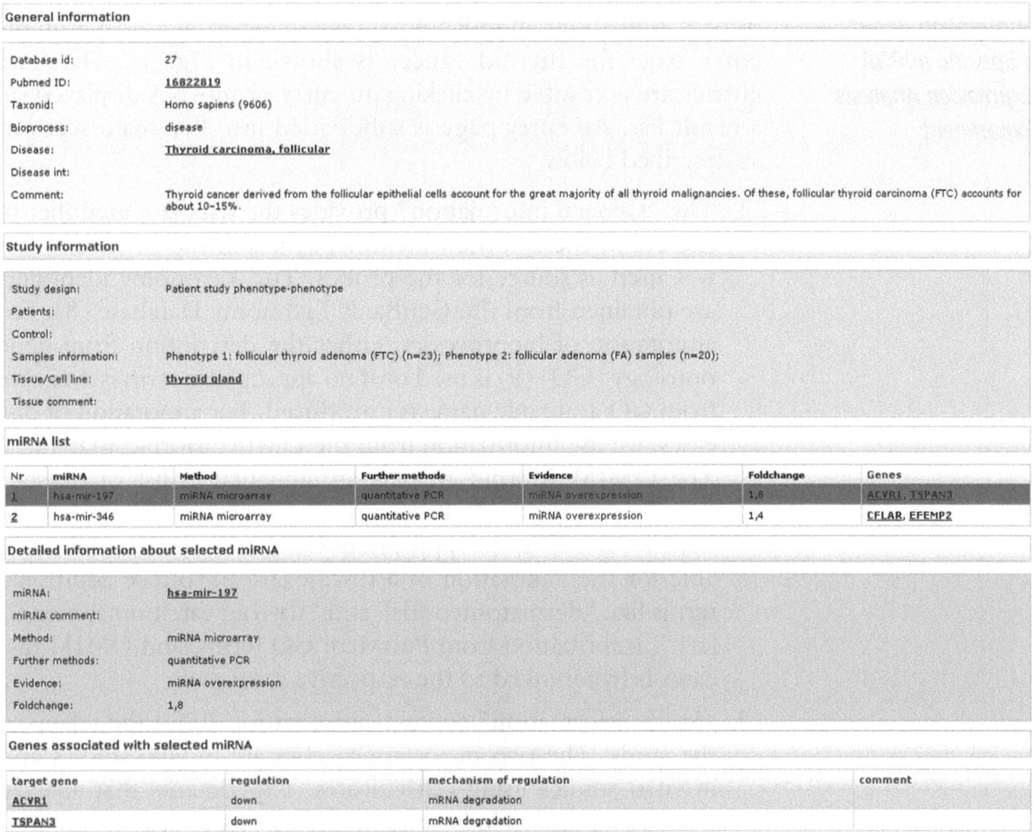

Fig. 1. (continued) (**c**) show all information that is annotated for an experiment.

analyses of lung development as well as of the diseases asthma and lung cancer. Results can be represented as list of entries or the list of miRNAs that were found to be differentially expressed in the respective experiments. The search results are hyperlinked to the respective annotation pages (Subheading 3.3). In conclusion, the "General search option" allows capturing a broad area of interest and is rather optimized for completeness than for specificity.

2. The "Specific search" allows the selection of individual annotated attributes shown in a pull-down menu. The list of attributes contains 18 topics, like miRNA name, disease, or study design. In addition to the "General search", the "Specific search" also offers the option to perform combined searches using "and", "or", and "and not" as operators. The specificity of the search using "Strict" or "Fuzzy" and the display of the search results are similar to the "General search".

3.3. Finding Information About a Specific miRNA Expression Analysis Experiment

The PhenomiR entry pages are the source of the annotated information about an miRNA expression experiment. An example entry page for thyroid cancer is shown in Fig. 1c. Database entries are accessible by clicking an entry or miRNA displayed on a result list. An entry page is subdivided into five main sections as described below.

1. The "General information" provides the database identifier of the entry, the PubMed identifier (7), and the organism that was used as source for the probes. The Taxonomy identifiers are obtained from the GenBank Taxonomy Database (8). For annotation of bioprocesses, either the description from gene ontology (GO) (9) is used or if no appropriate term is available from GO a suitable name is introduced. For annotation of diseases, we use information from the OMIM Morbid Map (10). The OMIM Morbid Map is an alphabetical list of diseases described in OMIM, including their corresponding cytogenetic locations. If no appropriate OMIM disease term is available for the annotation of a disease, we introduce additional terms like "dermatomyositis" and "thyroid carcinoma, medullary". Information from PubMed, GO terms, and OMIM diseases is hyperlinked to the respective resources.
2. "Study information" contains information about the set-up of the study. The two major approaches are in vivo studies and in vitro studies using cell cultures. Experiments that analyze miRNA expression in human beings are subdivided into studies, where patients are compared to healthy individuals, and studies, where for example different stages of a disease are being compared (e.g. PhenomiR database ID 141). There, same subdivision is applied for cell culture studies.

   In addition, PhenomiR provides cohort information that specifies the number of patients analyzed in a study and thus determines the statistical significance of the data. Peculiarities concerning the probes, such as preparation of specimens or treatment of cell lines with chemical compounds, are given in the "Samples information". Cell lines or tissues that were used as samples in the analyses are annotated using the BrendaTissueOntology (BTO) (11). Like sample information, tissues can also be supplemented with additional experimental details.
3. The "miRNA list" provides a list of all miRNAs that were up- or downregulated in the respective entry. The miRNA nomenclature is used according to miRBase that includes encoded information about the organism, e.g. hsa-miR-144 for the human miR-144. This has to be considered for searches using the "exact" mode. In addition, PhenomiR provides the experimental methods used for miRNA expression analyses. Information about differential expression of miRNAs in PhenomiR is given as the

qualitative attributes "miRNA overexpression" or "miRNA downregulation". This information is the minimal requirement for the annotation of an experiment in PhenomiR. If the authors provided additional quantitative data about differential miRNA expression, the information is given in a separated column. Finally, if a study extends research of differential miRNA expression by investigating target genes for selected miRNAs, this information is also given in the "miRNA list" and is linked to additional information (see below). If the user selects an miRNA by clicking on it, detailed information about expression of this miRNA is presented in the "Detailed information about selected miRNA" box below the miRNA list.

4. Details about target genes of miRNAs are given in "Detailed information about selected miRNA" section. Information in this section is rather concise and restricted to the mode of regulation target gene and the mechanism of regulation, e.g. "translational repression" or "mRNA degradation and translational repression".
5. "Comments" include particularly interesting or noteworthy information that does not fit into other sections. This might concern relations between different cell types (entry 494), effector proteins of cellular processes (entry 510), or gender-specific observations (entry 564).

3.4. Installing a Local Copy of the MySQL Database

For local access to the phenotypic data available in PhenomiR, all data is available for download at http://mips.helmholtz-muenchen.de/phenomir/download/phenomir.zip and can be used for complex analyses. The following demonstration shows how to install a local copy of PhenomiR in a local MySQL relational database.

This protocol assumes that the user has installed MySQL and has permission to upload data and create databases and tables. The commands shown here are for demonstration use only – for the complete database documentation, see http://www.mysql.com/ for more information.

1. Download the current PhenomiR release files on the Web page.
2. Unzip the file.
3. Start the MySQL client using the command:

   mysql -u <user> -h <host> -p <password> <phenomir.sql>.

The data is now available on a local database and can be used for local queries, including all queries that the Web resource allows.

3.5. Analysis of miRNA Expression Data

Several studies have shown that deregulated miRNAs can be used as biomarkers or to classify human cancers. PhenomiR provides a comprehensive data from studies that investigate deregulation of miRNA expression in diseases and biological processes. The following example illustrates how PhenomiR can be used to identify

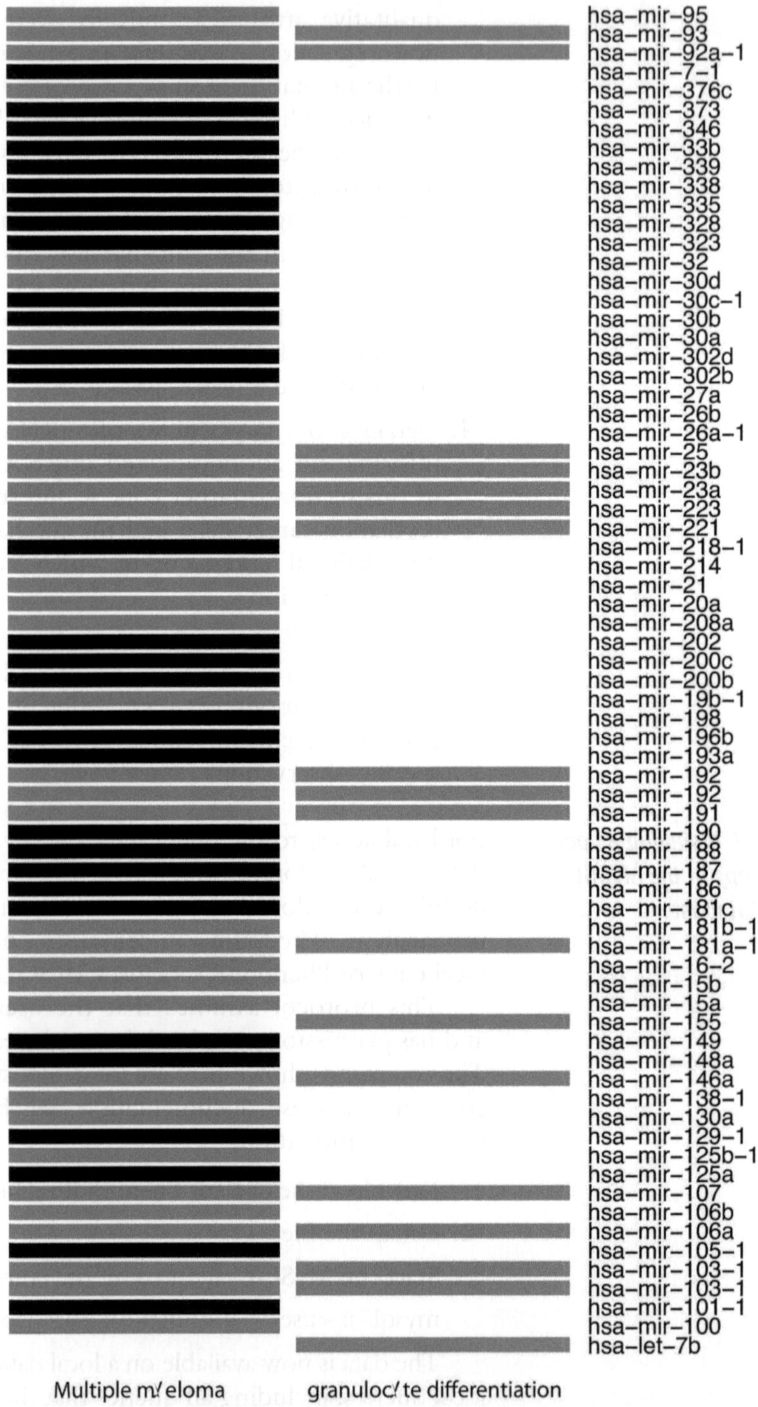

Fig. 2. Heatmap representation of miRNA expression analysis. The heatmap shows the microRNA expression profiles for multiple myeloma and granulocyte differentiation. Black bars indicate downregulated microRNAs and grey bars upregulated microRNAs. The result indicates that all downregulated microRNAs are rather disease state specific, whereas almost half of the upregulated microRNAs are also found to be expressed in the biological process of granulocyte differentiation.

a core set of myeloma-specific miRNAs by comparing disease and biological process-related miRNAs. This example assumes that the user has installed the R software environment for statistical computing. For documentation, see http://www.r-project.org/. The PhenomiR data used in this example can be downloaded at http://mips.helmholtz-muenchen.de/phenomir/download/phenomir.zip.

For access to the PhenomiR data in the R environment, prompt the command:

phenomir=read.delim('/DIR/phenomir.txt'),

whereas DIR should be replaced with the full path to the downloaded file. In the next step, one extract multiples myeloma-specific data using the command:

myeloma=phenomir[phenomir$id=='442'].

The corresponding biological process of granulocytes differentiation can be extracted using:

granulo=phenomir[phenomir$id=='545'].

In order to analyze both expression profiles, the data is combined:

df=merge(myeloma,granulo,by='miRNA',all = T).

In order to illustrate the expression profiles of multiples myeloma and granulocytes differentiation, the colour bar is set to red for downregulated miRNAs and green for upregulated:

hmcol=colorRampPalette(c('red','black','green'))(256).

Now the heatmap can be generated using the following command:

heatmap(as.matrix(df[,c('expr.x','expr.y')]),scale='none', Colv=NA,Rowv=NA,labRow=df[['miRNA']],margins=c(8,6), labCol=c('myeloma','granulo'), cexCol=2,col = hmcol).

The resulting heatmap is shown in Fig. 2 and can be used to distinguish between deregulated miRNAs in the disease state and the biological process. By comparing the expression profile, one obtains that all downregulated miRNAs are disease state specific, whereas almost half of the upregulated miRNAs are also found to be expressed in the biological process of granulocyte differentiation.

4. Notes

1. PhenomiR data structure and concepts: A large body of studies about miRNA expression in diseases and bioprocesses was published in recent years. The presentation of the results shows a large heterogeneity with respect to the completeness and the format. There are publications that show all data, including also miRNAs that are only marginally different in

their expression, and others limit the presented data according to an arbitrary threshold or top-ranked miRNAs. Downregulation of miRNA expression, e.g. for fivefold lower expression, is sometimes described as "–5"; other authors use "0.2" for the same result. In order to provide a high quality of the results, all information in PhenomiR is manually curated. Manual curation allows including all data in a correct format and also to include additional valuable information as described above (Subheading 3.3).

Publications may contain only one experiment, but 38% of studies investigate tissues from different diseases or a variety of cell lines. In PhenomiR, different experiments presented in one publication are annotated separately. Thus, each individual entry of the database refers to an instance of a publication describing a specific disease or bioprocess.

A design principle of PhenomiR is to use well-established ontologies and resources. This approach facilitates convenient post-processing of the data and allows to access additional information about the topic of interest via hyperlink. For annotation of miRNAs, we use miRBase as primary resource. In order to enable easy analysis of the dataset, miRNA designations from previous nomenclature releases were mapped to miRBase (12). Annotation of diseases is performed using the terms of the OMIM Morbid Map (10). OMIM is edited by experts in the field and contains information about the disease, such as clinical features, population genetics, and genes that are experimentally shown to be involved in the respective disease. If no appropriate disease name is available from OMIM, we introduce disease terms like "dermatomyositis" and "thyroid carcinoma, medullary". In order to enable analyses of the human disease network, Goh et al. introduced 22 high-level disease classes for OMIM diseases, such as cancer or cardiovascular (13). This annotation scheme is applied for the annotation of diseases in PhenomiR. Bioprocesses, such as developmental processes or response to environmental conditions, were also found to influence the expression of miRNAs. For the annotation of bioprocesses, we use terms of Gene Ontology or if required introduce additional terms like "monocytopoiesis" and 'tumor angiogenesis'.

For the evaluation of miRNA expression analyses, it important to know the used specimen. As was shown (6), results from patient studies and cell culture studies to some extent reveal different results for the same disease. In PhenomiR, cell lines and tissues are assigned according to the BTO.

A short description about the database content can be found on the PhenomiR home page via hyperlink of "About PhenomiR". Documentation of the search capabilities of the General search and the Specific Search and the data types that are queried by the General search are available via respective help buttons on the home page.

Search results in PhenomiR are not only presented as lists, but also as tag clouds to allow for visual and quick narrowing down of search results (Fig. 1b). Tag clouds display the relative abundance of terms that are associated with the query by a variable font size. A search for breast cancer shows that in this disease miR-21 and miR-9 are most frequently regulated, whereas in lung cancer this is the case for miR-1 and miR-7a. The terms of the tag cloud are hyperlinked to the respective list of results, e.g. the list of experiments, where miR-21 was deregulated in breast cancer.

2. PhenomiR release files and content: Every release of PhenomiR contains several database tables. These files are available from the Web page: http://mips.helmholtz-muenchen.de/phenomir/download/phenomir.zip.

   The README file contains important updated information regarding format changes and differences between releases.

   To obtain phenotypic annotation, you need at least the following four database tables: paper_mir, paper, mirs, disease. Mirs contains all data about miRNAs, such as name, sequence, and miRBase accession ID. The disease annotation contains disease names and Morbid Map identifiers. The paper table links the disease annotation and the corresponding miRNA phenotypic data, which are stored in the paper_mir relation. The paper_mir table contains all information about the deregulated miRNAs, such as expression, fold-change values, and detection method like microarray or QT-PCR.

   For more information about the different files and database content, please see the PhenomiR README file.

3. Contact information: The PhenomiR authors are keen to hear from users with suggestions for improvement of both the data and the Web interface. This feedback and any questions should be directed to andreas.ruepp@helmholtz-muenchen.de.

Acknowledgements

This work is funded in parts by the Federal Ministry of Education and Research (BMBF), its MedSys initiative (projects LungSys and SysMBo), the Helmholtz Alliance on Systems Biology (project CoReNe), and the German Research Foundation (InKoMBio, SPP1395).

References

1. Friedman, R. C., Farh, K. K., Burge, C. B., and Bartel, D. P. (2009) Most mammalian mRNAs are conserved targets of microRNAs. *Genome Res* **19**, 92–105.
2. Bishop, T. and Sham, P. (2001) *Analysis of Multifactorial Diseases* ELSEVIER.
3. Kim, Y. K., Yu, J., Han, T. S., Park, S. Y., Namkoong, B., Kim, D. H., et al. (2009) Functional links between clustered microRNAs: suppression of cell-cycle inhibitors by microRNA clusters in gastric cancer. *Nucleic Acids Res* **37**, 1672–81.
4. Mendell, J. T. (2008) miRiad roles for the miR-17-92 cluster in development and disease. *Cell* **133**, 217–22.
5. Baskerville, S. and Bartel, D. P. (2005) Microarray profiling of microRNAs reveals frequent coexpression with neighboring miRNAs and host genes. *RNA* **11**, 241–7.
6. Ruepp, A., Kowarsch, A., Schmidl, D., Buggenthin, F., Brauner, B., Dunger, I., et al. (2010) PhenomiR: a knowledgebase for microRNA expression in diseases and biological processes. *Genome Biol* **11**, R6.
7. Sayers, E. W., Barrett, T., Benson, D. A., Bolton, E., Bryant, S. H., Canese, K., et al. (2010) Database resources of the National Center for Biotechnology Information. *Nucleic Acids Res* **38**, D5–16.
8. Benson, D. A., Karsch-Mizrachi, I., Lipman, D. J., Ostell, J., and Sayers, E. W. (2010) GenBank. *Nucleic Acids Res* **38**, D46–51.
9. Ashburner, M., Ball, C. A., Blake, J. A., Botstein, D., Butler, H., Cherry, J. M., et al. (2000) Gene Ontology: tool for the unification of biology. *Nat Genet* **25**, 25–9.
10. Amberger, J., Bocchini, C. A., Scott, A. F., and Hamosh, A. (2009) McKusick's Online Mendelian Inheritance in Man (OMIM). *Nucleic Acids Res* **37**, D793–6.
11. Chang, A., Scheer, M., Grote, A., Schomburg, I., and Schomburg, D. (2009) BRENDA, AMENDA and FRENDA the enzyme information system: new content and tools in 2009. *Nucleic Acids Res* **37**, D588–92.
12. Griffiths-Jones, S., Saini, H. K., van, D. S., and Enright, A. J. (2008) miRBase: tools for microRNA genomics. *Nucleic Acids Res* **36**, D154–8.
13. Goh, K. I., Cusick, M. E., Valle, D., Childs, B., Vidal, M., and Barabasi, A. L. (2007) The human disease network. *Proc Natl Acad Sci USA* **104**, 8685–90.

Chapter 18

miRNA Expression Profiling: From Reference Genes to Global Mean Normalization

Barbara D'haene, Pieter Mestdagh, Jan Hellemans, and Jo Vandesompele

Abstract

MicroRNAs (miRNAs) are an important class of gene regulators, acting on several aspects of cellular function such as differentiation, cell cycle control, and stemness. These master regulators constitute an invaluable source of biomarkers, and several miRNA signatures correlating with patient diagnosis, prognosis, and response to treatment have been identified. Within this exciting field of research, whole-genome RT-qPCR-based miRNA profiling in combination with a global mean normalization strategy has proven to be the most sensitive and accurate approach for high-throughput miRNA profiling (Mestdagh et al., Genome Biol 10:R64, 2009). In this chapter, we summarize the power of the previously described global mean normalization method in comparison to the multiple reference gene normalization method using the most stably expressed small RNA controls. In addition, we compare the original global mean method to a modified global mean normalization strategy based on the attribution of equal weight to each individual miRNA during normalization. This modified algorithm is implemented in Biogazelle's qbase^{PLUS} software and is presented here for the first time.

Key words: miRNA profiling, miRNA expression, RT-qPCR, Global mean normalization

1. Introduction

Accurate quantification of micoRNA (miRNA) gene expression is a major challenge in the field and largely depends on two factors, i.e., the technology used to measure miRNA expression and the choice of a proper normalization strategy. Several methods have been developed to quantify miRNA expression such as microarrays (1–4) bead-based flow-cytometry (5) and small-RNA sequencing (6–9). While these methods enable genome-wide miRNA

Jian-Bing Fan (ed.), *Next-Generation MicroRNA Expression Profiling Technology: Methods and Protocols*,
Methods in Molecular Biology, vol. 822, DOI 10.1007/978-1-61779-427-8_18, © Springer Science+Business Media, LLC 2012

expression profiling, they typically require substantial amounts of input RNA which precludes the use of single cells, small biopsies, or body fluids such as serum, plasma, urine, or sputum. While reverse transcription quantitative PCR (RT-qPCR) intrinsically has a much higher specificity and sensitivity, down to a single molecule, it requires some adjustments to enable quantification of small RNA molecules such as miRNAs. Mature miRNAs consist of 21–25 nt and are too short to serve as templates in a RT-qPCR reaction. Therefore, different modifications of the classical RT-qPCR workflow have been developed in order to allow RT-qPCR-based miRNA expression profiling. One approach relies on the use of a miRNA-specific stem-loop primer that hybridizes to the 3 end of the mature miRNA (10). The loop unfolds upon denaturation, providing an elongated template that can be used in a subsequent qPCR-reaction. Alternatively, the mature miRNA is polyadenylated, and a poly-T primer is used to initiate the RT-reaction (11). Next to sensitivity, RT-qPCR-based approaches have a superior specificity, linear dynamic range of quantification, and a high level of flexibility making RT-qPCR the gold standard for small RNA expression profiling. Importantly, the accuracy of the results obtained through RT-qPCR miRNA expression profiling is largely dependent on a proper normalization strategy (12, 13). Different variables, inherent to the RT-qPCR workflow need to be controlled for in order to distinguish true biological changes from technical variation. These include the amount of starting material, enzymatic efficiencies, and overall transcriptional activity (14). The use of multiple stable reference genes is generally accepted as the method of choice for RT-qPCR data normalization (14). These stable reference genes can be identified from a set of candidate reference genes in a pilot experiment on a selection of samples that are representative for the experimental conditions under investigation. Different algorithms, such as geNorm, allow ranking the candidate reference genes according to their stability and indicating the optimal number of reference genes required for accurate normalization of gene expression. In the case of miRNA expression profiling, only few candidate reference miRNAs have been identified (15). Typically, other small endogenous noncoding RNAs such as small nuclear (U6) and small nucleolar (U24, U26) RNAs have been used. In 2009, Mestdagh et al. (13) introduced the global mean normalization method to normalize data from RT-qPCR miRNA profiling studies in which a large number of miRNAs are tested per sample (e.g., whole miRNome). This method outperformed other normalization strategies commonly used at the time (e.g., multiple target reference normalization using endogenous small RNA controls) and it has been considered the gold standard method since.

2. Normalizing Genome-Wide miRNA Expression Data: The Global Mean

The global mean normalization method consists of three successive steps. First, all Cq values above a certain threshold are considered noise and are discarded from further analysis (based on an 12 cycle sample preamplification procedure following mageplex reverse transcription, we routinely use 32 as cut-off value). The arithmetic average Cq value is then calculated for each individual sample and subsequently subtracted from each individual Cq value for that sample. The procedure results in normalized expression values in logE scale (E being the base of the exponential amplification function, with 2 being a good estimate); the more negative, the higher a particular miRNA is expressed (Table 1). In this chapter, we summarize the power of the previously described global mean normalization method in comparison to the multiple reference gene normalization method using the most stably expressed small RNA controls (Table 2). In addition, we compare the original global

Table 1
Normalized expression values of a particular miRNA in two samples (log scale, global mean normalization)

	Sample 1	Sample 2
hsa-let-7a	−1.50	−2.32

- Sample 1: the fold difference of let-7a compared to the mean Cq = $2^{-(-1.50)}$ = 2.83
  - Sample 1: Let-7a is 2.83 times higher expressed than the mean
- Fold change for let-7a of sample 1 relative to sample 2 = $2^{-(-1.507-(-2.32))}$ = 0.57
- Fold change for let-7a of sample 2 relative to sample 1 = $2^{-(-2.32-(-1.50))}$ = 1.76

* Assuming equal Cq values for equal transcript numbers

Table 2
Stably expressed small RNA controls used as reference genes

Neuroblastoma	T-ALL	Sputum (non-)smokers
RNU24	RNU24	U6[a]
RNU44	RNU44	
RNU58A	RNU48	
RNU6B	RNU58A	
	U18	
	Z30	

[a]Only U6 was found to be expressed in all sputum samples

mean method to a modified global mean normalization strategy based on the attribution of equal weight to each individual miRNA during normalization. This modified algorithm is implemented in Biogazelle's qbase^{PLUS} software as of version 2.0 (http://www.qbaseplus.com) and is presented here for the first time.

The modified global mean normalization strategy conveniently generates normalized relative quantities (NRQ values) in a linear scale. These values can be obtained using four simple steps in qbase^{PLUS} (1) auto-exclusion of miRNAs below a certain expression level, (2) conversion of Cq values into relative quantities (RQs), (3) calculation of sample specific normalization factor (NF) as the geometric mean of the RQs of all expressed targets per sample, and (4) conversion of RQs into normalized RQs (NRQs) by dividing the RQs by the sample specific NF. Steps 2–4 are simultaneously performed if the user selects one of the two available global mean normalization strategies. The NF can be calculated based on the RQs of all expressed targets in the sample for which the normalization is determined without taking into account the other samples ("modified global mean normalization"). Alternatively, the normalization factor is calculated based on the RQs of the targets that are expressed in all samples ("modified global mean normalization on common targets"). The different strategies to normalize large-scale RT-qPCR miRNA profiling data are compared in a comprehensive manner by reanalysis of published datasets (13, 16). These datasets include expression profiles for 430 miRNAs and 18 controls in 61 neuroblastoma (NB) tumor samples, 366 miRNAs and 18 controls in 49T-cell acute lymphoblastic leukemia (T-ALL) samples, and 636 miRNAs and 19 controls in 32 cell-free sputum samples from never smokers, smokers with COPD, and smokers without airflow limitation. The performance of the different normalization strategies is assessed by (1) evaluating their ability to reduce the overall variation, (2) determining their power to extract true biological variation, and (3) estimating the ability to reduce the number of false-positive and false-negative calls.

2.1. Evaluation of the Ability to Reduce the Overall Experimental Variation

Any variation in gene expression levels is composed of both true biological and experimentally induced (technical) variations. The purpose of normalization is to reduce the technical variation within a dataset, enabling a better appreciation of the biological variation.

To measure normalization performance, we calculated the standard deviation (SD) for each individual miRNA across all samples within a given dataset upon applying different normalization procedures. Lower standard deviations denote better removal of experimentally induced noise. Cumulative distributions of the SDs for not normalized and normalized data using different methods allows comparison of different normalization methods (Fig. 1).

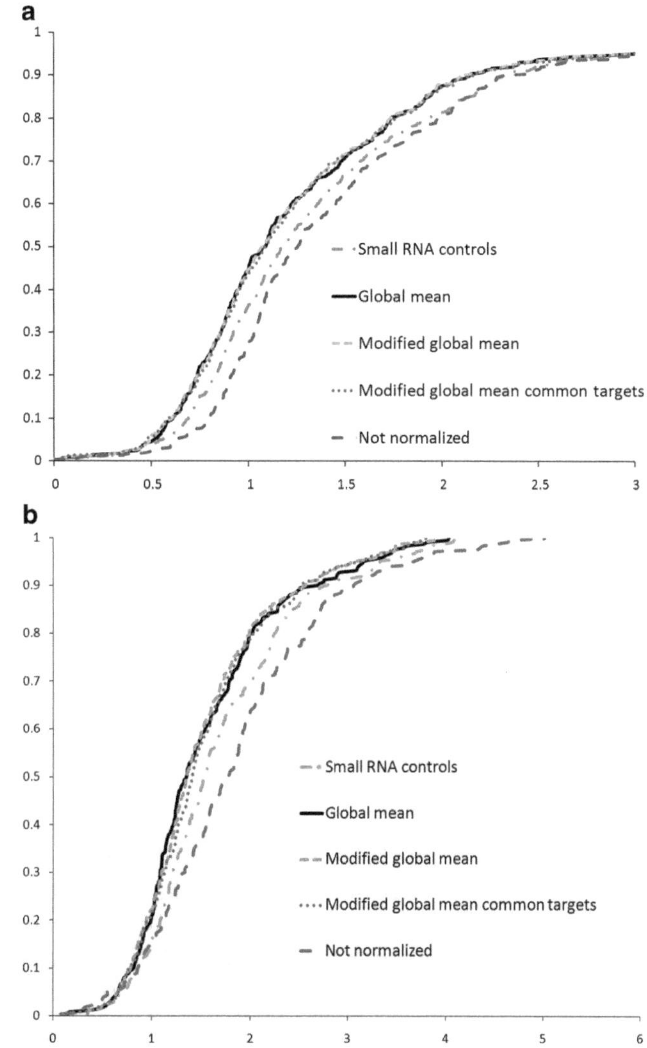

Fig. 1. Cumulative distribution of the SD values. (**a**) Neuroblastoma. (**b**) T-cell acute lymphoblastic leukemia (T-ALL). (**c**) Sputum (smokers and nonsmokers). Standard deviations (SDs) for each individual miRNA were calculated on $\log_2$ transformed relative expression levels normalized with (1) the original global mean normalization approach, (2) the modified global mean normalization, (3) the modified global mean normalization based on common targets, or (4) the multiple reference gene normalization using stably expressed small RNA controls. Standard deviations are also presented for not normalized expression data.

A shift to the left is discernable for all four normalization methods compared to not-normalized data, pointing at the intended reduction of the overall variation. The decrease in variation is least pronounced for multiple reference gene normalization using stable small RNA controls. The latter approach results in a small decrease of the SD values in the NB sample set for the 80% least variable

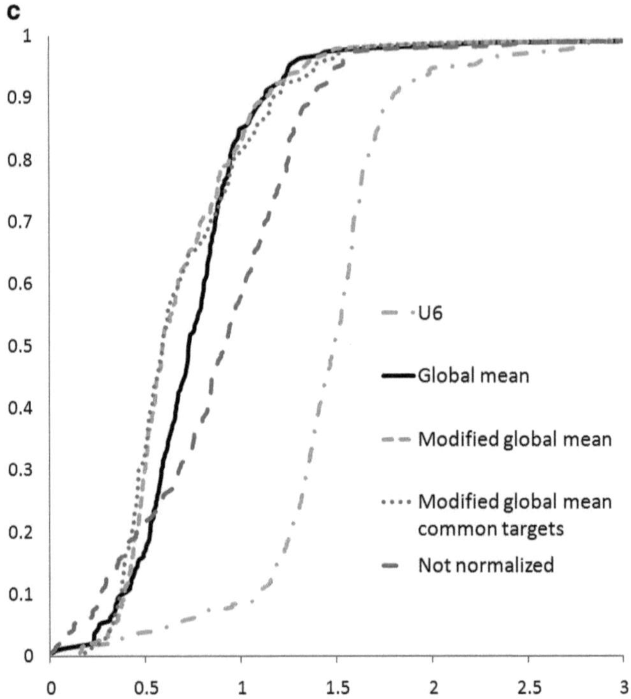

Fig. 1. (continued)

miRNAs. In the T-ALL sample set, it results in a pronounced decrease of the SD values for the 80% of most variable miRNAs. These observations indicate that elimination of technical variation is not effective for all miRNAs. Normalization using a single small RNA control (U6) in the sputum sample set even results in an increase of overall variability. In contrast, all three global mean normalization methods result in an overall decrease in variation that is (1) more pronounced compared to stable small RNA control normalization and (2) effective for all miRNAs that are measured.

In conclusion, all three global mean normalization-based methods are equally well suited to reduce the technical variation and outperform multiple reference gene normalization using stable small RNA controls.

2.2. Determination of the Power to Extract True Biological Variation

Good normalization approaches should not only reduce the technical variation as much as possible, but they should also accentuate true-biological differences. To assess the impact on appreciation of true biological differences, we evaluated the differential expression of the miRNAs belonging to the oncogenic mir-17-92 cluster in the NB data set. This cluster contains six miRNAs (miR-17, miR-18a, miR-19a, miR-20a, miR-19b, and miR-92) that are known to

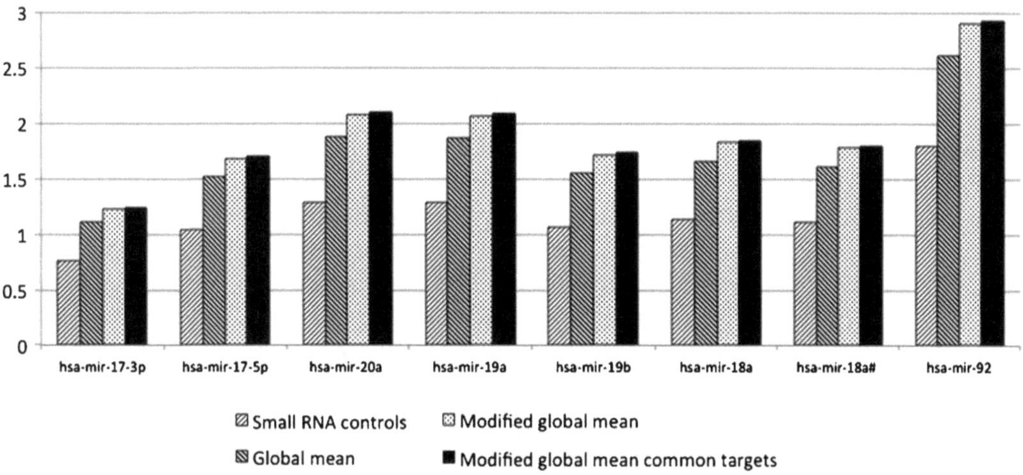

Fig. 2. miR-17-92 upregulation in MYCN-amplified neuroblastoma. Average fold change expression (linear scale) difference of six miRNAs residing within the miR-17-92 cluster in MYCN amplified neuroblastoma samples compared to MYCN single copy (MNSC) neuroblastoma samples. Fold changes were calculated upon data normalization with (1) the original global mean normalization approach, (2) the modified global mean normalization, (3) the modified global mean normalization based on common targets, (4) the multiple reference gene normalization using small RNA controls.

be upregulated in NB tumors with MYCN amplification (MNA) in comparison to samples with a normal MYCN copy number (MYCN single copy, MNSC).

The average fold change of the mir-17-92 cluster elements in MNA ($n=22$) compared to MNSC ($n=39$) samples was calculated upon normalization using (1) the original global mean normalization approach, (2) the modified global mean normalization, (3) the modified global mean normalization based on common targets, or (4) the multiple reference gene normalization using stable small RNA controls (Fig. 2). Normalized results were first log transformed, followed by calculation of the difference between the mean MNA and MNSC group values and exponentiation (anti-log) of the difference. When the data are normalized using the small RNA controls, only one out of eight miRNA genes within the miR-17-92 cluster reaches a 1.5-fold expression difference. In contrast, when the data are normalized using one of the global mean normalization methods seven out of eight miRNA transcripts reach at least a 1.5-fold expression difference. Both modified global mean normalization approaches result in true expression differences that are more pronounced in comparison to the original global mean normalization method; three out of eight miRNA transcripts reach a twofold expression difference as opposed to one out of eight, respectively. In conclusion, both the modified global mean normalization and the modified global mean normalization based on common targets perform slightly better than the original global mean normalization and clearly outperform the multiple reference gene normalization strategy using small RNA controls in appreciating true biological differences.

2.3. Balancing Up- and Down-regulated Genes

A fair assumption in most transcriptome-wide gene expression studies is that the number of up- and downregulated genes is approximately equal. To date, there is no biological evidence that the number of downregulated miRNAs is different from the number of upregulated miRNAs in MNA neuroblastoma samples compared to MNSC neuroblastoma samples and vice versa. It is hence acceptable to assume the existence of a balanced situation with an equal number of up- and downregulated miRNAs.

The overall differential miRNA expression in the two NB tumor sample subsets may give us an estimate of the ability of the normalization methods to reduce the number of false positives and negatives (Fig. 3, Table 3).

Normalization with small RNA controls suggests that most miRNAs are downregulated in MNA neuroblastoma samples. The global mean normalization-based methods result in a more balanced situation with an approximately equal number of up and downregulated miRNAs. Normalization with the original global mean normalization methods suggests that most miRNAs are

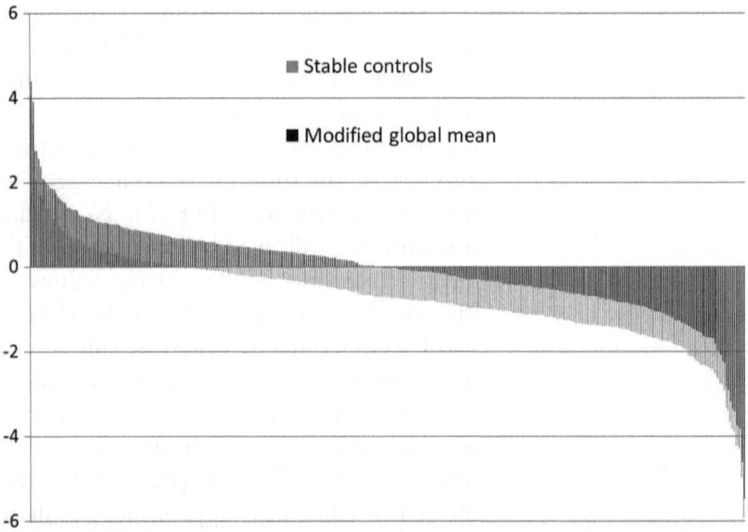

Fig. 3. Average fold change expression difference of each miRNA in neuroblastoma with respect to the MYCN amplification status. Average fold change expression difference ($\log_2$ scale) of each miRNA with a Cq value below 32 cycles in MYCN amplified neuroblastoma samples compared to MNSC neuroblastoma samples (differences calculated as explained for the miR-17-92 cluster in Fig. 2). Fold changes were calculated upon small RNA control normalization (*Light grey*) or modified global mean normalization (*Dark grey*). Fold changes are plotted in a $\log_2$ scale and sorted from positive (upregulated in MYCN amplified tumor samples) to negative (downregulated in MYCN amplified tumor samples).

Table 3
Number of up- and downregulated miRNAs in neuroblatoma with respect to the MYCN amplification status

Normalization	# Upregulated	# Downregulated
Small RNA controls	80	285
Original global mean	165	200
Modified global mean	180	185
Modified global mean common targets	181	184

Table 4
Number of up and downregulated miRNAs in sputum sample set (smokers and nonsmokers) with respect to the smoker status

Normalization	# Upregulated	# Downregulated
Small RNA controls (U6)	75	85
Original global mean	41	119
Modified global mean	70	90
Modified global mean common targets	72	88

downregulated in sputum samples from smokers (Table 4). The modified global mean normalization-based methods result in a more balanced situation with an approximately equal number of up and downregulated miRNAs. In conclusion, the modified global mean normalization-based methods result in a more balanced situation in which the number of downregulated miRNAs equals the number of upregulated miRNAs in neuroblastoma with respect to the MNA status. This clearly suggests that they have more power to reduce the number of false positives and false negatives in comparison to the multiple reference gene normalization using small RNA controls.

Considering all the results, it is fair to state that the modified global mean normalization methods are slightly better in terms of better appreciation of true biological changes. As biologically relevant miRNA expression differences can be quite small, we highly recommend applying the modified normalization method to obtain more accurate results for RT-qPCR miRNA profiling studies.

The impact of nonrandom missing data on calculating a global mean is reduced by removing the effect of differential expression between genes when calculating the modified global mean on relative quantities.

3. Multiple Reference Gene Normalization Using Small RNAs that Resemble the Global Mean

The global mean normalization methods are only valid for miRNA profiling studies in which a large number and unbiased set of genes are measured. This is because these methods are based on two assumptions (1) only a minority of miRNAs is differentially expressed and (2) the number of downregulated miRNAs is balanced against the number of upregulated miRNAs. Whole miRNome studies (using RT-qPCR, microarrays, or next-generation sequencing) often serve as a starting point in a pilot screen to identify differentially expressed miRNAs in certain subsets of samples. Subsequent studies on much larger sample groups are required to validate the statistical significance and to assess the biological relevance of the regulated miRNAs. In such RT-qPCR validation studies in which the expression of only a handful of miRNAs is measured, it is not valid to use a global mean normalization method.

To overcome this problem, Mestdagh et al. (13) proposed to use multiple stably expressed miRNAs or small RNA controls, identified by an expression pattern similar to the global mean level, referred to as genes resembling the mean expression value. Such reference genes can be identified from prior whole miRNome expression profiling studies or from a pilot experiment specifically performed to identify stably expressed genes resembling the mean. The results from Mestdagh et al. (13) indicate that a normalization factor based on the selection of miRNAs/small RNA controls resembling the mean expression value performs equally well compared to the mean expression value itself. Identification of such stably expressed reference miRNA/small RNA genes consists of two steps. First, candidate miRNAs/small RNA controls that resemble the mean expression value are identified, followed by selection of the most stably expressed reference from this group in a second step. In brief, the standard deviation (SD) for each individual miRNA across all samples is calculated after global mean normalization. The optimal set and number of miRNAs/small RNA controls for normalization is then determined through genorm^{PLUS} analysis of the ten best ranked candidate reference genes, being miRNAs or small RNA controls with the smallest standard deviation. To avoid the possibility of including co-regulated miRNAs in the genorm^{PLUS} analysis, miRNAs residing within the same gene cluster should be excluded, hereby retaining only one miRNA per cluster. For publically available RT-qPCR data, the miRNA body map enables selection of stably expressed miRNAs/small RNAs reference genes using the strategy outlined above (http://www.mirnabodymap.org).

4. Multiple Reference Gene Normalization Using Stably Expressed Small RNAs

If no prior whole miRNome expression profiling can be performed to identify genes whose expression pattern resembles the mean expression value, then a careful selection of the most stable small RNA controls should be performed in a typical geNorm pilot experiment (14).

Such a pilot experiment involves the analysis of a set of candidate reference small RNAs (preferentially more than 8, each belonging to a different family) (e.g., RNU X is not a good choice if RNU Y is already selected as candidate reference RNA) on a representative sample set (at least ten independent samples). A geNorm analysis determines the expression stability value for each gene (M value) and it calculates normalization factor V values. Both values are subsequently used to determine the optimal number and set of reference RNAs to be used in further studies. A useful tool is the genorm^{PLUS} module incorporated into qbase^{PLUS}, a substantial improvement over the old geNorm version running in Microsoft Excel, in terms of handling missing data, subranking of the best reference gene pair, fully automating the calculations, and providing interpretation of the results. The genorm^{PLUS} expert report enables straightforward interpretation as it provides recommendations on the number and nature of genes to be used for optimal normalization. In addition, it compares the stability of the proposed reference genes against empirically determined reference gene values (17).

Of note, Peltier and Latham (15) reported 2 miRNAs, miR-103, and miR-191 that were stably expressed in different normal and cancer tissues, suggesting that they may serve as universal reference miRNAs. In contrast, the reference miRNAs identified by Mestdagh et al. (13) varied substantially between different datasets. While miRNAs that are stably expressed across different tissues may exist, it is highly recommended to evaluate their stability in a selection of samples, representative for the entire sample set in the study. Please note that the results obtained through normalization with small RNAs should be interpreted with care. The side-by-side comparison of small RNA normalization with global mean normalization indicated that small RNAs are less efficient in reducing the technical variation and do not result in balanced expression differences. As biologically relevant miRNA expression differences can be quite small, it is not unconceivable that interesting expression differences will be missed.

Acknowledgment

This work was supported by the European Union Framework 7 project SysKid; Grant Number: 241544 (B. D'haene).

References

1. Castoldi M, Schmidt S, Benes V, Noerholm M, Kulozik AE, Hentze MW, et al. (2006) A sensitive array for microRNA expression profiling (miChip) based on locked nucleic acids (LNA). *RNA* **12**, 913–20.
2. Liu CG, Calin GA, Meloon B, Gamliel N, Sevignani C, Ferracin M, et al. (2004) An oligonucleotide microchip for genome-wide microRNA profiling in human and mouse tissues. *Proc Natl Acad Sci USA* **101**, 9740–4.
3. Nelson PT, Baldwin DA, Scearce LM, Oberholtzer JC, Tobias JW, and Mourelatos Z. (2004) Microarray-based, high-throughput gene expression profiling of microRNAs. *Nat Methods* **1**, 155–61.
4. Sioud M, and Rosok O. (2004) Profiling microRNA expression using sensitive cDNA probes and filter arrays. *Biotechniques* **37**, 574–6, 8–80.
5. Lu J, Getz G, Miska EA, Alvarez-Saavedra E, Lamb J, Peck D, et al. (2005) MicroRNA expression profiles classify human cancers. *Nature* **435**, 834–8.
6. Kuchenbauer F, Morin RD, Argiropoulos B, Petriv OI, Griffith M, Heuser M, et al. (2008) In-depth characterization of the microRNA transcriptome in a leukemia progression model. *Genome Res* **18**, 1787–97.
7. Morin RD, O'Connor MD, Griffith M, Kuchenbauer F, Delaney A, Prabhu AL, et al. (2008) Application of massively parallel sequencing to microRNA profiling and discovery in human embryonic stem cells. *Genome Res* **18**, 610–21.
8. Ramsingh G, Koboldt DC, Trissal M, Chiappinelli KB, Wylie T, Koul S, et al. (2010) Complete characterization of the microRNAome in a patient with acute myeloid leukemia. *Blood* **116**, 5316–26.
9. Schulte JH, Marschall T, Martin M, Rosenstiel P, Mestdagh P, Schlierf S, et al. (2010) Deep sequencing reveals differential expression of microRNAs in favorable versus unfavorable neuroblastoma. *Nucleic Acids Res* **38**, 5919–28.
10. Chen C, Ridzon DA, Broomer AJ, Zhou Z, Lee DH, Nguyen JT, et al. (2005) Real-time quantification of microRNAs by stem-loop RT-PCR. *Nucleic Acids Res* **33**, e179.
11. Shi R, and Chiang VL. (2005) Facile means for quantifying microRNA expression by real-time PCR. *Biotechniques* **39**, 519–25.
12. Brattelid T, Aarnes EK, Helgeland E, Guvag S, Eichele H, and Jonassen AK. (2010) The Normalization Strategy is Critical for the Outcome of miRNA Expression Analyses in the Rat Heart. *Physiol Genomics* doi: 10.1152/physiolgenomics.00131.2010.
13. Mestdagh P, Van Vlierberghe P, De Weer A, Muth D, Westermann F, Speleman F, et al. (2009) A novel and universal method for microRNA RT-qPCR data normalization. *Genome Biol* **10**, R64.
14. Vandesompele J, De Preter K, Pattyn F, Poppe B, Van Roy N, De Paepe A, et al. (2003) Accurate normalization of real-time quantitative RT-PCR data by geometric averaging of multiple internal control genes. *Genome Biol* **3**, RESEARCH0034.
15. Peltier HJ, and Latham GJ. (2008) Normalization of microRNA expression levels in quantitative RT-PCR assays: identification of suitable reference RNA targets in normal and cancerous human solid tissues. *RNA* **14**, 844–52.
16. Van Pottelberge GR, Mestdagh P, Bracke KR, Thas O, van Durme YM, Joos GF, et al. (2010) MicroRNA Expression in Induced Sputum of Smokers and Patients with Chronic Obstructive Pulmonary Disease. *Am J Respir Crit Care Med* 2010; 0: 201002–0304OCv1.
17. Hellemans J, Mortier G, De Paepe A, Speleman F, and Vandesompele J. (2007) qBase relative quantification framework and software for management and automated analysis of real-time quantitative PCR data. *Genome Biol* **8**, R19.

Chapter 19

miRNA Data Analysis: Next-Gen Sequencing

Preethi H. Gunaratne, Cristian Coarfa, Benjamin Soibam, and Arpit Tandon

Abstract

MicroRNAs (miRNAs) are short, noncoding RNAs that have the capacity to bind, capture, and silence hundreds of genes within and across diverse signaling pathways 1(Bartel, Cell 136:215–33, 2009) Specific sets of miRNAs characterize specific cell lineages of normal organisms and an increasing number of diseases have been shown to be associated with the dysregulation of specific miRNAs. Deep sequencing platforms have revealed unexpected complexity in relation to miRNAs, including 5′ and 3′-end-length heterogeneity and RNA editing. These insights not uncovered by previous microarray-based studies underscore the importance of data analysis tools that enable users to rapidly and easily analyze the unprecedented amounts of small RNA sequencing data that is emerging from next-generation sequencing platforms, such as Illumina/Solexa, SOLiD, and 454. In this chapter, we summarize the increasing number of analysis platforms that are available for miRNA discovery and profiling and the identification of functional miRNA–mRNA pairs in the context of biology and disease. We also discuss in greater detail our contributions to this effort.

Key words: miRDeep, miRNAkey, UEAsRNAtoolkit, miRanalyzer, SeqBuster, DSAP, mirTools, E-miR, SigTerms

1. Introduction

MicroRNAs (miRNAs) are small ~22-nt noncoding RNAs that can integrate multiple genes across and within signaling pathways to establish gene networks that underlie specific cellular contexts and biological processes (1). Historically, genome-wide analysis of miRNA expression has been performed through microarrays (2, 3). Following the advent of next-generation sequencing technologies, this effort is now largely sequencing based (4, 5). An increasing number of databases and analysis tools that can support both high-throughput analysis of millions of sequences from small RNA

Jian-Bing Fan (ed.), *Next-Generation MicroRNA Expression Profiling Technology: Methods and Protocols*, Methods in Molecular Biology, vol. 822, DOI 10.1007/978-1-61779-427-8_19, © Springer Science+Business Media, LLC 2012

(sRNA)-seq experiments as well as cross-correlational platforms that can integrate miRNA expression data with genome-wide gene expression data from the same treatments, cell types, or disease samples have been developed and published. This chapter discusses a number of these platforms with a focus on our own miRNA analysis and novel miRNA discovery platform (6) as well as miRNA–mRNA integration platform (7).

An optimal data analysis platform for miRNAs require the following functionalities.

1. Cleanup – Remove adaptor, contaminating *Escherichia coli*, <10 nt, >9 homopolymer runs, etc.
2. Determine nonredundant set of sequences.
3. Mapping to focused transcriptomes.
   (a) miRBase – 5′, 3′-end heterogeneity and RNA editing.
   (b) piRNA databases.
4. Mapping to the genome and intersecting with known annotations.
   (a) Problem of multiply mapped reads.
5. Differential expression analysis and graphic interface.
6. miRNA–mRNA enrichment.
7. Integrate into pathways/GO.

After the determination of miRNA profiles, a number of databases are now available for understanding and comparing miRNA profiles across a broad spectrum of tissues and diseases. miRBase (http://www.mirbase.org/) is currently the repository for miRNAs. This database which is updated regularly stores information on the mature miRNA sequences, precursor sequences, map locations, and overlapping annotations as well as predicted targets and a complete list of all publications that support each of the miRNA entries. An increasing number of species are included in every new miRBase version which is updated frequently (8). An increasing number of databases are being developed for predicting targets for each of the miRNAs (9). The commonly used ones include TargetScan (10), PicTar (11), and miRanda (12). The Mir2Disease-Base (13) is a database that extracts information from published literature and lists all the current disease associations for a given miRNA (13). The expression patterns of miRNAs across tissues can be obtained through GEO, miRGator (14), microRNA.org. Databases, such as miRSigDB (15) and miRGator, allow high-level integration to understanding how miRNAs expressed in a given sample relate to their putative targets in the context of signaling pathways. miRSigDB, miRGator, and recently published micro-RNA Expression and Sequence Analysis Database (mESAdb) allow miRNA expression and target gene expression to be linked with human diseases (16).

2. Materials

2.1. miRDeep

Link: http://www.mdc-berlin.de/en/research/research_teams/systems_biology_of_gene_regulatory_elements/projects/miRDeep).

Reference: Friedländer, M. R., Chen, W., Adamidi, C., Maaskola, J., Einspanier, R., Knespel, S., and Rajewsky, N. (2008) Discovering microRNAs from deep sequencing data using miRDeep. *Nat Biotechnol* **26**, 407–15.

2.2. miRNAkey

Link: http://ibis.tau.ac.il/miRNAkey/.

Reference: Ronen, R., Gan, I, Modai, S., Sukacheov, A., Dror, G., Halperin, E., and Shomron, N. (2010) miRNAkey: a software for microRNA Deep Sequencing analysis. *Bioinformatics* **26**, 2615–6.

2.3. UEAsRNAtoolkit

Link: http://srna-tools.cmp.uea.ac.uk/.

Reference: Moxon, S., Schwach, F., Dalmay, T., Maclean, D., Studholme, D. J., and Moulton, V. (2008) A toolkit for analyzing large-scale plant small RNA datasets. *Bioinformatics* **24**, 2252–3.

2.4. miRanalyzer

Link: http://web.bioinformatics.cicbiogune.es/microRNA/miRanalyser.php.

Reference: Hackenberg, M., Sturm, M., Langenberger, D., Manuel, J., Falcón-Pérez, and Aransay, A. M. (2009) miRanalyzer: a microRNA detection and analysis tool for next-generation sequencing experiments. *Nucleic Acids Res* **37**, W68–76.

2.5. SeqBuster

Link: http://estivill_lab.crg.es/seqbuster/.

Reference: Pantano, L., Estivill, X., Martí, E. (2010) SeqBuster, a bioinformatic tool for the processing and analysis of small RNAs datasets, reveals ubiquitous miRNA modifications in human embryonic cells. *Nucleic Acids Res* **38**, e34.

2.6. DSAP

Link: http://dsap.cgu.edu.tw/.

Huang, P. J., Liu, Y. C., Lee, C. C., Lin, W. C., Gan, R. R. C., Lyu, P. C., Tang, P. (2010) DSAP: Deep-Sequencing Small RNA Analysis Pipeline. *Nucleic Acids Res* **38**, W385–91.

2.7. mirTools

Link: http://centre.bioinformatics.zj.cn/mirtools/simple.php.

Reference: Zhu, E. L., Zhao, F. Q., Zhou, L. L., Hou, H. B., Xu G., Li, X. K., Bao, Q. Y, Sun, Z. S., and Wu, J. Y. (2010) mirTools: microRNA profiling and discovery based on high-throughput sequencing. *Nucleic Acids Research* **38**, W392–7.

2.8. E-miR

Link: http://www.lgtc.nl/services/general_services/software_tools/emir.php.

2.9. SigTerms

Link: http://SigTerms.sourceforge.net/.

Reference: Creighton, C. J., Nagaraja, A. K., Hanash, S. M., Matzuk, M. M., Gunaratne, P. H. (2008) A bioinformatics tool for linking gene expression profiling results with public databases of microRNA target predictions. *RNA* **14**, 2290–6.

3. Methods

3.1. miRDeep

The miRDeep package allows discovery of known or novel miRNAs from deep sequencing data. The package comprises a number of scripts to preprocess the mapped data, and a core algorithm that analyzes and scores these data. It requires Vienna package, Randfold application, and NCBI Blast package (which are all freely available). The input parameters can be varied with a single command line argument for custom trade-offs between sensitivity and specificity. The user can choose which potential precursor sequences to input to the core algorithm. These can be either sequences excised from the genome by miRDeep using the aligned reads as guidelines or custom sequences. After aligning reads to the genome, only a few hours on a standard Linux box are needed for genome-wide prediction using miRDeep.

miRDeep processes the deep sequencing data through a series of steps. The reads are aligned to the genome; genomic DNAs, bracketing the alignment, are extracted to compute their secondary structure. Potential precursors which are inconsistent with miRNA biogenesis are discarded. Then, it uses a simple probabilistic model of miRNA biogenesis to score each potential miRNA precursor for the combined compatibility of energy stability, positions, and frequencies of reads with Dicer processing. A number of factors, such as number of reads corresponding to the mature star sequence, the stability of the secondary structure, and conservation of the 5′ ends of the potential mature sequence, contribute to the score. Reads which do not map to rRNA, scRNA, snRNA, snoRNA, tRNA, 21U-RNAs, or protein-coding regions are passed through the miRDeep scoring algorithm. The output of miRDeep consists of a scored list of known and novel miRNA precursors and mature miRNAs in the deep sequencing data, as well as the estimates for the number of false positives.

Hence, miRDeep provides the users the capability to analyze their own deep sequencing data to detect known and novel miRNAs and at the same time estimate the quality of their results. miRDeep directly detects miRNAs by analyzing how sequenced RNAs are compatible with how miRNA precursors are processed in the cell.

3.2. miReduce

miReduce is a program that measures the relationship between gene expression fold change and regulatory motifs associated with these genes. For microRNA knockdown and overexpression studies, this program uses a linear regression model to determine the impact of each of these treatments on the target genes via the binding motifs in the 3′ UTRs of these genes. The model is based on linear regression, and is intended for genome-wide studies.

3.3. miRNAkey

miRNAkey is a software package designed as a base station for the analysis of miRNA sequencing data. miRNAkey takes FASTQ or FASTA format input files and generates a tabular output containing general and detailed reports on the sequence reads. miRNAkey also generates a table that compares each pair of samples. This table contains the combined observed miRNAs in the two samples, ranked from most differentially expressed to the least. However, miRNAkey does not provide tools for predicting new microRNAs. miRNAkey can run on a local Unix/Linux or Mac computer with 64-bit architecture. It has the following features.

1. Mapping to known miRNAs:

   The reads are mapped to known miRNA databases, such as miRBase, and mapped reads are counted for each miRNA species in each sample. These counts are converted into the normalized RPKM expression index (reads per kilobase per million mapped reads) to allow comparison across experiments. miRNAkey uses the SEQ-EM algorithm (17) to optimize the distribution of multiply aligned reads among the observed miRNAs, rather than discarding them, as is commonly done in this type of analysis.

2. Determining differentially expressed miRNAs:

   miRNAkey also supports quantification of differential expression for miRNAs between paired samples using chi-squared analysis, thus obtaining *P*-values for the differential expression of miRNAs. Other additional information regarding the input data, such as multiple mapping levels and post-clipping read lengths, can be generated through miRNAkey.

3.4. UEA sRNAtoolkit

This package provides different tools for the analysis of high-throughput, sRNA data. There are separate versions for plant and animal sRNA sequencing data. The UEA Web server consists of links to separate interfaces for different tools. The interface contains information on input file formats and allows the user to adjust parameters. Both plant and animal versions have the following tools.

1. Sequence file preprocessing tool: This tool reads in a FASTA or FASTQ format file along with 5′ (optional) and 3′ adaptor sequences and creates a FASTA format file with adaptors removed which can then be used with other tools on the site.

2. Filter tool: This tool filters sRNA sequences in FASTA files.
3. miRProf: It generates expression profiles of known miRNAs. PatMaN program is used to perform the searches of sRNAs against sequence databases. The user has an option of setting different parameters that would indicate matches to a known miRNA, such as mismatches, overhangs, etc.
4. miRCat: It inputs a FASTA file of sRNA reads, maps the reads to a reference genome, and identifies miRNAs. Genomic hit distribution patterns and secondary structure of genomic regions corresponding to sRNA hits are used for the prediction.
5. FiRePat: The sRNA expression dataset generated by the SiLoCo tool and gene expression levels from a gene expression assay, such as a microarray or digital tag expression profiling assay, are used to extract patterns in gene and sRNA expression datasets by this tool. These patterns may reveal functional targets of sRNAs.
6. SiLoCo: SiLoCo uses PatMaN program for the genome mapping of sRNA datasets (in FASTA format). By the relative position of sRNAs and their abundance, SiLoCo finds sRNA-producing loci. Only full-length perfect matches to the genome are considered. It can give output images that contain $\log_2$ sRNA expression ratio and the expression average for each locus. Multiple samples can be compared to identify differentially expressed loci.
7. SioMa: SiLoMa allows users to upload sRNAs data sets to be mapped to regions in the supported genomes or a user-provided sequence. The tool outputs text and image files, showing which sRNAs match and where the matches are on the reference sequence.
8. RNAfold with annotation: Vienna Package is used to generate the secondary structure of a long (up to 1 kb) RNA sequence. This structure is annotated by highlighting up to 20 short sequences. miRNA candidate sequences can be shown on a precursor hairpin. User can download the output in a PDF format.

The Plant Version Has Two Additional Tools

1. Targets tool: It takes as input a plain text FASTA format file of up to 50 miRNAs and runs target predictions against a chosen transcriptome dataset. The rules used for target prediction are based on those suggested by Allen et al. (18).
2. ta-siRNA prediction tool: This tool uses an algorithm proposed by Chen et al. (19) for calculating the probability of obtaining the observed percentage (or more) of phased sRNA matches by chance. An adjustable *P*-value cutoff is used to filter for loci with a significant degree of 21 nt phasing. In addition, a Web tool is provided to find target transcripts of sRNAs based on published rules for miRNAs and for highlighting sRNA locations on a folded RNA sequence.

3.5. miRanalyzer

This Web server tool analyzes sRNA sequencing data and requires a simple input file containing a list of unique reads and its copy numbers. The input file can be in two kinds of format: a tab-separated file with the read sequences and its counts and a multifasta file with read count as the description in the header. The output page includes multiple boxes that summarize information on the current state of the process, summary of the analysis of known microRNAs, matching of reads to several sets of transcribed sequences, summary of the detection of new microRNAs, and summary of the filtered and unmapped reads. There are links on each of the boxes that lead the user to more detailed results of each module. A Perl script for producing proper input files using the Genome Analyzer (Illumina Inc.) pipeline results is also available. The user can choose from a set of species to map the raw reads. The different modules available with miRanalyzer are described as follows.

1. Determine known miRNA sequences annotated in miRBase:
   Input reads which aligned to mature(miR), star (miR*), and hairpin sequences (pre-miR) in the miRBase repository are reported as known miRNA reads. The order of mapping against known miRNAs is mature, mature-star, unknown mature-star, and precursors/hairpin. Both unique matches and nonunique matches are allowed. Apart from the known mature-star sequences, a library with all other theoretically possible mature-star sequences is reported. To obtain unknown star sequences, the secondary structures for all hairpins using RNAfold is obtained. By means of location of mature miRNAs in the pre-miRNA hairpin, the information of the secondary structure and the characteristic "2-nt 3′ overhang" caused by Dicer, the corresponding sequence pairing with the mature miRNA, is extracted.
2. Ontological analysis:
   Annotation-Modules (20) are used to precompute the significant annotations of all target gene lists for all miRNAs from the miRanda software or TargetScan. In this way, precomputed ontological analyses are made available.
3. Determine other noncoding RNAs present:
   miRanalyzer maps the remaining reads to databases of transcribed sequences as mRNA, noncoding RNA, and (retro)-transposons.
4. Identify putative novel miRNAs:
   To detect new miRNAs, a machine learning algorithm is used. It is based on the WEKA (21) implementation of the *random forest* learning scheme (22) with the number of trees set to 100.

3.6. SeqBuster

SeqBuster is a Web-based toolkit to process and analyze large-scale sRNA datasets. The Web interface is based on Dynamic HTML (DHTML) and Common Gateway Interface (CGI) architecture. It consists of a preanalysis and an analysis component. The Perl language and R statistical package have been used for preanalysis and analysis modules, respectively. The users can upload R/perl-based packages to the server. In the preanalysis module, raw data are annotated to known miRNA and miRNA precursors, obtained from a miRNA and miRNA precursor databases or from a custom database. The processed and annotated data are stored in a MySQL database. The Web interface permits the analysis of the data using several R-based packages, such as comparative expression levels of individual sRNA loci between different samples and characterization of qualitative and quantitative miRNA variability (isomiRs). SeqBuster supports the visual display of results of each analysis through a DHTML format. The result can be stored in the server or downloaded to the local machine. Results data consist in a single table containing the name of every output file resulting from a specific analysis and may be saved permanently by the user. There is no functionality provided in SeqBuster for estimating novel miRNAs from the sequencing data. It was first applied to sRNA datasets of human embryonic stem cells. The available functions are described briefly below.

1. Precursor miRNA, miRNA, miRNA*, mRNA mapping:
   The precursor miRNA annotation allows one mismatch, three nucleotides in the 3′ addition variants, and the priority degree equal to 3. For miRNAs and miRNA*annotation, the following parameters are used: one mismatch, three nucleotides in the 3′ or 5′ trimming variants, three nucleotides in the 3′ addition variants, a priority degree equal to 1 and 2 for the miRNA and the miRNA* databases, respectively, and the parental database was the precursor miRNA database. The parameters for the alignment in mRNA and genome databases allows as much as one mismatch and up to three nucleotide additions in the 3′-terminus. The priority parameters are equal to 4 and 5 for the mRNA and genome databases, respectively.
2. IsomiR analysis of sequencing data:
   To characterize variants of miRNA, an IsomiR should have a frequency above 3 and should contribute in more than 10% to the total number of variants annotated in the same miRNA locus. A *Z*-score option is used to exclude sequencing errors as the possible cause of the nucleotide changes observed in some variants (23).
3. Determining differential expression:
   The Willcoxon test is used to determine statistically significant differences in the frequency distribution between samples.

For identifying differential expression profiles, a *Z*-test (24) is applied to show statistical significance in the differential expression; *P*-values are corrected by the Benjamini and Hochberg (25).

4. Function enrichment analysis:
   A TargetScan custom option (http://www.targetscan.org) is used to predict mRNA targets of the seed region isomiRs and miRNAs, which are differently expressed between libraries. Targets predicted by more than one enriched isomiRs or reference miRNAs are considered. Then, ingenuity pathway analysis (IPA) is used for the subsets of genes exclusively targeted by enriched isomiRs and those affected by the corresponding reference miRNAs. The *P*-value associated with a biological process is calculated with the right-tailed Fisher's exact test.

3.7. DSAP

DSAP is a Web server which provides tools to analyze sRNA sequencing data generated by SOLEXA. The dynamic Web interface is generated using the Perl CGI library, ChartDirector for Perl and Matrix2png. Data processing is performed using Perl and Linux shell scripts. The input to DSAP is a tab-delimited file consisting of all the unique sequence reads (tags) and their corresponding number of copies generated by the Solexa sequencing platform. The user can upload the input file through the DSAP Web interface and monitor the status of their jobs through a job status bar. The functions of DSAP suite include adaptor removal, clustering of tags, and classification of noncoding sRNAs and miRNAs basis on sequencing homology search against the Rfam and miRBase databases, respectively. DSAP also provides comparative miRNA expression profile analysis for up to five datasets. DSAP does not provide tools to predict unknown miRNAs. DSAP outputs a summary of the expression levels corresponding to matched ncRNAs and miRNAs in multicolor clickable bar charts linked to external databases. Other forms of output include $\log_2$-scaled color matrix display format for miRNA expression levels from different jobs and a cross-species comparative function to show the distribution of identified miRNAs in different species as deposited in miRBase. The different functionalities of DSAP are briefly described below.

1. Clustering:
   Sequence tags retained after the cleanup step with 100% sequence identity and identical sequence length are grouped as nonredundant sequence clusters. Each sequence cluster has a representative Cluster ID and its total read count. Supermatcher, based on the Smith–Waterman algorithm, from the EMBOSS (26) analysis package is used for the entire sequence alignment.
2. ncRNA matching:
   miRNA precursors from the Rfam database are used as a reference database to separate ncRNAs other than miRNAs in the

nonredundant sequence tag clusters. BLAST is applied to identify representative sequence clusters originating from rRNAs, tRNAs, snRNAs, snoRNAs, or other annotated ncRNAs (27).

3. Known miRNA matching:
   Representative sequence clusters remaining after ncRNA matching are compared with known mature miRNA sequences from miRBase with BLAST (hits with perfect alignments). Representative sequence clusters that show low sequence homology with known miRNAs are grouped as putative novel miRNAs.
4. Comparative miRNA analysis and cross-species distribution of miRNAs:
   DSAP displays non-normalized miRNA expression levels from different jobs or experimental results (in tab-delimited format) from other miRNA expression analyses, such as stem–loop real-time PCR, microarray, or SOLiD sequencing using a $\log_2$-transformed color matrix. DSAP can also show the distribution of identified miRNAs in different species from miRBase which provides a global view on the convergence and divergence of the identified miRNAs.

3.8. mirTools

Currently available tools, except mirTools, do not provide Web-based approaches to analyze multiple transcriptomes. miRTools was developed to characterize the sRNA transcriptome from deep sequencing data. mirTools is programmed in Perl and the Web server is hosted on an Apache 2.0 HTTP server under a Linux operating system. Users can submit raw data, a trimmed FASTA file obtained from preprocessed raw data, and access the results of the analysis through a front end which is implemented in PHP language scripts. mirTools also makes scripts available which are meant to remove adapters from raw reads. Its functionalities include classification of reads to known miRNAs, noncoding RNA, genomic repeats, or coding sequences; detailed annotation of known miRNAs, such as miRNA/miRNA*, absolute/relative reads count, and the most abundant tag; discovery of the novel miRNAs and identification of the differentially expressed miRNAs can be determined. These functionalities of miRTools are briefly described below.

1. Read filter:
   Low-quality reads are filtered out to exclude those most likely to represent sequencing errors and 3/5′ adaptor sequences. Reads are trimmed into clean, full-length reads and formatted into a nonredundant FASTA file. Each unique read is given a sequence tag and an expression profile (the number of counts).
2. sRNA annotation:
   All unique sequence tags are mapped onto the reference genome using the SOAP program (28). These unique sequence

tags are also aligned against miRBase (29), Rfam (8), repeat database produced by RepeatMasker (30), and the coding genes of the reference genome. This kind of mapping classifies the sequences into the following categories: known miRNA, degradation fragments of noncoding RNA, genomic repeats, and mRNA. In case of conflict, a hierarchy is conducted to assign the tag into a unique category, which starts with noncoding RNA, then known miRNA, followed by repeat-associated RNA and mRNA.

3. Differential expression detection:
The statistical significance (*P*-value) is inferred based on a Bayesian method (31). In default, a specific miRNA is deemed to be significantly differentially expressed when the *P*-value given by this method is ≤0.01 and there is at least a twofold change in normalized sequence counts.

4. Novel miRNA prediction:
Sequences that do not fall into above annotation categories but matched on the reference genome are used to detect candidate novel miRNA genes. In default, 100 nucleotides of genomic sequence flanking each side of these sequences are extracted and their RNA secondary structures are predicted using RNAfold (32). Novel miRNAs are identified by folding the flanking genomic sequence using the miRDeep program.

3.9. E-miR

E-miR provides scripts to generate an expression matrix for all know noncoding RNAs detected in the input data and reports transcripts per million and square root transformed expression levels. It generates Bed and Wig files for data visualization in the UCSC browser. E-miR does not include modules for predicting novel miRNA transcripts or other novel noncoding RNA transcripts and statistical inference of differential expression between (groups of) samples. The program can be downloaded from the Web page; E-miR does not host a Web server for the users to upload their data to be processed. E-miR offers the option of choosing between two different short read aligners, namely, Eland and Bowtie.

3.10. SigTerms

The public miRNA target prediction databases enable users to search the targets of one gene per query. Expression profiling can give to hundreds of differentially expressed genes; it is extremely cumbersome for users to manually associate hundreds of genes with the corresponding miRNAs targeting them. SigTerms integrates miRNA–mRNA target predictions with gene expression data on a large scale. It is desktop software application which, for a given target prediction database, retrieves all miRNA:mRNA functional pairs represented by an experimentally derived set of genes. Furthermore, for each miRNA, the software computes an enrichment statistic for overrepresentation of predicted targets within the gene set using Fisher's exact test. This can help to show roles for specific miRNAs and miRNA-regulated genes in the system under study. Currently, the software supports searching of results from PicTar, TargetScan, and miRanda algorithms. In addition, the software can accept any user-defined set of gene-to-class associations for searching, which can include the results of other target prediction algorithms, as well as gene annotation or gene-to-pathway associations. The software is a set of excel macros and very easy to use. Predicted targets by the three different algorithms are precompiled in the required format and are made available from the Web site. A more detailed section on how to use SigTerms is described in Subheading 4.

3.11. Small RNA Workbench

We are developing the Small RNA Workbench (http://genboree.org), a complex online system, powered by the Genboree discovery system, that enables the users to perform both primary and secondary sRNA analysis. The system enables users to formally characterize their own samples and upload the corresponding sequence files. In the first step, adapters are removed and a quality filter is applied; the result is one of representative compressed sRNA tags. Users can specify the adapter and control the quality filtering parameters. The resulting sequences are mapped against the target genome using software, such as Pash (33) and bwa (34), allowing up to 100 matching locations for each tag, with up to two mismatches. The mapping can be performed against a known assembled genome; a user has also the choice to upload his/her own genome definition. The following step is that of basic accounting of the sequencing tags, e.g., how many reads overlap with each category of features. By default, the Small RNA Workbench accounts for known sRNAs, such as miRNAs, piRNAs, sno/sca RNAs, and for well-defined genomic features, such as genes, CpG islands, and extensive categories of repeats. The users can also upload their own regions of interest and get a customized accounting of the sRNA tags. The results are made available in Excel spreadsheets for selected genomic feature sets and in annotation tracks. The tag density maps are computed and exposed for visualization in both the Genboree browser and also to the UCSC genome browser. The above steps conclude the primary analysis.

Often, however, users have biologically relevant questions that go beyond primary analysis. The Small RNA Workbench attempts to bridge the gap between primary analysis and testable hypothesis and publication quality figures in several ways. Typically, users would design an experiment combining multiple samples, with a collection of phenotypical labels. To enable impact analysis and leveraging of previous studies, the workbench enables users to import miRNA profiles from other studies. As a quick path to discovery, the workbench can generate basic heatmaps using R, using miRNA profiles data, and applying normalizations, such as *Z*-score normalization. In addition, it can generate PCA plots using a wide variety of clustering metrics by employing the QIIME package (35). To gain further insight, the workbench provides basic facilities for machine learning algorithms, such as *random forest* (22) and selection of discriminating miRNAs (36). Users can also perform general comparison of miRNA profiles between groups of samples, computing fold changes per miRNA, employing user-directed normalization schemes. It implements a key part of SigTerms, the miRNA–mRNA correlation. Specifically, using miRNA profiles, user-supplied gene lists, and known databases, such as PicTar, Miranda, and TargetScan, the workbench computes enrichment of miRNA–mRNA interactions. Finally, the workbench employs a state-of-the-art novel miRNA detection methods (6).

4. Notes

4.1. How to Use SigTerms?

The SigTerms software tool consists of the following Microsoft Excel macros for use in Excel.

1. FindSignificantTerms – Finds the significantly enriched term classes within a set of genes of interest.
2. CountTermToGene – Generates the Counts worksheet by adding up the term-to-gene associations from the Annotation worksheet.
3. DoSimulationTesting – Runs a number of random simulations in order to measure the true significance of a low-enrichment *P*-value for a term-to-gene set association.

To start up SigTerms, open the "SigTerms.xls" workbook in Excel. From the main menu, click "Tools," then "Macro," and "Macros." Select the name of the macro you wish to run. Click the "Run" button. A step-by-step procedure of using different macros is described below.

4.2. FindSignificantTerms

1. Open "SigTerms.xls" spreadsheet in Excel.
2. Open the appropriate Annotation workbook (see Subheading 4.1.1 for details) for your array.
3. Create a new blank spreadsheet in the Annotation workbook. Put your set of genes (the Entrez gene identifier) of interest in the first column of the new sheet.
4. Place the worksheet with the selected gene list at the front of any other spreadsheets. Run the "FindSignificantTerms" macro. On the displayed form, specify the total population

of genes for the purposes of computing enrichment *P*-values (see Subheading 4.1.2 for more details). The number of genes that are to be written out for each term can be limited by specifying a maximum *P*-value.

5. Hit the OK button. Two sheets are produced in the Annotation workbook: (a) the "Enriched Terms" sheet includes for each term the number of occurrences for the term in the gene set of interest and the probability (by one-sided Fisher's exact test) of finding the same number of occurrences or more of the term by chance and (b) the "Terms with Genes" sheet includes the same information as the "Enriched Terms" sheet, but in addition lists each gene that fell under a given term with that term, there being one row for each gene–term pair.

4.2.1. Annotation Workbook

The Annotation workbook contains two worksheets named "Counts" and the Annotation worksheet. The former lists each gene class term, along with the total number of times the term occurred in the Annotation worksheet.

The Annotation worksheet contains the gene-to-class associations in the following format.

1. Rows: Genes are listed starting from the second row, one gene per row.
2. Columns: The first and second columns may include any relevant information pertaining to the genes. The third and fourth columns list the Entrez identifier and the gene symbol (title description), respectively. Gene class associations for each gene are listed beginning from the fifth column. Starting from the fifth column, the top row lists the gene class type (e.g., "GO" or "TargetScan_pred").

4.2.2. Total Population of Genes for Computing Enrichment P-Values

The total number of genes of the population from which the user-specified gene set was selected can be obtained in the following ways.

1. If the user-specified gene set was collected from a set of profiling experiments, use the total number of unique-named genes (no duplicates, no genes, or ESTs without an Entrez identifier, and no RNA probes with ambiguous gene mappings) present in the array platform.
2. Use the total number of genes listed in the Annotation worksheet.

4.3. CountTermToGene

The SigTerms Web page hosts a set of precompiled Annotation workbooks for several types of gene class associations of potential interest (e.g., Gene Ontology annotations, miRNA targeting predictions, oncogenic signatures, etc.). If users create a new Annotation sheet, the "CountTermToGene" macro needs to be run in order to generate the Counts worksheet.

4.4. DoSimulation-Testing

1. Repeat steps 1–3 under Subheading 4.1.
2. Run the "DoSimulationTesting" macro. In the form displayed, enter the number of simulations you want to run and the number of genes you want to sample from your population in each test (the same number of genes that you had input into the "FindSignificantTerms" macro).
3. Hit the OK button on the form. One or more new sheets are generated in the current workbook. Each of the columns in these new sheets contains a set of *P*-values generated from a single simulation test (*P*-values greater than 0.05 for a simulation are not listed). These simulation results can be used in order to estimate the true significance of a *P*-value obtained from the set of genes of interest.

References

1. Bartel, D. P. (2009) MicroRNAs: Target recognition and regulatory functions. *Cell* **136**, 215–33.
2. Thomson, J. M., Parker, J., Perou, C. M., and Hammond, S. M. (2004) A custom microarray platform for analysis of microRNA gene expression. *Nat Methods* **1**, 47–53.
3. Miska, E. A., Alvarez-Saavedra, E., Townsend, M., Yoshii, A., Sestan, N., Rakic, P., et al. (2004) Microarray analysis of microRNA expression in the developing mammalian brain. *Genome Biol* **5**, R68.
4. Mardis, E. R. (2008) The impact of next-generation sequencing technology on genetics. *Trends Genet* **24**, 133–41.
5. Morozova, O. and Marra, M. A. (2008) Applications of next-generation sequencing technologies in functional genomics. *Genomics* **92**, 255–64.
6. Creighton, C. J., Reid, J. G., Gunaratne, P. H. (2009) Expression profiling of microRNAs by deep sequencing. *Brief Bioinformatics* **10**, 490–7.
7. Creighton, C. J., Nagaraja, A. K., Hanash, S. M., Matzuk, M. M., Gunaratne, P. H. (2008) A bioinformatics tool for linking gene expression profiling results with public databases of microRNA target predictions. *RNA* **14**, 2290–6.
8. Griffiths-Jones, S., Moxon, S., Marshall, M., Khanna, A., Eddy, S. R, and Bateman, A. (2005) Rfam: annotating non-coding RNAs in complete genomes. *Nucleic Acids Res* **33**, D121–4.
9. Sethupathy, P., Megraw, M., and Hatzigeorgiou, A. (2006) A guide through present computational approaches for the identification of mammalian microRNA targets. *Nat Methods* **3**, 881–6.
10. Lewis, B. P., Shih, I. H., Jones-Rhoades, M. W., Bartel, D. P., and Burge, C. B. (2003) Prediction of mammalian microRNA targets. *Cell* **26**, 787–98.
11. Krek, A., Grün, D., Poy, M., Wolf, R., Rosenberg, L., Epstein, E., et al. (2005) Combinatorial microRNA target predictions. *Nat Genet* **37**, 495–500.
12. Betel, D., Wilson, M., Gabow, A., Marks, D., and Sander, C. (2008) The microRNA.org resource: Targets and expression. *Nucleic Acids Res* **36**, D149–53.
13. Jiang, Q., Wang, Y., Hao, Y., Juan, L., Teng, M., Zhang, X., et al. (2009) miR2Disease: a manually curated database for microRNA deregulation in human disease. *Nucleic Acids Res* **37**, D98–104.
14. Nam, S., Kim, B., Shin, S., and Lee, S. (2008) miRGator: an integrated system for functional annotation of microRNAs. *Nucleic Acids Res* **36**, D159–64.
15. Subramanian, A., Tamayo, P., Mootha, V. K., Mukherjee, S., Ebert, B. L., Gillette, M. A., et al. (2005) Gene set enrichment analysis: a knowledge-based approach for interpreting genome-wide expression profiles. *Proc Natl Acad Sci USA* **102**, 15545–50.
16. Kaya, K. D., Karakülah, G., Yakicier, C. M., Acar, A. C., Konu, O. (2011) mESAdb: microRNA expression and sequence analysis database *Nucleic Acids Res* **39**, D170–80.

17. Pasaniuc, B., Zaitlen, N., and Halperin, E. (2010) Accurate estimation of expression levels of homologous genes in RNA-seq experiments. *Proceedings of the Fourteenth International Conference on Research in Computational Biology* 397–409.
18. Allen, E., Xie, Z., Gustafson, A. M., and Carrington, J. C. (2005) microRNA-directed phasing during trans-acting siRNA biogenesis in plants. *Cell* **121**, 207–21.
19. Chen, H. M., Li, Y. H., Wu, S. H. (2007) Bioinformatic prediction and experimental validation of a microRNA-directed tandem trans-acting siRNA cascade in Arabidopsis. *Proc Natl Acad Sci USA* **104**, 3318–23.
20. Hackenberg, M. and Matthiesen, R. (2008) Annotation-Modules: a tool for finding significant combinations of multisource annotations for gene lists. *Bioinformatics* **24**, 1386–93.
21. Witten, I. H. and Frank, E. (2005) Data Mining: practical machine learning tools and techniques. Morgan Kaufmann Publishers, San Francisco.
22. Breiman, L. (2001) Random forests. *Machine Learning* **45**, 28.
23. Dohm, J. C., Lottaz, C., Borodina, T., and Himmelbauer, H. (2008) Substantial biases in ultra-short read data sets from high-throughput DNA sequencing. *Nucleic Acids Res* **36**, e105.
24. Reinartz, J., Bruyns, E., Lin, J. Z., Burcham, T., Brenner, S., Bowen, B., et al. (2002) Massively parallel signature sequencing (MPSS) as a tool for in-depth quantitative gene expression profiling in all organisms. *Brief Funct Genomic Proteomic* **1**, 95–104.
25. Benjamini, Y. and Hochberg, Y. (1995) Controlling the false discovery rate: a practical and powerful approach to multiple testing. *J Roy Stat Soc Ser B* **57**, 289–300.
26. Mullan, L. J. and Bleasby, A. J. (2002) Short EMBOSS user guide. European molecular biology open software suite. *Brief Bioinformatics* **3**, 92–4.
27. Altschul, S. F., Madden, T. L., Schaffer, A. A., Zhang, J., Zhang, Z., Miller, W., et al. (1997) Gapped BLAST and PSI-BLAST: a new generation of protein database search programs. *Nucleic Acids Res* **25**, 3389–402.
28. Li, R., Yu, C., Li, Y., Lam, T. W, Yiu, S. M, Kristiansen, K., et al. (2009) SOAP2: an improved ultrafast tool for short read alignment. *Bioinformatics* **25**, 1966–7.
29. Griffiths-Jones, S., Saini, H. K., van Dongen, S., and Enright, A. J. (2008) miRBase: tools for microRNA genomics. *Nucleic Acids Res* **36**, D154–8.
30. Chen, N. (2004) Using RepeatMasker to identify repetitive elements in genomic sequences. *Curr Protoc Bioinform* **4**, 108.
31. Audic, S., and Claverie, J. M. (1997) The significance of digital gene expression profiles. *Genome Res* 7, 986–95.
32. Hofacker, I. L. (2003) Vienna RNA secondary structure server. *Nucleic Acids Res* **31**, 3429–31.
33. Coarfa, C., Yu, F., Miller, C. A., Chen, Z., Harris, R. A, Milosavljevic, A. (2010) Pash 3.0: A versatile software package for read mapping and integrative analysis of genomic and epigenomic variation using massively parallel DNA sequencing. *BMC Bioinformatics* **23**, 572.
34. Li, H. and Durbin, R. (2009) Fast and accurate short read alignment with Burrows-Wheeler transform. *Bioinformatics* **25**, 1754–60.
35. Caporaso, J. G., Kuczynski, J., Stombaugh, J., Bittinger, K., Bushman, F. D., Costello, E. K., et al. (2010) QIIME allows analysis of high-throughput community sequencing data. *Nat Methods* 7, 335–6.
36. Kursa, M. B. and Rudnicki, W. R. (2010) Feature selection with the Boruta package. *J Stat Softw* **36**, 1–13.

Chapter 20

Integrated miRNA Expression Analysis and Target Prediction

William Ritchie and John E.J. Rasko

Abstract

The accurate prediction of microRNA targets is essential to understanding their function. Commonly used software produces a prohibitive number of predicted targets for each microRNA. Here, we describe procedures that refine these predictions by integrating available software and expression data from experiments available online. These procedures are tailored to experiments, where predicting true targets is more important than detecting all putative targets. Our approach is tailored to the experimental biologist who seeks to identify a workable set of putative microRNA target genes for further characterization.

Key words: MicroRNA, Target genes, MicroRNA expression, Gene ontology, mimiRNA, Non-coding RNA, mRNA

1. Introduction

Bioinformatics prediction of microRNA targets has provided substantial insights into microRNA functions and their impact on gene regulation. Because the most efficient algorithms produce prohibitively large lists comprising hundreds of putative microRNA targets, complementary strategies can be adopted to reduce this list for time-consuming wet-lab testing.

Here, we present two strategies that increase the accuracy of predicted targets in exchange for sensitivity. The first technique called multitargeting (Fig. 1, 3.1) takes advantage of the fact that microRNAs that target the same gene multiple times can be detected with higher signal-to-noise ratios than those that target the same gene once (1). The second technique uses readily available expression data to identify mRNAs for which the expression appears to be dependent on changes in microRNA expression (Fig. 1, 3.2). The expression of microRNAs that inhibit mRNAs should be negatively correlated with the expression of their bona fide targets.

Jian-Bing Fan (ed.), *Next-Generation MicroRNA Expression Profiling Technology: Methods and Protocols*,
Methods in Molecular Biology, vol. 822, DOI 10.1007/978-1-61779-427-8_20, © Springer Science+Business Media, LLC 2012

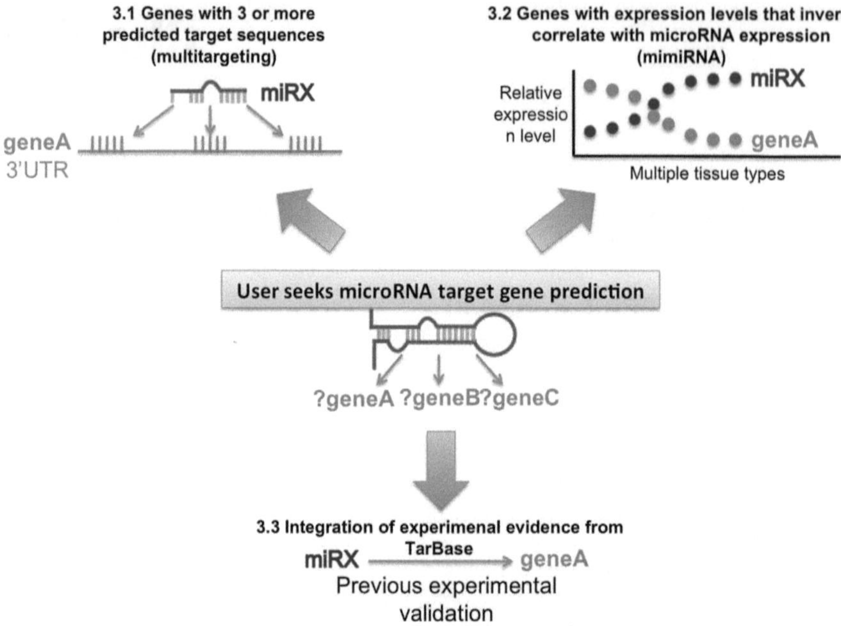

Fig. 1. Three independent approaches to discovering microRNA targets. This figure summarizes and cross references the three approaches designed to assist the experimental biologist in refining putative microRNA target gene predictions (detailed in the text). For a given microRNA "miRX," the investigator can take advantage of the fact that bone fide target gene UTRs often contain multiple predicted target sequences (3.1 multitargeting). Additionally, our online Web tool "mimiRNA" can be used to identify mRNAs for which the expression is inversely correlated with changes in microRNA expression (3.2 mimiRNA). Integration of search results with experimentally confirmed target genes is the final means of distilling data (3.3 integration).

In this guide, we use the TargetScan (2) algorithm and the miRBase version of the miRanda (3) algorithm. Because they use different criteria to predict targets, we consider these algorithms to be complementary and results from both should be considered for subsequent investigation. The principles underlying these two algorithms are explained in Subheading 4 (see Note 1). We also use the TarBase (4) Web site that compiles a list of experimental evidence supporting microRNA targets (Fig. 1, 3.3). This evidence ranges from the "gold standard" Luciferase assay to much weaker evidence, such as changes in mRNA expression subsequent to microRNA knockdown or knock-in. Finally, we use the mimiRNA (5) server that provides useful tools for the analysis of microRNA targets, such as expression profiling and multitargeting.

2. Materials

A computer with an Internet connection is required. The investigator begins with the official gene symbol of an mRNA gene of interest ("A1bg," for example) or the official miRBase (http://www.mirbase.org/)

name of a microRNA of interest ("has-mir-99a," for example). Microsoft Excel or its equivalent in free software, such as OpenOffice or StarOffice, is also required.

3. Methods

3.1. Using Multitargeting to Discover MicroRNA Targets (Miranda and TargetScan)

With TargetScan

1. Go to TargetScan (http://www.targetscan.org/). Select the species, paste or type your microRNA identifier in the section "Enter a microRNA name," and click "Submit."
2. In the results page, in the "Conserved sites" column, in the subcolumn named "total," search for the number 3 or above. This corresponds to genes with at least three conserved predicted sites for the inputted microRNA. The gene symbols corresponding to these rows in the left column are the targets of interest. If no such genes are found, the threshold should not be reduced to select genes with less sites because there is insufficient proof that this increases prediction accuracy. Users should instead use the same approach with the miRanda algorithm or use the approach described in Subheading 3.2.

With miRanda

1. Go to the miRBase implementation of miRanda (http://www.ebi.ac.uk/enright-srv/microcosm). Click on the Search button.
2. Select the genome, paste your microRNA identifier in the section "Enter microRNA id," and click "Search."
3. At the top of the Target Listing page, next to "Download Table," click on the "TXT" link. This allows you to download the results in text format. Your browser should prompt you to save this file. Save it in a convenient place on your computer.
4. Right click (Ctrl+ click on Mac) the saved folder and select "Open With," and then Excel. Make sure that Excel recognizes the individual columns (there should be 13 columns) of this file. Select the column with gene names (column "M"). Copy this column.
5. Go to mimiRNA's multitool (http://mimiRNA.centenary.org.au/mep/MultiT.htm) and paste the list in the text area next to "Input list of microRNAs or Genes." Click "Find multiple occurrences." The results page shows you genes with three or more targets for your microRNA. This function of mimiRNA can be used for lists of genes or microRNAs from any source.

3.2. Using Expression Data to Discover Targets

1. Go to mimiRNA (http://mimiRNA.centenary.org.au). Click option 5 "Which genes does my microRNA target."
2. Select your microRNA in the scroll-down menu. In the "Integrate with data from" section, select "targetScan" or "miRanda." This allows you to integrate predictions of these algorithms with those made by mimiRNA. Click "Find correlated genes."
3. The right column is a list of gene symbols for which the expression is negatively correlated with the input microRNA and are therefore potential targets. Gene symbols with a ">>>" symbol were also predicted targets according to TargetScan (or miRanda). Each gene can be clicked to go to its Entrez Gene description at the NCBI Web site.

3.3. Combining Experimental Evidence from TarBase

1. Go to mimiRNA's experimental evidence Web tool (http://mimiRNA.centenary.org.au/mep/expEv.htm). This tool combines an input list of targets with experimental evidence taken from the TarBase.
2. Select an organism and a microRNA of interest from the pull-down menus. If the microRNA of interest is not in the pull-down menu, this means that there is no experimental evidence supporting it from TarBase. In the text box next to "Input list of microRNA targets," paste a list of microRNA targets, each target separated by a new line. The input can be a list of targets found by miRanda as described in Subheading 3.1 or by any other target prediction program.
3. Click "Combine experimental evidence." This displays the common targets between the input list and targets with experimental evidence compiled in TarBase.

Investigators can further explore the type of evidence supporting targets by selecting the microRNA, its target gene, and organism at http://diana.cslab.ece.ntua.gr/tarbase/. Clicking on the "+" sign next to "Experimental Conditions" on the results page shows if the evidence is strong (existence of Direct Support) or weak (Indirect Support).

4. Notes

1. There are many available programs dedicated to microRNA target prediction. In this chapter, we use TargetScan and the miRBase version of miRanda for three main reasons. They have relatively low false-positive rates (<30%), they are hosted on reliable servers, but also the two approaches (and therefore the set of targets they predict) are different.

   The miRBase version of miRanda searches for exact matches between the 3′UTR of mRNA genes and the seed region (generally, nucleotide positions 2–9) of the microRNA (6). It then uses a thermodynamic model to evaluate whether the duplex formed between the whole microRNA sequence and the mRNA gene is stable. The microRNA/mRNA pairs for which these conditions are true in at least two other species are retained as predictions.

TargetScan v5 relies on the complementarity between the mRNA and the microRNA seed sequences. Each seed that binds to an mRNA sequence is attributed a score based on the degree of conservation of the region to which the seed binds, whether the complementarity to the seed region is 8 or 7 nucleotides long, and whether it has an A at position 1. TargetScan also calculates a context score, which depends on the AU composition of the flanking region and the distance of the target site to the 3′UTR end.

TargetScan is, therefore, biased toward conserved microRNA targets, whereas miRanda is biased toward microRNAs that bind with higher calculated affinity to their target.

2. mimiRNA is an online resource that integrates expression data at the time of writing from over 1,483 samples and permits visualization of the expression of 635 human microRNAs across 188 different tissues or cell types. mimiRNA uses a natural language classification algorithm to automatically pair microRNA and mRNA experiments from separate sources. This enables it to provide reliable expression profiles and to discover functional relations between microRNAs and mRNAs, such as microRNA targets. All the expression data can be downloaded from the mimiRNA Web site by clicking on the link next to "Download normalised microRNA expression data" and "Download normalised mRNA expression data." mimiRNA also incorporates a decision tree algorithm to discover distinguishing microRNA features between two tissue or cell types.

References

1. Ritchie, W., Flamant, S. and Rasko, J. E. (2009) Predicting microRNA targets and functions: traps for the unwary. *Nat Methods* **6**, 397–8.
2. Friedman, R. C., Farh, K. K., Burge, C. B. and Bartel, D. P. (2009) Most mammalian mRNAs are conserved targets of microRNAs. *Genome Res* **19**, 92–105.
3. John, B., Enright, A. J., Aravin, A., Tuschl, T., Sander, C. and Marks, D. S. (2004) Human MicroRNA targets. *PLoS Biol* **2**, e363.
4. Sethupathy, P., Corda, B. and Hatzigeorgiou, A. G. (2006) TarBase: A comprehensive database of experimentally supported animal microRNA targets. *RNA* **12**, 192–7.
5. Ritchie, W., Flamant, S. and Rasko, J. E. (2010) mimiRNA: a microRNA expression profiler and classification resource designed to identify functional correlations between microRNAs and their targets. *Bioinformatics* **26**, 223–7.
6. Lewis, B. P., Shih, I. H., Jones-Rhoades, M. W., Bartel, D. P. and Burge, C. B. (2003) Prediction of mammalian microRNA targets. *Cell* **115**, 787–98.

Chapter 21

miRNAs in Human Cancer

Xiaomin Zhong, George Coukos, and Lin Zhang

Abstract

MicroRNAs (miRNAs) are small (~18–25 nucleotides), endogenous, noncoding RNAs that regulate gene expression in a sequence-specific manner via the degradation of target mRNAs or the inhibition of protein translation. miRNAs are predicted to target up to one-third of all human mRNAs. Each miRNA can target hundreds of transcripts and proteins directly or indirectly, and more than one miRNA can converge on a single target transcript; thus, the potential regulatory circuitry afforded by miRNAs is enormous. Increasing evidence is revealing that the expression of miRNAs is deregulated in cancer. High-throughput miRNA quantification technologies provide powerful tools to study global miRNA profiles. It has become progressively more apparent that, although the number of miRNAs (~1,000) is much smaller than the number of protein-coding genes (~22,000), miRNA expression signatures more accurately reflect the developmental lineage and tissue origin of human cancers. Large-scale studies in human cancer have further demonstrated that miRNA expression signatures are associated not only with specific tumor subtypes but also with clinical outcomes.

Key words: MicroRNA, Noncoding RNA, Cancer

1. Introduction

Cancer is a disease involving multistep changes in the genome (1). Recent studies have focused mainly on protein-coding genes, and little is known about the alterations of functional noncoding sequences in cancer (2–4). MicroRNAs (miRNAs) are small (~18–25 nucleotide), endogenous, noncoding RNAs that regulate gene expression in a sequence-specific manner (5–8). Rapidly accumulating evidence indicates that miRNAs are involved in the initiation and progression of cancer in several ways. First, miRNAs act as key regulators of various fundamental biological processes that share common pathways with cancer, such as development, differentiation, apoptosis, and cell proliferation (5–7). Second, increasing evidence shows that the expression of miRNAs is markedly deregulated in cancer due to multiple genomic and epigenetic alterations (2, 9–22), and third, several miRNAs have been shown to serve as tumor

Jian-Bing Fan (ed.), *Next-Generation MicroRNA Expression Profiling Technology: Methods and Protocols*, Methods in Molecular Biology, vol. 822, DOI 10.1007/978-1-61779-427-8_21, © Springer Science+Business Media, LLC 2012

suppressor genes or oncogenes (3, 12, 19, 20, 23–34). The investigation of miRNAs in cancer may provide novel strategies for both the diagnosis and treatment of this disease.

2. Biogenesis of miRNAs

With the exception of those from the Alu repeat regions, which are transcribed by RNA polymerase III (Pol III) (35), most miRNA genes are derived from primary miRNA transcripts (pri-miRNAs) which are produced by Pol II and contain a 5′ cap and a poly(A) tail (36, 37). The pri-miRNA is cleaved within the nucleus by a multiprotein complex called Microprocessor, which is composed of the RNase III enzyme Drosha and the double-stranded RNA-binding domain (dsRBD) protein DGCR8/Pasha (38–42), into a ~70-nt hairpin precursor known as pre-miRNA. Next, the pre-miRNA is exported into the cytoplasm by Exportin-5 via a Ran-GTP-dependent mechanism (43–45). The pre-miRNA is further cleaved into a mature ~22-nt miRNA:miRNA* duplex by an RNase III enzyme, Dicer, in association with its partners, TRBP/Loquacious and PACT in human cells (46, 47). Subsequently, an RNA-induced silencing complex called RISC is assembled with the protein Argonaute (Ago) 2 (48, 49). The miRNA strand is selectively incorporated into the RISC complex (50, 51) and guides the complex specifically to its mRNA targets through base-pairing interactions (Fig. 1).

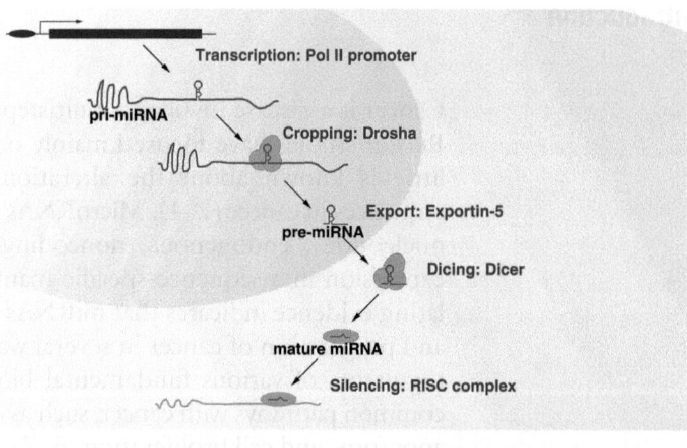

Fig. 1. Biogenesis of miRNAs. miRNA genes are derived from primary miRNA transcripts which are produced by Pol II and contain a 5′ cap and a poly(A) tail. The pri-miRNA is cleaved within the nucleus by a multiprotein complex called Microprocessor into a ~70-nt hairpin precursor known as pre-miRNA. Next, the pre-miRNA is exported into the cytoplasm by Exportin-5. The pre-miRNA is further cleaved into a mature ~22-nt miRNA:miRNA* duplex by an RNase III enzyme, Dicer. Subsequently, an RNA-induced silencing complex called RISC is assembled with the protein Argonaute (Ago) 2. The miRNA strand is selectively incorporated into the RISC complex and guides the complex specifically to its mRNA targets through base-pairing interactions.

3. miRNA Silencing Mechanisms

miRNAs can downregulate the expression of their target genes via two different mechanisms; the mechanism used depends on the complementarity between the miRNA and its target. miRNAs with perfect or near-perfect complementarity to the target sequence induce the cleavage and degradation of the transcript by initiating deadenylation and decapping of the mRNA (52). However, most miRNAs bind imperfectly to their target sequences and function by repressing protein translation. The underlying molecular mechanisms resulting in this repression have been studied intensively using in vitro cell-free systems (53). Recently, Wakiyama et al. established a cell-free system derived from human embryonic kidney (HEK) 293 cells and demonstrated that efficient miRNA-guided translational repression requires a m^7G-cap as well as a poly(A) tail (54); this is consistent with a previous report using a rabbit reticulocyte lysate system (55). In addition, a study utilizing extracts from mouse Krebs-2 ascites cells (56) showed that inhibition of translation initiation can be due to changes in ribosome recruitment to the mRNA as well as targeting of the mRNA cap structure. The dilemma of how the miRNA ribonucleoprotein complex (miRNPs) that is bound to the 3′ UTR of a target mRNA interferes with the initiation of translation was resolved by Kiriakidou et al. (57). They identified a motif (MC) within the Mid domain of Ago proteins, which bears significant similarity to the m^7G cap-binding domain of eIF4E, an essential translation initiation factor. In their model, the Ago proteins compete with the eIF4E for cap binding and thus repress the initiation of translation. Inhibition of translation initiation by *mir-2* was similarly observed in a cell-free system from *Drosophila* embryos. Interestingly, *mir-2* induced the formation of structures (heavier than 80S) known as "pseudo-polysomes," which resemble cytoplasmic processing bodies (P bodies) (58).

4. Oncomirs in Cancer

Changes in miRNA expression and function may contribute to the initiation and maintenance of tumors. Such miRNAs have been referred to as "oncomirs," and they may serve as both tumor suppressors and oncogenes (3). The first indication of miRNAs as tumor suppressors came from a report by Calin et al. where they found that patients diagnosed with B-cell chronic lymphocytic leukemia (B-CLL) have frequent deletions or downregulation of the *mir-15a* and *mir-16-1* genes on chromosome 13q14.3 (23).

A follow-up study demonstrated that *mir-15a* and *mir-16-1* negatively regulate the antiapoptotic protein BCL2 at a posttranscriptional level (59), and their function in leukaemogenesis and lymphomagenesis has been supported by additional studies (9, 10, 60). Some other miRNAs have also been shown to function as tumor suppressor genes, for example, the *let-7* family, which are negative regulators of RAS (26). Other miRNAs can function as oncogenes. The *mir-17-92* cluster was found by He et al. (25) to be upregulated in 65% of B-cell lymphomas, and its overexpression accelerated the development of malignant lymphomas in a transplantation mouse model. Upregulation of *mir-21* has been reported in glioblastomas (24) and breast cancer (61), where it exerts an antiapoptotic function. *mir-155* is remarkably overexpressed and linked to tumorigenesis (likely in cooperation with MYC) in pediatric Burkitt, Hodgkin, primary mediastinal, and diffuse large-B-cell lymphomas (62–65) as well as in breast cancer (61). In addition, *mir-372* and *mir-373* have been implicated as oncogenes in testicular germ cell tumors (28).

5. Deregulation of miRNAs in Cancer

The underlying mechanisms of miRNA deregulation in human cancer are not well understood, however, recent findings indicate that multiple processes are involved (21). It has been well documented that most primary miRNAs are transcribed from Pol II promoters that are regulated by transcription factors (5–8, 36, 37), and several examples of miRNA deregulation in cancer due to transcriptional deregulation have been reported (27, 29, 66–68). Recent studies also suggest that epigenetic alterations play a critical role in deregulating miRNA expression in human cancers (12, 21, 69), and that mutations may contribute to the downregulation of mature miRNAs (10). Over 50% of miRNAs are aligned to genomic fragile sites or regions associated with cancers (13), and several groups have provided evidence that DNA copy number abnormalities are involved in miRNA deregulation (14, 23, 25). Finally, the key proteins in the miRNA biogenesis pathway may be dysfunctional (70) or deregulated in cancer (71–74), thereby enhancing tumorigenesis (75). Thus, transcriptional deregulation, epigenetic alterations, mutations, DNA copy number abnormalities, and defects in the miRNA biogenesis machinery might contribute either alone, but more likely together, to the deregulation of miRNAs in human cancer.

Global expression of miRNAs is seemingly deregulated in most cancer types, according to reports from recent high-throughput studies (2, 9–11, 21, 76, 77). Interestingly, some studies suggest that miRNA expression may be widely downregulated in human tumors relative to normal tissues, as revealed by bead-based flow cytometry (9) and miRNA microarrays (21). However, other microarray studies have reported a tumor-specific pattern of down- and upregulation of miRNA genes (10, 11, 76). The selection of control samples may therefore be critical in the interpretation of these results. For example, normal ovaries are composed mainly of stroma, with smaller amounts of surface epithelium, thus the whole ovary may not serve as an optimal control for epithelial ovarian cancer studies (78). Recently, miRNA profiles in ovarian cancer were reported by several independent groups, using different control samples (17, 20, 21, 79). Although the overall conclusion of these studies is consistent, i.e., the expression of miRNAs is highly deregulated in ovarian cancer, the detailed expression patterns of each study are distinct, highlighting the challenge of determining an appropriate control tissue sample for profiling studies.

Transcriptional regulation is one of the key steps controlling the expression of miRNA, and several studies have demonstrated that miRNA deregulation in cancer can be due to changes in miRNA transcription. The expression of several miRNAs, including the oncogenic miRNA *mir-17-92* cluster, is regulated by the transcriptional factor c-Myc (27, 66). It has been shown that *c-Myc* is amplified and overexpressed in several types of human tumors, which suggests that this may contribute to the upregulation of *mir-17-92* in cancer (27, 66). In addition, the miRNA tumor suppressor *mir-34* is regulated by the transcription factor p53 (29, 80–84), and p53 inactivation is believed to decrease *mir-34* expression in human cancers (85). Finally, activation of the transcriptional factor HIF may be important to the upregulation of *mir-210* expression in human cancer (67, 68, 86). Taken together, an increasing amount of evidence indicates that transcriptional deregulation plays an important role in the deregulation of miRNAs in human cancer.

Epigenetic changes such as DNA methylation and histone modification play an important role in chromatin remodeling and in the general regulation of protein-coding gene expression in human cancer (87). Likewise, such mechanisms may also function to affect miRNA expression in cancer. To test this hypothesis, several groups treated cancer cell lines with DNA-demethylating reagents and/or histone deacetylase inhibitors in vitro and monitored miRNA expression by microarray analysis (12, 76, 88, 89). The results suggest that epigenetic alterations may play a critical role in regulating miRNA expression in human cancers, and therefore epigenetic treatments may provide novel strategies for cancer therapy.

Alterations in DNA copy number is one mechanism that can modify gene expression and function, and DNA dosage alterations in somatic cells are frequent contributors to cancer (90). The first example of a miRNA gene with an alteration in DNA copy number in cancer was reported in CLL patients. The genes *mir-16-1* and *mir-15a* on chromosome 13q14 were deleted in more than 50% of the CLL patients studied, with concurrent reduced expression in ~65% patients (23). Additional studies demonstrated that these two miRNAs suppress *BCL2* expression and may serve as tumor suppressor genes in this disease (59). Deletions of *mir-16-1* and *mir-15a* were later identified in epithelial tumors, such as pituitary adenomas (91), and in ovarian and breast cancers (14). In 2004, the amplification of *C13orf25* on chromosome 13q31-32 was first reported in lymphoma patients (92). Interestingly, this amplified region contains seven miRNAs as a polycistronic cluster, and there was an increased expression of primary and mature miRNAs derived from this locus in this type of lymphoma (25, 93). We now know that this miRNA cluster actually serves as an oncogene in human cancer (25, 94, 95) by altering the balance between cell death and proliferation via a *c-Myc* mediated pathway (27, 95). Using a bioinformatics based approach on data obtained from public databases, Calin et al. (13) compared 186 miRNA loci to the sequences of previously reported nonrandom genetic alterations and found that miRNA genes frequently reside in fragile sites, as well as in minimal regions of loss of heterozygosity, minimal regions of amplification, and common breakpoint regions. Recently, this finding was experimentally confirmed in an array-based comparative genomic hybridization (aCGH) study in 227 human tumors (14). A more recent study has suggested that a loss in genomic copy number may account for the downregulation of approximately 15% of miRNAs in advanced ovarian tumors (21). These findings support the notion that alterations in the DNA copy number of miRNA genes are highly prevalent in cancer and may account in part for miRNA gene deregulation.

6. miRNA-Based Cancer Therapy

The binding of miRNAs to their targets are governed by the rules of Watson–Crick base pairing. Therefore, an obvious molecule that could be used to inhibit an miRNA is an anti-miRNA oligonucleotide (AMO), which competitively blocks the interaction between the miRNA and its target (96). AMOs can be chemically modified in a variety of ways to improve their stability. One example is a locked nucleic acid (LNA), often referred to as inaccessible RNAs, which is a bicyclic high-affinity RNA analogue where the ribose moiety is chemically locked in an RNA-mimicking N-type (C3′-endo) conformation by the introduction of an extra 2′-O, 4′-C methylene bridge (97). The locked ribose conformation enhances base stacking and backbone pre-organization and significantly increases the thermal stability upon hybridization with complementary single-stranded RNA target molecules. In addition, LNAs are compatible with

RNase H cleavage and display high aqueous solubility and low toxicity in vivo (98). Other oligonucleotide analogues, such as morpholinos (99), 2′-*O*-methyl- (100), and 2′-*O*-methoxyethyl-modified (2′-MOE) oligonucleotides (101) have also been shown to be efficient in functionally inhibiting miRNAs. Besides chemical modifications, some improvements in inhibitor potency have been observed by increasing the length of the AMOs (102). Optimized secondary structural elements that flank the antisense core have also been shown to be highly potent and specifically block RISC activity in vitro for extended periods of time, thus suggesting structures surrounding or adjacent to the antisense core sequence are major determinants of inhibitor potency (103). In summary, a combination of optimization of sequences, structures, and/or chemical modifications may be required to produce a potent AMO.

Since protein-coding tumor suppressor genes can inhibit tumor growth, it has been proposed that restoring tumor suppressive miRNAs may also have an antitumorigenic effect. An example of miRNA replacement therapy is with *mir-15* and *mir-16*, which target BCL2 (10) and are often deleted in CLL patients (23). It has been reported that the transfection of *mir-15/16* expressing constructs resulted in the reduction of BCL2 protein levels and increased apoptosis in cancer cell lines. This study highlights the possibility of treating tumors displaying BCL2 overexpression by restoring *mir-15a* and *mir-16-1* expression. Another therapeutic candidate is *mir-124a*, whose expression is downregulated in acute lymphoblastic leukemia (ALL) due to hypermethylation of the promoter as well as histone modifications, resulting in an upregulation of the expression of target genes, including CDK6, and the phosphorylation of retinoblastoma (Rb). Accordingly, forced expression of *pre-mir-124a* led to decreased tumorigenicity in a xenogeneic mouse model of ALL (104). Mendell et al. have recently demonstrated that *mir-26a* in hepatocellular carcinoma (HCC) represents an additional example of a tumor suppressing miRNA, and that systemic administration of this miRNA using adeno-associated virus (AAV) in an animal model of HCC results in inhibition of cancer cell proliferation, induction of tumor-specific apoptosis, and significant protection from disease progression without toxicity (105).

Acknowledgments

This work was supported in part by grants from the Breast Cancer Alliance, the Ovarian Cancer Research Fund (Liz Tilberis Scholar), the Mary Kay Ash Charitable Foundation, the National Cancer Institute (R01CA142776 and Ovarian Cancer SPORE P50-CA83638-7951 project 3), and the US Department of Defense (W81XWH-10-1-0082).

References

1. Hanahan D, and Weinberg R A. (2000) The hallmarks of cancer, *Cell* **100**, 57–70.
2. Calin G A, and Croce C M. (2006) MicroRNA signatures in human cancers, *Nat Rev Cancer* **6**, 857–66.
3. Esquela-Kerscher A, and Slack F J. (2006) Oncomirs – microRNAs with a role in cancer, *Nat Rev Cancer* **6**, 259–69.
4. Calin G A, Liu C G, Ferracin M, Hyslop T, Spizzo R, Sevignani C., et al. (2007) Ultraconserved regions encoding ncRNAs are altered in human leukemias and carcinomas, *Cancer Cell* **12**, 215–29.
5. Bartel D P. (2004) MicroRNAs: genomics, biogenesis, mechanism, and function, *Cell* **116**, 281–97.
6. Ambros V. (2004) The functions of animal microRNAs, *Nature* **431**, 350–355.
7. He L, and Hannon G J. (2004) MicroRNAs: small RNAs with a big role in gene regulation, *Nat Rev Genet* **5**, 522–31.
8. Zamore P D, and Haley B. (2005) Ribo-gnome: the big world of small RNAs, *Science* **309**, 1519–24.
9. Lu J, Getz G, Miska E A, Alvarez-Saavedra E, Lamb J, Peck D., et al. (2005) MicroRNA expression profiles classify human cancers, *Nature* **435**, 834–8.
10. Calin G A, Ferracin M, Cimmino A, Di Leva G, Shimizu M, Wojcik S E., et al. (2005) A MicroRNA signature associated with prognosis and progression in chronic lymphocytic leukemia, *N Engl J Med* **353**, 1793–801.
11. Volinia S, Calin G A, Liu C G, Ambs S, Cimmino A, Petrocca F., et al. (2006) A microRNA expression signature of human solid tumors defines cancer gene targets, *Proc Natl Acad Sci USA* **103**, 2257–61.
12. Saito Y, Liang G, Egger G, Friedman J M, Chuang J C, Coetzee G A, et al. (2006) Specific activation of microRNA-127 with downregulation of the proto-oncogene BCL6 by chromatin-modifying drugs in human cancer cells, *Cancer Cell* **9**, 435–43.
13. Calin G A, Sevignani C, Dumitru C D, Hyslop T, Noch E, Yendamuri S, et al. (2004) Human microRNA genes are frequently located at fragile sites and genomic regions involved in cancers, *Proc Natl Acad Sci USA* **101**, 2999–3004.
14. Zhang L, Huang J, Yang N, Greshock J, Megraw M S, Giannakakis A, et al. (2006) microRNAs exhibit high frequency genomic alterations in human cancer, *Proc Natl Acad Sci USA* **103**, 9136–41.
15. Fazi F, Racanicchi S, Zardo G, Starnes L M, Mancini M, Travaglini L, et al. (2007) Epigenetic silencing of the myelopoiesis regulator microRNA-223 by the AML1/ETO oncoprotein, *Cancer Cell* **12**, 457–66.
16. Sevignani C, Calin G A, Nnadi S C, Shimizu M, Davuluri R V, Hyslop T, et al. (2007) MicroRNA genes are frequently located near mouse cancer susceptibility loci, *Proc Natl Acad Sci USA* **104**, 8017–22.
17. Iorio M V, Visone R, Di Leva G, Donati V, Petrocca F, Casalini P, et al. (2007) MicroRNA signatures in human ovarian cancer, *Cancer Res* **67**, 8699–707.
18. Lu L, Katsaros D, de la Longrais I A, Sochirca O, and Yu H. (2007) Hypermethylation of let-7a-3 in epithelial ovarian cancer is associated with low insulin-like growth factor-II expression and favorable prognosis, *Cancer Res* **67**, 10117–22.
19. Shell S, Park S M, Radjabi A R, Schickel R, Kistner E O, Jewell D A, et al. (2007) Let-7 expression defines two differentiation stages of cancer, *Proc Natl Acad Sci USA* **104**, 11400–5.
20. Yang H, Kong W, He L, Zhao J J, O'Donnell J D, Wang J, et al. (2008) MicroRNA expression profiling in human ovarian cancer: miR-214 induces cell survival and cisplatin resistance by targeting PTEN, *Cancer Res* **68**, 425–33.
21. Zhang L, Volinia S, Bonome T, Calin G A, Greshock J, Yang N, et al. (2008) Genomic and epigenetic alterations deregulate microRNA expression in human epithelial ovarian cancer, *Proc Natl Acad Sci USA* **105**, 7004–9.
22. Yang N, Kaur S, Volinia S, Greshock J, Lassus H, Hasegawa K, et al. (2008) MicroRNA microarray identifies Let-7i as a novel biomarker and therapeutic target in human epithelial ovarian cancer, *Cancer Res* **68**, 10307–14.
23. Calin G A, Dumitru C D, Shimizu M, Bichi R, Zupo S, Noch E, et al. (2002) Frequent deletions and down-regulation of micro-RNA genes miR15 and miR16 at 13q14 in chronic lymphocytic leukemia, *Proc Natl Acad Sci USA* **99**, 15524–9.
24. Chan J A, Krichevsky A M, and Kosik K S. (2005) MicroRNA-21 is an antiapoptotic factor in human glioblastoma cells, *Cancer Res* **65**, 6029–33.
25. He L, Thomson J M, Hemann M T, Hernando-Monge E, Mu D, et al. (2005) A microRNA polycistron as a potential human oncogene, *Nature* **435**, 828–33.

26. Johnson S M, Grosshans H, Shingara J, Byrom M, Jarvis R, Cheng A, et al. (2005) RAS is regulated by the let-7 microRNA family, *Cell* **120**, 635–47.
27. O'Donnell K A, Wentzel E A, Zeller K I, Dang C V, and Mendell J T. (2005) c-Myc-regulated microRNAs modulate E2F1 expression, *Nature* **435**, 839–43.
28. Voorhoeve P M, le Sage C, Schrier M, Gillis A J, Stoop H, Nagel R, et al. (2006) A genetic screen implicates miRNA-372 and miRNA-373 as oncogenes in testicular germ cell tumors, *Cell* **124**, 1169–81.
29. He L, He X, Lim L P, de Stanchina E, Xuan Z, Liang Y, et al. (2007) A microRNA component of the p53 tumour suppressor network, *Nature* **447**, 1130–4.
30. Mayr C, Hemann M T, and Bartel D P. (2007) Disrupting the pairing between let-7 and Hmga2 enhances oncogenic transformation, *Science* **315**, 1576–9.
31. Corney D C, Flesken-Nikitin A, Godwin A K, Wang W, and Nikitin A Y. (2007) MicroRNA-34b and MicroRNA-34c are targets of p53 and cooperate in control of cell proliferation and adhesion-independent growth, *Cancer Res* **67**, 8433–8.
32. Ma L, Teruya-Feldstein J, and Weinberg R A. (2007) Tumour invasion and metastasis initiated by microRNA-10b in breast cancer, *Nature* **449**, 682–8.
33. Tavazoie S F, Alarcon C, Oskarsson T, Padua D, Wang Q, Bos P D, et al. (2008) Endogenous human microRNAs that suppress breast cancer metastasis, *Nature* **451**, 147–52.
34. Varambally S, Cao Q, Mani R S, Shankar S, Wang X, Ateeq B, et al. (2008) Genomic loss of microRNA-101 leads to overexpression of histone methyltransferase EZH2 in cancer, *Science* **322**, 1695–9.
35. Borchert G M, Lanier W, and Davidson B L. (2006) RNA polymerase III transcribes human microRNAs, *Nat Struct Mol Biol* **13**, 1097–101.
36. Lee Y, Kim M, Han J, Yeom K H, Lee S, Baek S H, et al. (2004) MicroRNA genes are transcribed by RNA polymerase II, *EMBO J* **23**, 4051–60.
37. Cai X, Hagedorn C H, and Cullen B R. (2004) Human microRNAs are processed from capped, polyadenylated transcripts that can also function as mRNAs, *RNA* **10**, 1957–66.
38. Denli A M, Tops B B, Plasterk R H, Ketting R F, and Hannon G J. (2004) Processing of primary microRNAs by the Microprocessor complex, *Nature* **432**, 231–5.
39. Gregory R I, Yan K P, Amuthan G, Chendrimada T, Doratotaj B, Cooch N, et al. (2004) The Microprocessor complex mediates the genesis of microRNAs, *Nature* **432**, 235–40.
40. Curiel T J, Coukos G, Zou L, Alvarez X, Cheng P, Mottram P, et al. (2004) Specific recruitment of regulatory T cells in ovarian carcinoma fosters immune privilege and predicts reduced survival, *Nat Med* **10**, 942–9.
41. Landthaler M, Yalcin A, and Tuschl T. (2004) The human DiGeorge syndrome critical region gene 8 and Its D. melanogaster homolog are required for miRNA biogenesis, *Curr Biol* **14**, 2162–7.
42. Lee Y, Ahn C, Han J, Choi H, Kim J, Yim J, et al. (2003) The nuclear RNase III Drosha initiates microRNA processing, *Nature* **425**, 415–9.
43. Yi R, Qin Y, Macara I G, and Cullen B R. (2003) Exportin-5 mediates the nuclear export of pre-microRNAs and short hairpin RNAs, *Genes Dev* **17**, 3011–6.
44. Lund E, Guttinger S, Calado A, Dahlberg J E, and Kutay U. (2004) Nuclear export of microRNA precursors, *Science* **303**, 95–8.
45. Bohnsack M T, Czaplinski K, and Gorlich D. (2004) Exportin 5 is a RanGTP-dependent dsRNA-binding protein that mediates nuclear export of pre-miRNAs, *RNA* **10**, 185–91.
46. Hutvagner G, McLachlan J, Pasquinelli A E, Balint E, Tuschl T, and Zamore P D. (2001) A cellular function for the RNA-interference enzyme Dicer in the maturation of the let-7 small temporal RNA, *Science* **293**, 834–8.
47. Ketting R F, Fischer S E, Bernstein E, Sijen T, Hannon G J, and Plasterk R H. (2001) Dicer functions in RNA interference and in synthesis of small RNA involved in developmental timing in C. elegans, *Genes Dev* **15**, 2654–9.
48. Gregory R I, Chendrimada T P, Cooch N, and Shiekhattar R. (2005) Human RISC couples microRNA biogenesis and posttranscriptional gene silencing, *Cell* **123**, 631–40.
49. Maniataki E, and Mourelatos Z. (2005) A human, ATP-independent, RISC assembly machine fueled by pre-miRNA, *Genes Dev* **19**, 2979–90.
50. Schwarz D S, Hutvagner G, Du T, Xu Z, Aronin N, and Zamore P D. (2003) Asymmetry in the assembly of the RNAi enzyme complex, *Cell* **115**, 199–208.
51. Du T, and Zamore P D. (2005) microPrimer: the biogenesis and function of microRNA, *Development* **132**, 4645–52.
52. Pillai R S, Bhattacharyya S N, and Filipowicz W. (2007) Repression of protein synthesis by

miRNAs: how many mechanisms? *Trends Cell Biol* **17**, 118–26.

53. Standart N, and Jackson R J. (2007) MicroRNAs repress translation of m7Gppp-capped target mRNAs in vitro by inhibiting initiation and promoting deadenylation, *Genes Dev* **21**, 1975–82.
54. Wakiyama M, Takimoto K, Ohara O, and Yokoyama S. (2007) Let-7 microRNA-mediated mRNA deadenylation and translational repression in a mammalian cell-free system, *Genes Dev* **21**, 1857–62.
55. Wang B, Love T M, Call M E, Doench J G, and Novina C D. (2006) Recapitulation of short RNA-directed translational gene silencing in vitro, *Mol Cell* **22**, 553–60.
56. Mathonnet G, Fabian M R, Svitkin Y V, Parsyan A, Huck L, Murata T, et al. (2007) MicroRNA inhibition of translation initiation in vitro by targeting the cap-binding complex eIF4F, *Science* **317**, 1764–7.
57. Kiriakidou M, Tan G S, Lamprinaki S, De Planell-Saguer M, Nelson P T, and Mourelatos Z. (2007) An mRNA m7G cap binding-like motif within human Ago2 represses translation, *Cell* **129**, 1141–51.
58. Thermann R, and Hentze M W. (2007) Drosophila miR2 induces pseudo-polysomes and inhibits translation initiation, *Nature* **447**, 875–8.
59. Cimmino A, Calin G A, Fabbri M, Iorio M V, Ferracin M, Shimizu M, et al. (2005) miR-15 and miR-16 induce apoptosis by targeting BCL2, *Proc Natl Acad Sci USA* **102**, 13944–9.
60. Michael M Z, SM O C, van Holst Pellekaan N G, Young G P, and James R J. (2003) Reduced accumulation of specific microRNAs in colorectal neoplasia, *Mol Cancer Res* **1**, 882–91.
61. Iorio M V, Ferracin M, Liu C G, Veronese A, Spizzo R, Sabbioni S, et al. (2005) MicroRNA gene expression deregulation in human breast cancer, *Cancer Res* **65**, 7065–70.
62. van den Berg A, Kroesen B J, Kooistra K, de Jong D, Briggs J, Blokzijl T, et al. (2003) High expression of B-cell receptor inducible gene BIC in all subtypes of Hodgkin lymphoma, *Genes Chromosomes Cancer* **37**, 20–8.
63. Metzler M, Wilda M, Busch K, Viehmann S, and Borkhardt A. (2004) High expression of precursor microRNA-155/BIC RNA in children with Burkitt lymphoma, *Genes Chromosomes Cancer* **39**, 167–9.
64. Kluiver J, Poppema S, de Jong D, Blokzijl T, Harms G, Jacobs S, et al. (2005) BIC and miR-155 are highly expressed in Hodgkin, primary mediastinal and diffuse large B cell lymphomas, *J Pathol* **207**, 243–9.
65. Eis P S, Tam W, Sun L, Chadburn A, Li Z, Gomez M F, et al. (2005) Accumulation of miR-155 and BIC RNA in human B cell lymphomas, *Proc Natl Acad Sci USA* **102**, 3627–32.
66. Chang T C, Yu D, Lee Y S, Wentzel E A, Arking D E, West K M, et al. (2008) Widespread microRNA repression by Myc contributes to tumorigenesis, *Nat Genet* **40**, 43–50.
67. Giannakakis A, Sandaltzopoulos R, Greshock J, Liang S, Huang J, Hasegawa K, et al. (2007) miR-210 links hypoxia with cell cycle regulation and is deleted in human epithelial ovarian cancer, *Cancer Biol Ther* **7**.
68. Kulshreshtha R, Ferracin M, Wojcik S E, Garzon R, Alder H, Agosto-Perez F J, et al. (2007) A microRNA signature of hypoxia, *Mol Cell Biol* **27**, 1859–67.
69. Yang N, Coukos G, and Zhang L. (2008) MicroRNA epigenetic alterations in human cancer: one step forward in diagnosis and treatment, *Int J Cancer* **122**, 963–8.
70. Thomson J M, Newman M, Parker J S, Morin-Kensicki E M, Wright T, and Hammond S M. (2006) Extensive post-transcriptional regulation of microRNAs and its implications for cancer, *Genes Dev* **20**, 2202–7.
71. Muralidhar B, Goldstein L D, Ng G, Winder D M, Palmer R D, Gooding E L, et al. (2007) Global microRNA profiles in cervical squamous cell carcinoma depend on Drosha expression levels, *J Pathol* **212**, 368–77.
72. Chiosea S, Jelezcova E, Chandran U, Luo J, Mantha G, Sobol R W, et al. (2007) Overexpression of Dicer in precursor lesions of lung adenocarcinoma, *Cancer Res* **67**, 2345–50.
73. Chiosea S, Jelezcova E, Chandran U, Acquafondata M, McHale T, Sobol R W, et al. (2006) Up-regulation of dicer, a component of the MicroRNA machinery, in prostate adenocarcinoma, *Am J Pathol* **169**, 1812–20.
74. Karube Y, Tanaka H, Osada H, Tomida S, Tatematsu Y, Yanagisawa K, et al. (2005) Reduced expression of Dicer associated with poor prognosis in lung cancer patients, *Cancer Sci* **96**, 111–5.
75. Kumar M S, Lu J, Mercer K L, Golub T R, and Jacks T. (2007) Impaired microRNA processing enhances cellular transformation and tumorigenesis, *Nat Genet* **39**, 673–7.
76. Yanaihara N, Caplen N, Bowman E, Seike M, Kumamoto K, Yi M, et al. (2006) Unique

microRNA molecular profiles in lung cancer diagnosis and prognosis, *Cancer Cell* **9**, 189–98.

77. Cummins J M, He Y, Leary R J, Pagliarini R, Diaz L A, Jr., Sjoblom T, et al. (2006) The colorectal microRNAome, *Proc Natl Acad Sci USA* **103**, 3687–92.
78. Zorn K K, Jazaeri A A, Awtrey C S, Gardner G J, Mok S C, Boyd J, et al. (2003) Choice of normal ovarian control influences determination of differentially expressed genes in ovarian cancer expression profiling studies, *Clin Cancer Res* **9**, 4811–8.
79. Nam E J, Yoon H, Kim S W, Kim H, Kim Y T, Kim J H, et al. (2008) MicroRNA Expression Profiles in Serous Ovarian Carcinoma, *Clin Cancer Res* **14**, 2690–5.
80. Bommer G T, Gerin I, Feng Y, Kaczorowski A J, Kuick R, Love R E, et al. (2007) p53-Mediated Activation of miRNA34 Candidate Tumor-Suppressor Genes, *Curr Biol* **17**, 1298–307.
81. Chang T C, Wentzel E A, Kent O A, Ramachandran K, Mullendore M, Lee K H, et al. (2007) Transactivation of miR-34a by p53 broadly influences gene expression and promotes apoptosis, *Mol Cell* **26**, 745–52.
82. Raver-Shapira N, Marciano E, Meiri E, Spector Y, Rosenfeld N, Moskovits N, et al. (2007) Transcriptional activation of miR-34a contributes to p53-mediated apoptosis, *Mol Cell* **26**, 731–43.
83. Tarasov V, Jung P, Verdoodt B, Lodygin D, Epanchintsev A, Menssen A, et al. (2007) Differential regulation of microRNAs by p53 revealed by massively parallel sequencing: miR-34a is a p53 target that induces apoptosis and G1-arrest, *Cell Cycle* **6**, 1586–93.
84. Corney D C, Flesken-Nikitin A, Godwin A K, Wang W, and Nikitin A Y. (2007) MicroRNA-34b and MicroRNA-34c Are Targets of p53 and Cooperate in Control of Cell Proliferation and Adhesion-Independent Growth, *Cancer Res* **67**, 8433–8.
85. He L, He X, Lowe S W, and Hannon G J. (2007) microRNAs join the p53 network--another piece in the tumour-suppression puzzle, *Nat Rev Cancer* **7**, 819–22.
86. Camps C, Buffa F M, Colella S, Moore J, Sotiriou C, Sheldon H, et al. (2008) hsa-miR-210 Is induced by hypoxia and is an independent prognostic factor in breast cancer, *Clin Cancer Res* **14**, 1340–8.
87. Egger G, Liang G, Aparicio A, and Jones P A. (2004) Epigenetics in human disease and prospects for epigenetic therapy, *Nature* **429**, 457–63.
88. Scott G K, Mattie M D, Berger C E, Benz S C, and Benz C C. (2006) Rapid alteration of microRNA levels by histone deacetylase inhibition, *Cancer Res* **66**, 1277–81.
89. Diederichs S, and Haber D A. (2006) Sequence Variations of MicroRNAs in Human Cancer: Alterations in Predicted Secondary Structure Do Not Affect Processing, *Cancer Res* **66**, 6097–104.
90. Pinkel D, and Albertson D G. (2005) Array comparative genomic hybridization and its applications in cancer, *Nat Genet* **37** **Suppl**, S11–7.
91. Bottoni A, Piccin D, Tagliati F, Luchin A, Zatelli M C, and degli Uberti E C. (2005) miR-15a and miR-16-1 down-regulation in pituitary adenomas, *J Cell Physiol* **204**, 280–5.
92. Ota A, Tagawa H, Karnan S, Tsuzuki S, Karpas A, Kira S, et al. (2004) Identification and characterization of a novel gene, C13orf25, as a target for 13q31-q32 amplification in malignant lymphoma, *Cancer Res* **64**, 3087–95.
93. Tagawa H, and Seto M. (2005) A microRNA cluster as a target of genomic amplification in malignant lymphoma, *Leukemia* **19**, 2013–6.
94. Hayashita Y, Osada H, Tatematsu Y, Yamada H, Yanagisawa K, Tomida S, et al. (2005) A polycistronic microRNA cluster, miR-17-92, is overexpressed in human lung cancers and enhances cell proliferation, *Cancer Res* **65**, 9628–32.
95. Dews M, Homayouni A, Yu D, Murphy D, Sevignani C, Wentzel E, et al. (2006) Augmentation of tumor angiogenesis by a Myc-activated microRNA cluster, *Nat Genet.* **38**, 1060–5.
96. Weiler J, Hunziker J, and Hall J. (2006) Anti-miRNA oligonucleotides (AMOs): ammunition to target miRNAs implicated in human disease?, *Gene Ther* **13**, 496–502.
97. Vester B, and Wengel J. (2004) LNA (locked nucleic acid): high-affinity targeting of complementary RNA and DNA, *Biochemistry* **43**, 13233–41.
98. Elmen J, Lindow M, Schutz S, Lawrence M, Petri A, Obad S, et al. (2008) LNA-mediated microRNA silencing in non-human primates, *Nature* **452**, 896–9.
99. Flynt A S, Li N, Thatcher E J, Solnica-Krezel L, and Patton J G. (2007) Zebrafish miR-214 modulates Hedgehog signaling to specify muscle cell fate, *Nat Genet* **39**, 259–63.
100. Krutzfeldt J, Rajewsky N, Braich R, Rajeev K G, Tuschl T, Manoharan M, et al. (2005) Silencing of microRNAs in vivo with 'antagomirs', *Nature* **438**, 685–9.

101. Esau C, Davis S, Murray S F, Yu X X, Pandey S K, Pear M, et al. (2006) miR-122 regulation of lipid metabolism revealed by in vivo antisense targeting, *Cell Metab* **3**, 87–98.
102. Hutvagner G, Simard M J, Mello C C, and Zamore P D. (2004) Sequence-specific inhibition of small RNA function, *PLoS Biol* **2**, E98.
103. Vermeulen A, Robertson B, Dalby A B, Marshall W S, Karpilow J, Leake D, et al. (2007) Double-stranded regions are essential design components of potent inhibitors of RISC function, *RNA* **13**, 723–30.
104. Agirre X, Vilas-Zornoza A, Jimenez-Velasco A, Martin-Subero J I, Cordeu L, Garate L, et al. (2009) Epigenetic silencing of the tumor suppressor microRNA Hsa-miR-124a regulates CDK6 expression and confers a poor prognosis in acute lymphoblastic leukemia, *Cancer Res* **69**, 4443–53.
105. Kota J, Chivukula R R, O'Donnell K A, Wentzel E A, Montgomery C L, Hwang H-W, et al. (2009) Therapeutic microRNA Delivery Suppresses Tumorigenesis in a Murine Liver Cancer Model, *Cell* **137**, 1005–17.

Chapter 22

Blood-Based miRNA Preparation for Noninvasive Biomarker Development

Svenja Debey-Pascher, Jing Chen, Thorsten Voss, and Andrea Staratschek-Jox

Abstract

This chapter describes several methods for the isolation of miRNAs from peripheral whole blood samples or constituent fractions thereof, such as peripheral blood mononuclear cells, plasma, and serum. The methods described here are recently introduced protocols dedicated to the isolation of total RNAs including small RNAs, e.g., miRNeasy Kit and PAXgene Blood miRNA Kit, or alternatively for the enrichment of low-molecular-weight RNA (LMW RNA) fractions including small RNAs, e.g., using the miRNeasy Kit. Furthermore, modifications of classical RNA purification protocols to facilitate the recovery of small RNAs are highlighted.

Key words: miRNA, Peripheral blood mononuclear cells, Peripheral blood, RNA stabilization, Low-molecular-weight RNA, Small RNA enrichment

1. Introduction

Today, gene expression studies have been widely accepted as a powerful tool to investigate the transcriptome of a given source to describe disease-specific signatures as well as to identify pathogenetic relevant genes and its deregulated transcription. In peripheral blood, gene expression profiling is used to associate specific blood-based signatures with the occurrence of a variety of diseases such as infectious disease (1, 2), cardiovascular disease (3, 4), autoimmune disease (5), and cancer (reviewed in ref. 6, 7). Moreover, specific signatures are used to achieve differential diagnosis of hematological neoplasias such as leukemia (reviewed in ref. 8, 9).

In addition to the mRNA expression approach, the expression profiling of microRNAs (miRNAs) was established to further characterize tissue specific signatures. These miRNAs belong to the

Jian-Bing Fan (ed.), *Next-Generation MicroRNA Expression Profiling Technology: Methods and Protocols*, Methods in Molecular Biology, vol. 822, DOI 10.1007/978-1-61779-427-8_22, © Springer Science+Business Media, LLC 2012

small noncoding RNA fraction comprising RNA molecules of up to 24-nucleotide in length (10). miRNAs were found to play important roles as major switches in processes such as development, cell proliferation, hematopoietic differentiation, regulation of the development of a lymphoid lineage, oncogenic transformation, and apoptosis (reviewed in ref. 11, 12). miRNAs represent a family of functional RNAs of 19–23 nt cleaved from 60- to 110-nt hairpin precursor microRNAs (pre miRNA) by Dicer, which is a RNase III enzyme (13). Pre miRNAs in turn are processed from the primary transcript (pri RNA) by the RNase III enzyme, Drosha. The regulatory role of miRNAs is conducted by translational repression or degradation of specific target mRNAs (14, 15).

Many reports have already described altered expression of miRNAs in cancer samples compared with normal tissues including breast cancer (16), sarcomas (17) leukemias (18, 19), lymphomas (20), prostate cancer (21), or other types of diseases such as autoimmune diseases (22) and diabetes (23) or viral infections (24). These data indicate that investigating miRNA expression can be used for the identification of new and early diagnostic as well as prognostic and clinical markers, to define tumor subtypes and to classify human cancer entities. Interestingly, miRNAs were reported to be actively secreted by tumor cells through the formation of microvesicles (25, 26). Such microvesicles can be traced back to peripheral blood indicating that peripheral blood might be a perfect source to monitor tumor-associated miRNA expression signatures for early diagnosis and prediction of therapeutic outcome (27). Subsequently, the most recent blood-based disease specific miRNA signatures were identified in patients suffering from multiple sclerosis (28) rheumatoid arthritis (29), stroke (30), coronary artery disease (31) and different types of cancer (32–36) indicating that similar to mRNA profiles, miRNA profiles in peripheral whole blood or compartments thereof (e.g., cells, serum, plasma) also can reflect the disease specific alteration of other tissues and thus can be used as an easily accessible surrogate tissue in clinical settings. Such array based miRNA profiles can be established with high accuracy and reliability due to recent established protocols (37, 38).

In general, blood sampling for diagnostic and prognostic applications in most clinical trials is widely established and recently several protocols have been developed for the isolation of small RNAs from blood and serum samples as well as for other tissues. In this chapter we describe several options for the isolation of miRNAs either from stabilized whole blood samples or from isolated peripheral blood mononuclear cells (PBMC) and from plasma/serum samples. The protocols described here are either newly developed (miRNeasy Kit, PAXgene Blood miRNA System) or represent modifications of common RNA purification protocols to enable the recovery of small RNAs within these protocols. Of note is that miRNAs, in general, can be isolated along with the total RNA fraction. Alternatively, miRNA can be enriched using the low-molecular-weight RNA (LMW RNA) fraction containing all RNA molecules ≤200 nt including tRNA, 5S rRNA, and 5.8S rRNA, as well as small RNAs such as miRNA and short interfering RNA (siRNA). Depending on the selected isolation methods subsequent analyses can yield significantly different results (38). Therefore, the appropriate method has to be carefully chosen depending on the aim of the respective study.

2. Materials

2.1. Blood Collection

1. BD Vacutainer® CPT™ Cell Preparation Tube with Sodium Citrate (Becton Dickinson) for subsequent isolation of peripheral PBMC.
2. PAXgene Blood RNA collection tubes (Becton Dickinson) for direct stabilization of whole blood RNA.
3. Vacutainer tubes containing anticoagulant (e.g., BD Vacutainer plastic EDTA tube, 10 mL, lavender top (Becton Dickinson)) for separation of plasma.
4. Vacutainer tubes (containing either no additive or a clot activator) for separation of serum.

   Clot activator and silica gel:

   For example BD Vacutainer Plus plastic serum tube, 10 mL, red top (Becton Dickinson).

   For example BD Vacutainer Plus plastic serum tube (transport tube), 10 mL, mottled red/gray top (Becton Dickinson).

   No additive:

   For example BD Vacutainer Plus tube with clear BD Hemogard closure, 3 mL, clear top (Becton Dickinson).
5. BD Vacutainer™ holder (Becton Dickinson).
6. BD Safety-Lok™ blood collection set (Becton Dickinson).

2.2. Serum/Plasma Isolation

1. Serological pipettes of appropriate volumes (sterile).
2. Centrifuge tubes.
3. Cryovials.
4. Benchtop centrifuge (NOT refrigerated) with swing-out rotor and appropriate carriers.

2.3. PBMC Isolation

1. Serological pipettes of appropriate volumes (sterile).
2. Benchtop centrifuge with appropriate carriers.
3. 50-mL centrifuge tubes.

4. 1× PBS.
5. Trypan Blue.

2.4. RNA Isolation: General User Supplied Equipment Required

1. Pipettes (1 μL–4 mL).
2. Sterile, aerosol-barrier, RNase-free pipette tips.
3. RNase-free 1.5-mL or 2-mL microcentrifuge tubes.
4. Graduated cylinder.
5. Vortex mixer.
6. Crushed ice.
7. Permanent pen for labeling.
8. Variable-speed microcentrifuge capable of attaining 1,000–8,000 × *g*, and equipped with a rotor for 2-mL microcentrifuge tubes for centrifugation at 4°C and at room temperature (15–25°C).
9. Shaking incubator capable of incubating at 55°C and 65°C and shaking at 400 rpm, not exceeding 1,400 rpm (e.g., Eppendorf® Thermomixer Compact, or equivalent).
10. Disposable gloves

2.5. PAXgene Blood miRNA Kit (PreAnalytiX) Including the Following Components

1. PAXgene RNA Spin Columns (red).
2. PAXgene Shredder Spin Columns (lilac).
3. Microcentrifuge Tubes (1.5 mL).
4. Processing Tubes (2 mL).
5. Buffer BM1 (resuspension buffer).
6. Buffer BM2 (binding buffer).
7. Buffer BM3 (wash buffer concentrate); Note: Before using for the first time, add the required volume of ethanol (96–100%) as indicated on the bottle to obtain a working solution.
8. Buffer BM4 (wash buffer concentrate); Note: Before using for the first time, add the required volume of ethanol (96–100%) as indicated on the bottle to obtain a working solution.
9. Buffer BR5 (elution buffer).
10. RNase-Free Water.
11. Proteinase K.
12. RNase-Free DNase Set.
13. RNase-Free DNase I (lyophilized) 1,500 U.
14. Buffer RDD (DNA digestion buffer).
15. RNase-Free Water.
16. Secondary Hemogard Closures.

2.6. Blood miRNA: User Supplied Reagents/ Equipment

1. Ethanol (96–100%, purity grade p.a).
2. Isopropanol (100%, purity grade p.a).
3. Centrifuge capable of attaining 1,000–8,000 × *g*, and equipped with a swing-out rotor and buckets to hold PAXgene Blood RNA Tubes.

2.7. PAXgene Blood RNA Kit (PreAnalytiX)

1. BR1 Resuspension Buffer.
2. BR2 Binding Buffer.
3. BR3 Wash Buffer 1.
4. BR4 Wash Buffer 2 (concentrate), Note: Before using for the first time, add 4 volumes of ethanol (96–100%, purity grade p.a.) as indicated on the bottle to obtain a working solution.
5. BR5 Elution Buffer.
6. RNase-Free Water.
7. Proteinase K (green lid).
8. PAXgene RNA Spin Columns (red).
9. Processing Tubes (2 mL).
10. Secondary BD Hemogard™ Closures.
11. Microcentrifuge Tubes (1.5 mL).
12. DNase l, RNase-Free 1,500 U.
13. RDD DNA Digestion Buffer (white lid).
14. DNase Resuspension Buffer (tube, lilac lid).
15. PAXgene Shredder Spin Columns (lilac).

2.8. PAXgene Blood RNA: User Supplied Reagents/Equipment

1. Ethanol (96–100%, purity grade p.a.)
2. Centrifuge capable of attaining 3,000–5,000 × *g*, and equipped with a swing-out rotor and buckets to hold PAXgene Blood RNA Tubes.

2.9. miRNeasy® Mini Kit (QIAGEN) Including the Following Components

1. RNeasy® Mini Spin Columns (each packaged with a 2-mL Collection Tube).
2. Collection Tubes (1.5 mL).
3. Collection Tubes (2 mL).
4. QIAzol Lysis Reagent.
5. Buffer RWT.
6. Buffer RPE.
7. RNase-Free Water.

2.10. miRNeasy®: User Supplied Reagents

1. Chloroform (without added isoamyl alcohol).
2. Ethanol (70% and 96–100%).

2.11. RNeasy® MinElute Cleanup Kit (QIAGEN) Including the Following Components

1. RNeasy® MinElute Spin Columns (each in a 2-mL Collection Tube).
2. Collection Tubes (1.5 mL).
3. Collection Tubes (2 mL).
4. Buffer RLT.
5. Buffer RPE (concentrate); Note: Before using for the first time, add 4 volumes of ethanol (96–100%) as indicated on the bottle to obtain a working solution.
6. RNase-Free Water.

2.12. RNeasy® User Supplied Reagents

1. Ethanol (80% and 96–100%).

2.13. RNeasy Mini Kit (QIAGEN) Including the Following Components

1. RNeasy Mini Spin Columns (pink).
2. Collection Tubes (1.5 mL).
3. Collection Tubes (2 mL).
4. Buffer RLT.
5. Buffer RW1.
6. Buffer RPE (concentrate). Note: Before using for the first time, add 4 volumes of ethanol (96–100%) as indicated on the bottle to obtain a working solution.
7. RNase-Free Water.

2.14. RNeasy User Supplied Reagents

1. 14.3 M β-mercaptoethanol (β-ME).
2. 96–100% ethanol.
3. 70% ethanol.

2.15. TRIzol® RNA Isolation

1. TRIzol® Reagent (Invitrogen).

2.16. TRIzol® User Supplied Reagents

1. Chloroform.
2. Isopropyl alcohol.
3. 75% Ethanol (in DEPC-treated or RNAse-free water).
4. RNase-free water.

2.17. QIAamp Circulating Nucleic Acid Kit (QIAGEN) Including the Following Components

1. QIAGEN Mini columns.
2. Tube Extenders.
3. Collection Tubes (2.0 mL).
4. Elution Tubes (1.5 mL).
5. VacConnectors.
6. Buffer ACL.

7. Buffer ACB (concentrate). Note: Before use, add 200 mL isopropanol (100%) to 300 mL buffer ACB concentrate to obtain 500 mL Buffer ACB. Mix well after adding isopropanol.
8. Buffer ACW1 (concentrate). Note: Before use, add 25 mL ethanol (96–100%) to 19 mL buffer ACW1 concentrate to obtain 44 mL Buffer ACW1. Mix well after adding ethanol.
9. Buffer ACW2 (concentrate). Note: Before use, add 30 mL ethanol (96–100%) to 13 mL buffer ACW2 concentrate to obtain 43 mL Buffer ACW2. Mix well after adding ethanol.
10. Buffer AVE (purple caps).
11. QIAGEN Proteinase K.
12. Carrier RNA (red caps).

2.18. QIAamp: User Supplied Reagents/ Equipment

1. Pipette tips (adjustable).
2. Water bath or heating block capable of holding 50 mL centrifuge tubes at 60°C.
3. Heating block, or rocking platform capable of holding 2-mL collection tubes at 56°C.
4. Microcentrifuge with rotor for 2 mL tubes.
5. 50-mL centrifuge tubes.
6. Vacuum manifold (e.g., QIAvac 24 Plus, QIAGEN).
7. Vacuum regulator (QIAGEN) for easy monitoring of vacuum pressures and easy releasing of vacuum.
8. Vacuum pump capable of producing a vacuum of –800 to –900 mbar.
9. Additional Buffer ATL as lysis agent (QIAGEN).
10. Additional isopropanol (100%, 7 mL per extraction).
11. RNase-Free DNase set with Buffer RDD (QIAGEN for 50 on-column digestion).
12. One additional 2-mL collection tube per sample (QIAGEN, 1,000 collection tubes).
13. Optional: VacValves (QIAGEN).
14. Optional: QIAvac Connecting System (QIAGEN).

2.19. Pure Link miRNA Isolation Kit (Invitrogen) Including the Following Components

1. Binding Buffer (L3).
2. Wash Buffer (W5).
3. Sterile, RNase-free Water (pH > 7.0).
4. Spin Cartridge in Collection Tubes.
5. Wash Tubes and Recovery Tubes.

2.20. Pure Link miRNA: User Supplied Reagents/Equipment

1. 96–100% ethanol.
2. 70% ethanol.
3. Samples (total RNA).
4. Microcentrifuge capable of centrifuging >10,000 × *g*.

3. Methods

3.1. Preparation of Different Blood Based Fractions Prior to RNA Extraction

Peripheral blood consists of different components, e.g., the cellular fraction including the PBMC-fraction, the erythrocyte fraction, plasma and serum, respectively. Depending on the purpose of the underlying study, each of these fractions as well as the whole peripheral blood sample can be used as a primary source for RNA extraction. Next, the procedure of blood collection as well as the separation of the different blood components is described.

3.1.1. Blood Collection

Blood and plasma/serum collection should be performed according to standard clinical venipuncture techniques. The following protocol is according to the BD Vacutainer® Evacuated Blood Collection Tubes Product Insert (Form No 8363769, 11/2010) and the PAXgene Blood RNA Tube Product Circular (PAXgene blood collection, PreAnalytiX).

Since BD Vacutainer® Tubes and PAXgene tubes contain anticoagulants and chemical additives respectively, it is important to prevent possible backflow from the tubes during blood withdrawal. To prevent backflow, the following precautions should be taken when drawing blood into the tube:

1. Hold the patient's arm in a downward position during the blood collection procedure.
2. Hold the tube with the stopper upright to prevent possible backflow from the tube.
3. Loosen the tourniquet as soon as the blood starts to flow into the tube, or within 2 min of application.
4. Take care that the tube contents do not touch the stopper or the end of the needle during the collection procedure.
   (a) Prior to the blood withdrawal procedure select the appropriate number and type of tubes. Note: Ensure that the blood collection tubes are at room temperature (18–25°C) prior to use and properly labeled with patient identification.
   (b) Open the needle package and thread the needle onto holder. Do not remove the needle shield.
   (c) Insert the tube into the holder and leave in this position.

(d) Select a site for venipuncture.

(e) Apply the tourniquet. Disinfect the venous puncture site and allow the skin to dry. Note: Never touch the venipuncture area after disinfection.

(f) Remove the needle shield and perform the venipuncture by leaving the patient's arm in a downward position.

(g) Push the collection tube onto the needle thereby penetrating the diaphragm of the stopper.

(h) As soon as blood flows in the collection tube, release the tourniquet. Take care that no contents of the tube have contact with the stopper or the end of the needle during blood withdrawal.

(i) When multiple tubes are collected, remove the first tube from the holder when blood flow ceases and place a new tube into the holder. Note: Complete filling is important especially if PAXgene Blood RNA Tubes are used. Underfilled tubes are the main reason for reduced RNA yields. During filling of the succeeding tube, gently invert the previous tube 8–10 times to mix anticoagulant additive/chemical with the blood. Note: Do not shake the blood samples since this can cause hemolysis. Note: When using the BD Vacutainer® Safety-Lok™ blood collection set, a reduced draw (~ 0.5 mL) will occur on the first tube due to the trapped air in the blood collection set tubing. If only one PAXgene blood RNA tube is drawn, a "discard tube" should be added prior to drawing blood into the PAXgene blood RNA tube. Otherwise, the PAXgene blood RNA tube should be the last tube drawn in the blood collection procedure.

(j) After the last tube has been drawn and mixed as described above, withdraw the needle from the vein and apply pressure to the puncture site until bleeding stops using dry, sterile gauze. If desired, apply a bandage. Note: The top of the stopper may contain residual blood after phlebotomy. Appropriate precaution should be taken when handling tubes, due to blood contamination. Consider any needle holder that comes in contact with blood as hazardous.

(k) After collection, discard the needle into a sharps container. Note: Do not reshield the needle.

3.1.2. Stabilization and Storage of Whole Blood Based RNA Using PAXgene Blood RNA Tubes

1. According to the manufacturer, the use of PAXgene tubes enables the immediate stabilization of RNA in whole blood samples during blood collection. To further allow for stabilization PAXgene blood RNA tubes should be stored upright at room temperature (18–25°C) for at least 2 h and up to 72 h. After this incubation proceed to RNA isolation or store at −20 or −80°C for long-term storage.

2. To freeze PAXgene™ blood RNA tubes, deposit the tubes upright in a wire rack. Note: according to PreAnalytiX guidelines for Freezing PAXgene™ Blood RNA Tubes it is important not to deposit and freeze tubes in a styrofoam tray, because this may damage the tubes. If the tubes are to be stored below −20°C, first freeze the tubes −20°C for 24 h before transferring the tubes to −70 or −80°C.
3. To thaw PAXgene™ blood RNA tubes, incubate the tubes in a wire rack for approximately 2 h at ambient temperature (18–22°C) as recommended by the manufacturer, but do not incubate them at temperatures above 25°C. One suggestion is to remove the tubes from the freezer the afternoon before processing, thaw overnight, and process the tubes the next day. After thawing, carefully invert the tubes ten times before starting RNA preparation.

3.1.3. Plasma/Serum/PBMC Isolation from Whole Blood

Blood Plasma Preparation. The preparation of blood plasma and blood serum is according to a protocol established by ProImmune Limited (PR31 Version 1.0, 03/2009).

1. For blood collection use vacutainer tube(s) containing ~1.8 mg K_2EDTA per mL blood (may vary depending on manufacturer, for blood donation procedure see Subheading 3.1.1). Allow complete filling of the blood collection tubes to ensure the correct blood-to anticoagulant ratio.
2. Mix blood and anticoagulant by inverting Vacutainer tubes carefully 10 times and store at room temperature until centrifugation.
3. Perform centrifugation of samples immediately for 10 min at 1,000–2,000 RCF (generally 1,300 RCF) at room temperature (refer to speed recommended by the manufacturer). Note: Do not apply brake to stop centrifuge. After centrifugation the plasma fraction (top) is separated from the residual blood components (bottom).
4. Carefully collect the supernatant (plasma) at room temperature in a centrifuge tube. Take care not to disturb the bottom layer or transfer any additional blood components.
5. Check plasma for turbidity. Recentrifuge turbid samples and carefully collect the supernatant to remove remaining insoluble material.
6. Aliquot the plasma into labeled cryovials and store at −80°C if not used immediately. Label cryovials with the relevant information, including details of respective additives present in the blood.

Blood Serum Preparation. The serum preparation procedure is according to the protocol for Preparation of Blood Plasma and Serum described by ProImmune Limited (PR31 Version 1.0, 03/2009).

1. For blood collection use Vacutainer tube(s) without any addition of an anticoagulant (for blood donation procedure see Subheading 3.1.1).
2. To allow clotting, incubate the blood samples at room temperature for 30–45 min (maximum 60 min) in an upright position. In the case of using a clot-activator tube, carefully mix the blood and clot activator by inverting the Vacutainer tubes carefully 5–6 times before incubation.
3. Perform centrifugation for 15 min (usually 1,000–2,000 RCF, refer to the speed recommended by the manufacturer). Note: Do not apply brake to stop centrifuge. Carefully collect the supernatant (serum) at room temperature in a centrifuge tube. Take care not to disturb the bottom layer or to transfer any other blood components. Use a clean pipette for each tube.
4. Check serum for turbidity. Recentrifuge turbid samples and carefully collect the supernatant to remove remaining insoluble material.
5. Aliquot serum into cryovials and store at −80°C if not used immediately. Label cryovials with the relevant information, including details of additives present in the blood.

3.1.4. PBMC Preparation Using the BD Vacutainer® CPT™ Cell Preparation Tubes with Sodium Citrate

The PBMC fraction includes all blood cells with a round nucleus (e.g., lymphocytes, monocytes and macrophages) and can be isolated using a density gradient centrifugation. Here the use of the BD Vacutainer® CPT™ Tube with sodium citrate is described in accordance with the manufacturer's recommendations (Becton Dickinson: Vacutainer® CPT™ Cell Preparation Tubes with Sodium Citrate Product Insert).

1. Properly label the BD Vacutainer® CPT™ Tube with sodium citrate with the patient's identification. The tubes should be at room temperature (18–25°C).
2. Perform blood collection according to the standard technique for BD Vacutainer® Brand Blood Collection Tubes (for blood donation procedure see Subheading 3.1.1).
3. Store the tube upright at room temperature, until the centrifugation step. Note: For best results blood samples should be centrifuged within 2 h after blood collection.
4. Gently invert the blood sample (8–10 times) immediately prior to centrifugation. Make sure that the proper centrifuge carrier/adapter is used.
5. Centrifuge tubes in a horizontal rotor (swing-out) for 20–30 min at 1,500–1,800 RCF at room temperature (18–25°C).
6. After centrifugation, PBMC and platelets will be in a white fraction underneath the plasma fraction. Invert the unopened tube

gently 5–10 times to resuspend the cells into the plasma. To collect the PBMC transfer the entire contents of the tube above the gel barrier into a separate 50-mL centrifuge tube containing 35 mL PBS. Mix the cells by inverting the tube 5 times.

7. Centrifuge for 10 min at 300 RCF at room temperature. Carefully remove the supernatant without disturbing the cell pellet.
8. Fill up the volume to 10 mL with PBS. Close the tube and resuspend the cells by inverting the tube five times.
9. Count the cell number by trypan blue dye and centrifuge the cell suspension for 10 min at 300 RCF. Remove as much supernatant as possible without disturbing the cell pellet.
10. Cells could be either lysed directly by adding appropriate amount of TRIzol® Reagent (see RNA-Isolation with TRIzol® reagent, Subheading 3.2.1) or alternatively cryopreserved in DMSO containing medium before RNA isolation. Note: cryopreservation of viable PBMC has a negative impact on mRNA as well as on miRNA composition ((38) and SDP, ASJ, unpublished results). We therefore suggest to directly lyse cells in TRIzol® Reagent and store TRIzol® lysates at –80°C instead of cryopreserving PBMC in liquid nitrogen.

3.2. RNA Extraction

As described blood based RNA can be extracted from whole blood as well as from fractions thereof (e.g., PBMC, plasma or serum). Depending on the input source different extraction protocols have to be followed which are described in more detail below. Figure 1 gives a brief overview of all these different methods and the respective outcomes.

3.2.1. Isolation of Total RNA Including the LMW RNA Fraction from PMBC using TRIzol® and the RNeasy® MinElute Cleanup Kit

This protocol is for the isolation of total RNA based on Invitrogen's TRIzol® reagent according to the manufacturer's recommendations (Part No. 15596018.pps, Rev. date: 12 Jun 2007, Invitrogen).

RNA Isolation with TRIzol® Reagent

1. Isolate PBMC as described in Subheading 3.1.4, steps 1–9 (PBMC preparation with BD Vacutainer® CPT™ Cell Preparation Tubes with Sodium Citrate).
2. Add 1 mL TRIzol® Reagent/1×10^7 cells and lyse cells by repetitive pipetting. Allow for complete dissociation of nucleoprotein complexes by storing the samples at room temperature for 5 min. At this step TRIzol® lysates can be stored at –80°C until further processing. Otherwise continue with step 3.
3. Add 0.2 mL chloroform per initial mL TRIzol® used for the initial homogenization and shake vigorously by hand for 15 s.
4. Let rest at room temperature for 3 min.
5. Centrifuge at no more than $12{,}000 \times g$, for 10 min at 4°C.
6. Transfer the upper aqueous phase to a fresh tube. Note: avoid transfer of genomic DNA from the interphase.

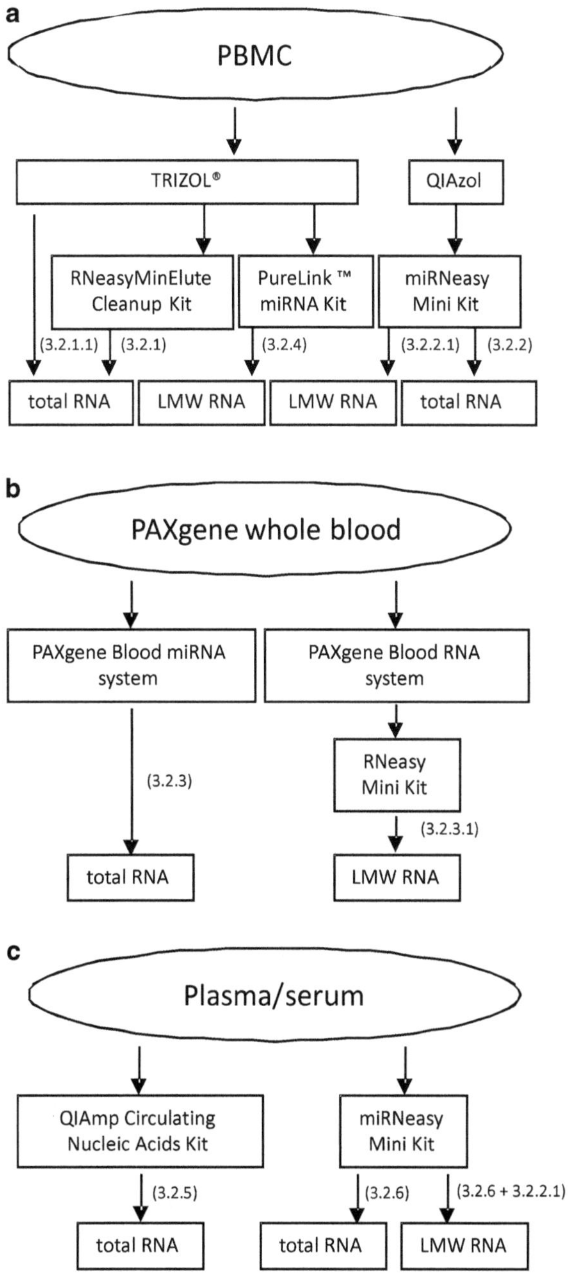

Fig. 1. Overview of different miRNA isolation options from PBMC (**a**), whole blood samples collected in PAXgene Blood RNA Tubes (**b**) and from serum or plasma samples, respectively (**c**). LMW RNA: low-molecular-weight fraction (RNA < 200 nt). When following the respective protocols as described in the text, total RNA including the LMW RNA fraction can be isolated from each source. Numbers in brackets refer to the relevant chapters.

7. Add 0.5 volumes of isopropanol per mL TRIzol® used for the initial homogenization, mix by inverting the tube and incubate at room temperature for 10 min.
8. Centrifuge at no more than 12,000 × *g*, for 30 min at 4°C.

9. Remove the supernatant and wash the RNA pellet twice with 1 mL 70–80% EtOH per mL TRIzol® used for the initial homogenization. Mix the sample by vortexing and centrifuge at no more than 7,500 × *g*, for 5 min at 4°C.
10. Remove the supernatant and let the RNA dry (5–10 min). Avoid complete drying of the RNA pellet since this will greatly decrease solubility.
11. Resuspend the RNA in RNase-free water by passing the RNA sample through a pipette tip a few times, and incubate for 5–10 min at 55°C. Then store on ice or at −80°C until further processing.

RNeasy® MinElute Cleanup

The RNeasy® MinElute Cleanup protocol (QIAGEN) was adapted to isolate total RNA, including small RNAs <200 nucleotides (miRNA).

Note: A maximum of 45 μg RNA in a maximum volume of 200 μL can be cleaned up in this protocol. This amount corresponds to the binding capacity of the RNeasy® MinElute spin column. Do not overload the column as this will significantly reduce RNA yield and purity.

1. Adjust the RNA sample to a volume of 100 μL with RNase-free water. Add 350 μL buffer RLT, and mix well.
2. Add 700 μL of 96–100% ethanol to the diluted RNA, and mix well by pipetting. Do not centrifuge. Proceed immediately to step 3.
3. Transfer 700 μL of the sample (1,150 μL total) to an RNeasy® MinElute spin column placed in a 2-mL collection tube (supplied). Close the lid gently, and centrifuge for 15 s at 8,000 × *g* (10,000 rpm). Discard the flowthrough. Transfer the remaining sample (450 μL) to the RNeasy® MinElute spin column and repeat the centrifugation. Discard the flowthrough.
4. Place the RNeasy® MinElute spin column in a new 2-mL collection tube (supplied). Add 500 μL Buffer RPE to the spin column. Close the lid gently, and centrifuge for 15 s at 8,000 × *g* (10,000 rpm) to wash the spin column membrane. Discard the flowthrough. Reuse the collection tube in step 5. Note: Buffer RPE is supplied as a concentrate. Ensure that ethanol is added to Buffer RPE before use.
5. Add 500 μL of 80% ethanol to the RNeasy® MinElute spin column. Close the lid gently, and centrifuge for 2 min at 8,000 × *g* (10,000 rpm) to wash the spin column membrane. Discard the flowthrough and collection tube.

Note: After centrifugation, carefully remove the RNeasy® MinElute spin column from the collection tube so that the column does not contact the flowthrough. Otherwise, carryover of ethanol will occur.

6. Place the RNeasy® MinElute spin column in a new 2-mL collection tube (supplied). Open the lid of the spin column, and centrifuge at full speed for 5 min. Discard the flowthrough and collection tube.
7. Place the RNeasy® MinElute spin column in a new 1.5-mL collection tube (supplied). Add 14 μL RNase-free water directly to the center of the spin column membrane. Close the lid gently, and centrifuge for 1 min at full speed to elute the RNA.

To avoid damage to their lids, place the spin columns into the centrifuge with at least one empty position between columns. Orient the lids so that they point in a direction opposite to the rotation of the rotor (e.g., if the rotor rotates clockwise, orient the lids counterclockwise). It is important to dry the spin column membrane since residual ethanol may interfere with downstream reactions. Centrifugation with the lids open ensures that no ethanol is carried over during RNA elution. Store RNA on ice or at −80°C until further processing.

3.2.2. Isolation of Total RNA Including the LMW RNA Fraction from PBMC using the miRNeasy Kit

The following procedure is according to the miRNeasy Mini Handbook (QIAGEN)

1. Prepare PBMC as described under Subheading 3.1.4, steps 1–9.
2. Loosen the cell pellet thoroughly by flicking the tube. Add 700 μL QIAzol Lysis Reagent to a maximum of 1×10^7 cells. Vortex or pipette to mix. Note: Incomplete loosening of the cell pellet may lead to inefficient lysis and reduced RNA yields.
3. If processing $\leq 3 \times 10^6$ cells, homogenize the cells by vortexing for 1 min.

   Note: Incomplete homogenization leads to significantly reduced RNA yields and can cause clogging of the RNeasy Mini spin column. Note: Homogenized cell lysates can be stored at −70°C for several months.

4. Place the tube containing the homogenate on the benchtop at room temperature (15–25°C) for 5 min. This step promotes dissociation of nucleoprotein complexes.
5. Add 140 μL chloroform to the tube containing the homogenate and cap it securely. Shake the tube vigorously for 15 s. Thorough mixing is important for subsequent phase separation.
6. Place the tube containing the homogenate on the benchtop at room temperature for 2–3 min.
7. Centrifuge for 15 min at 12,000 × *g* at 4°C. After centrifugation, heat the centrifuge up to room temperature (15–25°C) if the same centrifuge will be used for the next centrifugation steps. After centrifugation, the sample separates into 3 phases: an upper, colorless, aqueous phase containing RNA; a white

interphase; and a lower, red, organic phase. The volume of the aqueous phase should be approximately 350 μL.

Note: If you want to purify a separate miRNA-enriched fraction, follow the protocol described in Subheading 3.2.2 after performing this step.

Enrichment of the LMW RNA Fraction Using the miRNeasy Kit and RNeasy MinElute Cleanup Kit

8. Transfer the upper aqueous phase to a new collection tube (supplied). Add 1.5 volumes (usually 525 μL) of 100% ethanol and mix thoroughly by pipetting up and down several times. Do not centrifuge. Continue without delay with step 9. A precipitate may form after addition of ethanol, but this will not affect the procedure.
9. Pipette up to 700 μL of the sample, including any precipitate that may have formed, into an RNeasy Mini spin column in a 2 mL collection tube (supplied). Close the lid gently and centrifuge at ≥8,000 × g (10,000 rpm) for 15 s at room temperature (15–25°C). Discard the flowthrough. Reuse the collection tube in step 10.
10. Repeat step 9 using the remainder of the sample. Discard the flowthrough. Reuse the collection tube in step 11.
11. Add 700 μL Buffer RWT to the RNeasy Mini spin column. Close the lid gently and centrifuge for 15 s at 8,000 × g (10,000 rpm) to wash the column. Discard the flowthrough. Reuse the collection tube in step 12.
12. Pipette 500 μL Buffer RPE onto the RNeasy Mini spin column. Close the lid gently and centrifuge for 15 s at 8,000 × g (10,000 rpm) to wash the column. Discard the flowthrough. Reuse the collection tube in step 13.
13. Add another 500 μL Buffer RPE to the RNeasy Mini spin column. Close the lid gently and centrifuge for 2 min at 8,000 × g (10,000 rpm) to dry the RNeasy Mini spin column membrane. The long centrifugation dries the spin column membrane, ensuring that no ethanol is carried over during RNA elution. Residual ethanol may interfere with downstream reactions. Note: Following centrifugation, remove the RNeasy Mini spin column from the collection tube carefully so the column does not contact the flowthrough. Otherwise, carryover of ethanol will occur.
14. Optional: Place the RNeasy Mini spin column into a new 2-mL collection tube (not supplied) and discard the old collection tube with the flowthrough. Centrifuge in a microcentrifuge at full speed for 1 min. Perform this step to eliminate any possible carryover of Buffer RPE or if residual flowthrough remains on the outside of the RNeasy Mini spin column after step 13.
15. Transfer the RNeasy Mini spin column to a new 1.5-mL collection tube (supplied). Pipette 30–50 μL RNase-free water directly onto the RNeasy Mini spin column membrane.

Close the lid gently and centrifuge for 1 min at 8,000×*g* (10,000 rpm) to elute the RNA.

16. If the expected RNA yield is >30 μg, repeat step 15 with a second volume of 30–50 μL RNase-free water. Elute into the same collection tube. To obtain a higher total RNA concentration, this second elution step may be performed by using the first eluate (from step 15). The yield will be 15–30% less than the yield obtained using a second volume of RNase-free water, but the final concentration will be higher.

According to the miRNeasy Mini Handbook (QIAGEN) the LMW RNA fraction can be enriched from PBMC or serum/plasma in a separate fraction. For this protocol an additional RNeasy MinElute® Cleanup Kit is required.

1. Follow steps 1–7 of the protocol "Isolation of Total RNA Including the LMW RNA Fraction from PBMC Using the miRNeasy Kit" as described in Subheading 3.2.2 or steps 1–6 of the protocol "Isolation of total RNA, including the LMW RNA fraction from serum or plasma using the miRNeasy Mini Kit" (described in Subheading 3.2.6, step 6).
2. Transfer the upper aqueous phase to a new reaction tube (not supplied). Add 1 volume of 70% ethanol (usually 350 μL) and mix thoroughly by vortexing. Do not centrifuge. Proceed immediately to step 3.
3. Pipette the sample (approx. 700 μL), including any precipitate that may have formed, into an RNeasy Mini spin column placed in a 2-mL collection tube. Close the lid gently and centrifuge at 8,000×*g* (10,000 rpm) for 15 s at room temperature (15–25°C). Pipette the flowthrough (which contains miRNA) into a 2-mL reaction tube (not supplied). Discard the RNeasy Mini spin column and follow steps 4–10.
4. Add 450 μL of 100% ethanol (0.65 volumes) to the flowthrough from step 3 and mix thoroughly by vortexing. Do not centrifuge. Proceed immediately to step 5.
5. Pipette 700 μL of the sample into an RNeasy MinElute spin column placed in a 2-mL collection tube. Close the lid gently and centrifuge for 15 s at 8,000×*g* (10,000 rpm) at room temperature (15–25°C). Discard the flowthrough. Repeat this step until the whole sample has been pipetted into the spin column. Discard the flowthrough each time.
6. Optional: Add 700 μL Buffer RWT to the RNeasy MinElute spin column. Close the lid gently and centrifuge for 15 s at 8,000×*g* (10,000 rpm) to wash the column. Discard the flowthrough. This step is optional for the miRNA-enriched fraction because most impurities have already been removed on the first RNeasy Mini spin column.

7. Pipette 500 μL Buffer RPE into the RNeasy MinElute spin column. Close the lid gently and centrifuge for 15 s at 8,000 × *g* (10,000 rpm). Discard the flowthrough.
8. Add 500 μL of 80% ethanol to the RNeasy MinElute spin column. Close the lid gently and centrifuge for 2 min at 8,000 × *g* (10,000 rpm) to dry the spin column membrane. Discard the flowthrough and the collection tube. Note: After centrifugation, remove the RNeasy MinElute spin column from the collection tube carefully so that the column does not contact the flowthrough. Otherwise, carryover of ethanol will occur.
9. Place the RNeasy MinElute spin column into a new 2-mL collection tube. Open the lid and centrifuge for 5 min at 8,000 × *g* (10,000 rpm).
10. Place the RNeasy MinElute spin column into a 1.5-mL collection tube and pipette 14 μL RNase-free water onto the spin column membrane. Close the lid gently and centrifuge for 1 min at 8,000 × *g* (10,000 rpm) to elute the miRNA-enriched fraction.

3.2.3. Isolation of Total RNA Including the LMW RNA Fraction from Whole Blood Samples Using the PAXgene Blood miRNA System

The PAXgene Blood miRNA System consists of PAXgene Blood RNA Tubes for collection, stabilization and transport of blood, and the PAXgene Blood miRNA Kit for purification of total RNA, including RNA longer than approximately 18 nucleotides using well-established PAXgene silica-membrane technology. The purified RNA includes both mRNA and small RNAs such as miRNA for molecular analysis. This protocol is according to the manufacturer's recommendation described in the PAXgene Blood miRNA Kit Handbook (PreAnalytiX) and is also available as an automated version for the QIAcube instrument (QIAGEN).

1. Centrifuge the PAXgene Blood RNA Tube for 10 min at 3,000–5,000 × *g* using a swing-out rotor.

   Note: Ensure that the blood sample has been incubated in the PAXgene Blood RNA Tube for a minimum of 2 h at room temperature (15–25°C), to achieve complete lysis of blood cells. Note: The rotor must contain tube adapters for round-bottom tubes. If other types of tube adapter are used, the tubes may break during centrifugation.
2. Remove the supernatant by decanting or pipetting. Add 4 mL RNase-free water to the pellet and close the tube using a fresh secondary Hemogard closure. If decanting the supernatant, take care not to disturb the pellet, and dry the rim of the tube with a clean paper towel.
3. Vortex until the pellet is visibly dissolved, and centrifuge for 10 min at 3,000–5,000 × *g* using a swing-out rotor. Remove the entire supernatant by decanting or pipetting and discard. Small debris remaining in the supernatant after vortexing but before centrifugation will not affect the procedure.

Note: Incomplete removal of the supernatant will inhibit lysis and dilute the lysate, and therefore affect the conditions for binding RNA to the PAXgene membrane.

4. Add 350 μL Buffer BM1, and vortex until the pellet is visibly dissolved.
5. Pipette the sample into a 1.5-mL microcentrifuge tube. Add 300 μL Buffer BM2 and 40 μL proteinase K. Mix by vortexing for 5 s and incubate for 10 min at 55°C in a shaking incubator at 400–1,400 rpm. After incubation, set the temperature of the shaking incubator to 65°C for use in step 20. Note: Do not mix Buffer BM2 and proteinase K together before adding them to the sample.
6. Pipette the sample into a PAXgene Shredder spin column (lilac) placed in a 2-mL processing tube and centrifuge for 3 min at full speed (do not exceed 20,000 × *g*).
7. Carefully transfer the entire supernatant of the flowthrough from the PAXgene Shredder spin column to a new 1.5-mL microcentrifuge tube without disturbing the pellet in the processing tube.
8. Add 700 μL of isopropanol (100%, purity grade p.a.) and mix by vortexing.
9. Pipette 700 μL sample into the PAXgene RNA spin column (red) placed in a 2-mL processing tube. Close the lid gently and centrifuge for 1 min at 8,000–20,000 × *g*. Place the spin column in a new 2-mL processing tube, and discard the old processing tube containing flowthrough.
10. Pipette the remaining sample into the PAXgene RNA spin column (red). Close the lid gently, and centrifuge for 1 min at 8,000–20,000 × *g*. Place the spin column in a new 2-mL processing tube and discard the old processing tube containing flowthrough.
11. Add 350 μL Buffer BM3 to the PAXgene RNA spin column. Close the lid gently, and centrifuge for 15 s at 8,000–20,000 × *g*. Place the spin column in a new 2-mL processing tube, and discard the old processing tube containing flowthrough. Note: Buffer BM3 is supplied as a concentrate. Ensure that ethanol is added to Buffer BM3 before use.
12. Add 10 μL DNase I stock solution to 70 μL Buffer RDD in a 1.5 mL microcentrifuge tube. Mix by gently flicking the tube and centrifuge briefly to collect residual liquid from the sides of the tube. Note: DNase I is especially sensitive to physical denaturation. Mixing should only be carried out by gently flicking the tube. Do not vortex.
13. Pipette the DNase I incubation mix (80 μL) directly onto the PAXgene RNA spin column membrane, and incubate on the benchtop (20–30°C) for 15 min. Note: Ensure that the DNase

I incubation mix is placed directly onto the membrane. DNase digestion will be incomplete if part of the mix is applied to and remains on the walls or O-ring of the spin column.

14. Add 350 μL Buffer BM3 to the PAXgene RNA spin column. Close the lid gently, and centrifuge for 15 s at 8,000–20,000 × *g*. Place the spin column in a new 2-mL processing tube and discard the old processing tube containing flowthrough.
15. Add 500 μL Buffer BM4 to the PAXgene RNA spin column. Close the lid gently, and centrifuge for 15 s at 8,000–20,000 × *g*. Discard the flowthrough. Place the spin column in a new 2-mL processing tube and discard the old processing tube containing flowthrough. Note: Buffer BM4 is supplied as a concentrate. Ensure that ethanol is added to Buffer BM4 before use.
16. Add another 500 μL Buffer BM4 to the PAXgene RNA spin column. Close the lid gently, and centrifuge for 2 min at 8,000–20,000 × *g*. Note: After centrifugation, carefully remove the PAXgene RNA spin column from the processing tube so that the column does not contact the flowthrough. Otherwise, carryover of ethanol will occur.
17. Discard the processing tube containing flowthrough, and place the PAXgene RNA spin column in a new 2-mL processing tube (supplied). Centrifuge at 8,000–20,000 × *g* for 1 min. It is important to dry the spin column membrane, since residual ethanol may interfere with downstream reactions.
18. Discard the processing tube containing flowthrough. Place the PAXgene RNA spin column in a new 1.5-mL microcentrifuge tube and pipette 40 μL Buffer BR5 directly onto the spin column membrane. Close the lid gently, and centrifuge for 1 min at 8,000–20,000 × *g* to elute the RNA. Be sure to add Buffer BR5 directly to the spin column membrane. This wets the entire membrane, ensuring maximum elution efficiency.
19. Repeat the elution step (step 18) as described, using 40 μL Buffer BR5 and the same microcentrifuge tube.
20. Incubate the eluate for 5 min at 65°C in the shaking incubator without shaking. After incubation, chill immediately on ice. This incubation at 65°C denatures the RNA for downstream applications. Do not exceed the incubation time or temperature.
21. If the RNA eluate will not be used immediately, store at −80°C. Since the RNA remains denatured after freezing and thawing, it is not necessary to repeat the incubation at 65°C.

Isolation of LMW RNA Fraction from Whole Blood Samples Using the PAXgene Blood RNA System and RNeasy Mini Kit

A way to enrich the portion of small RNA species from the PAXgene Blood RNA tubes is to harvest the flowthrough of the PAXgene RNA column after the binding step of the cellular RNA and isolate the small RNA in a second purification using the RNeasy Mini Kit. This protocol is adapted from Kruhøffer et al. (39) for the manual isolation of small RNAs from single PAXgene Blood RNA tubes.

The original method in this paper describes a partially automated process that is more convenient for laboratories with higher sample throughput.

1. Centrifuge the PAXgene Blood RNA Tube for 10 min at 3,000–5,000 × *g* using a swing-out rotor. Ensure that the blood sample has been incubated in the PAXgene Blood RNA Tube for a minimum of 2 h at room temperature (15–25°C), to achieve complete lysis of blood cells. The rotor must contain tube adapters for round-bottom tubes. If other types of tube adapter are used, the tubes may break during centrifugation.
2. Remove the supernatant by decanting or pipetting. Add 4 mL RNase-free water to the pellet, and close the tube using a fresh secondary BD Hemogard closure (supplied with the kit). If the supernatant is decanted, take care not to disturb the pellet, and dry the rim of the tube with a clean paper towel.
3. Vortex until the pellet is visibly dissolved, and centrifuge for 10 min at 3,000–5,000 × *g* using a swing-out rotor. Remove and discard the entire supernatant. Small debris remaining in the supernatant after vortexing but before centrifugation will not affect the procedure. Incomplete removal of the supernatant will inhibit lysis and dilute the lysate, and therefore affect the conditions for binding RNA to the PAXgene membrane.
4. Add 350 μL resuspension buffer BR1 and vortex until the pellet is visibly dissolved.
5. Pipette the sample into a 1.5-mL microcentrifuge tube. Add 300 μL binding buffer BR2 and 40 μL proteinase K. Mix by vortexing for 5 s and incubate for 10 min at 55°C using a shaking incubator at 400–1,400 rpm. Do not mix binding buffer BR2 and proteinase K together before adding them to the sample.
6. Pipette the lysate directly into a PAXgene Shredder spin column (lilac) placed in a 2-mL processing tube, and centrifuge for 3 min at maximum speed (but not to exceed 20,000 × *g*). Carefully pipette the lysate into the spin column and visually check that the lysate is completely transferred to the spin column .To prevent damage to columns and tubes, do not exceed 20,000 × *g*. Some samples may flow through the PAXgene Shredder spin column without centrifugation. This is due to low viscosity of some samples and should not be taken as an indication of product failure.
7. Carefully transfer the entire supernatant of the flowthrough fraction to a fresh 1.5-mL microcentrifuge tube without disturbing the pellet in the processing tube.
8. Add 350 μL ethanol (96–100%, purity grade p.a). Mix by vortexing, and centrifuge briefly (1–2 s at 500–1,000 × *g*) to

remove drops from the inside of the tube lid. The length of the centrifugation must not exceed 1–2 s, as this may result in pelleting of nucleic acids and reduced yields of total RNA.

9. Pipette 700 μL sample into the PAXgene RNA spin column (red) placed in a 2-mL processing tube, and centrifuge for 1 min at 8,000–20,000 × *g*. Place the spin column (PRC) in a new 2-mL processing tube and save the old processing tube containing flowthrough for small RNA isolation.
10. Pipette the remaining sample into the PAXgene RNA spin column and centrifuge for 1 min at 8,000–20,000 × *g*. Throw away the spin column and save the processing tube containing flowthrough for small RNA isolation.
11. Add 700 μL ethanol (98%) per 500 μL flowthrough of the PAXgene Blood RNA binding step and mix carefully by pipetting. In order to process the whole flowthrough of steps 9 and 10 (~1 mL), add 1.4 mL ethanol.
12. Apply the sample in 700 μL portions to an RNeasy mini spin column placed in a 2-mL collection tube, centrifuge for 1 min at 8,000–20,000 × *g*, discard the flowthrough after each loading step. After the last loading step, discard the flowthrough and collection tube.
13. Place the RNeasy mini column into a new 2-mL collection tube and add 500 μL Buffer RPE, centrifuge for 15 s at 8,000–20,000 × *g*. Discard the flowthrough.
14. Repeat the washing step with 500 μL Buffer RPE and centrifuge for 2 min at 8,000–20,000 × *g* to dry the spin column membrane.
15. To elute, transfer the RNeasy column to a new 1.5-mL collection tube. Pipette 30–50 μL RNase free water directly onto the RNeasy silica-gel membrane and centrifuge for 1 min at 8,000–20,000 × *g*. Alternatively, Buffer BR5 can be used for elution.

3.2.4. Isolation of the LMW RNA Fraction from Total RNA Samples using the Pure Link miRNA Isolation Kit

This protocol is for the enrichment of miRNA from total RNA based on Invitrogen's "PureLink™ miRNA Isolation Kit" (Invitrogen #K1570-01, Instruction Manual Version C, 18 July 2005, 25-0753). Total RNAs are used as starting materials instead of cell lysates. Figure 2 depicts the enriched LMW RNA fraction after using the Pure Link miRNA Isolation Kit.

Start with total RNA (preferable 5 μg, minimal 1–2 μg).

1. Add 300 μL Binding Buffer to the tube (1.7 mL size) containing the RNA sample and mix well by pipetting or vortexing.
2. Add 300 μL 70% ethanol to the tube with well mixed binding buffer and total RNA sample, and mix well by pipetting or vortexing.

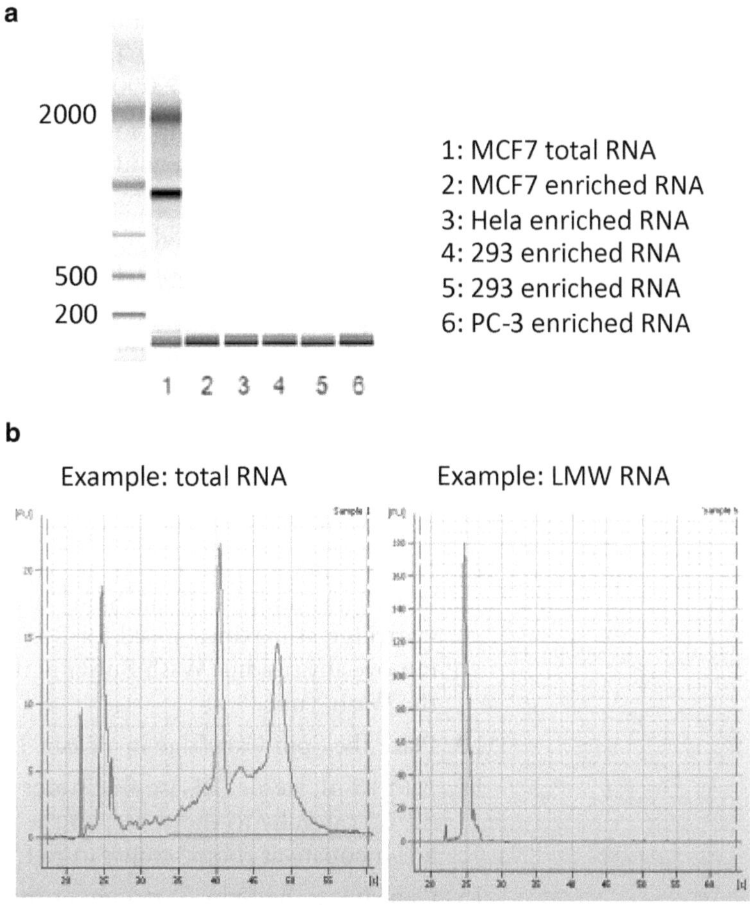

Fig. 2. Isolation of the LMW RNA fraction from total RNA samples using the Pure Link miRNA Isolation Kit. The LWM RNA fraction can be subsequently enriched from total RNA using Invitrogen's "PureLink™ miRNA Isolation Kit." Bioanalyzer gel analysis (**a**) and electropherograms (**b**) for the size distribution of enriched small RNA with Pure Link miRNA Isolation Kit in comparison to total RNA. The *y*-axis represents fluorescence units (FU) and the *x*-axis runtime in seconds (s).

3. Load the entire mixed contents (total volume is ~600 μL) into a clean Spin Cartridge with Collection Tube attached underneath. Close the lid.
4. Briefly centrifuge the Spin Cartridge with collection tube at 12,000 × *g* for 1 min at room temperature.
5. Throw away the Spin Cartridge; Keep the Collection Tube with flowthrough inside.
6. Add 700 μL of 100% ethanol to the Collection Tube containing flowthrough to obtain a final concentration of ~70% ethanol. Mix well by pipetting or vortexing.

7. Transfer 700 μL of the mixture from step 6 to a clean Spin Cartridge with Collecting Tube underneath. Briefly centrifuge the Spin Cartridge at 12,000 × *g* for 1 min at room temperature
8. Transfer the remaining mixture from step 6 to the Spin Cartridge in the step 7 and centrifuge again at 12,000 × *g* for 1 min at room temperature.
9. Discard the flowthrough in the Collection Tube. Attach the Collection Tube without flowthrough underneath the Spin Cartridge.
10. Wash the Spin Cartridge by adding 500 μL Wash Buffer (provided in the kit, ETOH preadded) and centrifuging at 12,000 × *g* for 1 min at room temperature. Discard the flowthrough in the Collection Tube/Wash Tube.
11. Repeat the wash step one more time (step 10) with 500 μL Wash Buffer.
12. Discard the flowthrough in the Wash Tube. Attach the same Wash Tube underneath the Spin Cartridge. Centrifuge the Spin Cartridge at maximal speed for 2 min at room temperature to remove any residual Wash Buffer in the Spin Cartridge. Discard the Wash Tube.
13. Place the Spin Cartridge in a clean 1.7-mL Recovery Tube.
14. Add 15 μL sterile, RNase-free water (pH >7.0) (prewarmed to 37°C) carefully to the center of the Spin Cartridge. Close the lid. Incubate at room temperature for 2 min.
15. Centrifuge the Spin Cartridge at maximal speed for 2 min at room temperature.
16. Store the Recovery Tube containing small RNA at −80°C for long term storage.

3.2.5. Isolation of Total RNA Including the LMW RNA Fraction from Serum or Plasma using the QIAamp® Circulating Nucleic Acid Kit

According to the QIAamp Circulating Nucleic Acid Handbook (QIAGEN) this protocol describes how to isolate circulating miRNAs and circulating total RNA from plasma or serum samples that are either fresh or frozen.

1. Pipette 400 μL QIAGEN Proteinase K into a 50-mL tube (not provided).
2. Add 3 mL of plasma into the 50-mL tube.
3. Add 3.2 mL of Buffer ACL (without carrier RNA) and 1.0 mL Buffer ATL; close the cap and mix by pulse-vortexing for 30 s. Note: Buffer ATL should be added to the lysis mixture as the last component. Make sure that a visible vortex forms in the tube. In order to ensure efficient lysis, it is essential that the sample and Buffers ACL and ATL are mixed thoroughly to yield a homogeneous solution.
4. Incubate at 60°C for 30 min.

5. Briefly spin the tube to remove drops from the inside of the lid.
6. Add 9.0 mL of Buffer ACB and 7.0 mL isopropanol to the lysate, close the cap, and mix thoroughly by pulse-vortexing for 15–30 s.
7. Incubate the lysate–Buffer ACB mixture for 5 min on ice.
8. Insert the QIAamp Mini column into the VacConnector on the QIAvac 24 Plus. Insert a 20 mL tube extender into the open QIAamp Mini column. Make sure that the tube extender is firmly inserted into the QIAamp Mini column to avoid leakage of sample. Note: Keep the collection tube for the dry spin in step 16.
9. Carefully apply the lysate from step 7 into the tube extender of the QIAamp Mini column. Switch on the vacuum pump. When all lysates have been drawn through the columns completely, switch off the vacuum pump and release the pressure to 0 mbar. Carefully remove and discard the tube extender. If removal of DNA is not required for the analysis of miRNA, skip directly to step 13. For fast and convenient release of the vacuum pressure, use the Vacuum Regulator (see Ordering Information). Note: To avoid cross-contaminations, be careful not to cross neighboring QIAamp columns while tube extenders are removed.
10. Optional: Transfer the column from the vacuum manifold into the 2-mL collection tube (not provided) and centrifuge the column 1 min at 14,000 rpm. Note: This step will remove traces of lysate that could interfere with the subsequent DNase digestion.
11. Optional: Per each sample, add 10 μL DNase I stock solution (see above) to 70 μL Buffer RDD. Mix by gently inverting the tube. Buffer RDD is supplied with the RNase-Free DNase Set. Note: DNase I is especially sensitive to physical denaturation. Mixing should only be carried out by gently inverting the tube. Do not vortex.
12. Optional: Place the QIAamp Mini columns back onto their original positions on the QIAvac 24 Plus vacuum manifold. Pipette the DNase I incubation mix (80 μL) directly onto the QIAamp Mini column membrane, and incubate at ambient temperature (20–30°C) for 15 min. Note: Make sure to pipette the DNase I incubation mix directly onto the QIAamp Mini column membrane. DNase digestion will be incomplete if part of the mix sticks to the walls or the O-ring of the QIAamp Mini column.
13. Apply 600 μL of Buffer ACW1 to the QIAamp Mini column. Leave the lid of the column open and switch on the vacuum pump. After all of Buffer ACW1 has been drawn through the QIAamp Mini column, switch off the vacuum pump and release the pressure to 0 mbar.

14. Apply 750 μL of Buffer ACW2 to the QIAamp Mini column. Leave the lid of the column open and switch on the vacuum pump. After all of Buffer ACW2 has been drawn through the QIAamp Mini column, switch off the vacuum pump and release the pressure to 0 mbar.
15. Apply 750 μL of ethanol (96–100%) to the QIAamp Mini column. Leave the lid of the column open and switch on the vacuum pump. After all of the ethanol has been drawn through the QIAamp Mini column, switch off the vacuum pump and release the pressure to 0 mbar.
16. Close the lid of the QIAamp Mini column, remove it from the vacuum manifold and discard the VacConnector. Place the QIAamp Mini column in a clean 2-mL collection tube (saved from step 8) and centrifuge at full speed (20,000 × *g*; 14,000 rpm) for 3 min.
17. Place the QIAamp Mini column into a new 2-mL collection tube, open the lid, and incubate the assembly at 56°C for 10 min to dry the membrane completely. If using the same heating block used in step 4, reduce the temperature by 4°C.
18. Place the QIAamp Mini column in a clean 1.5 mL elution tube and discard the collection tube from step 16. Carefully apply 20–150 μL of Buffer AVE to the center of the QIAamp Mini column membrane. Close the lid and incubate at room temperature for 3 min.
19. Centrifuge at full speed (20,000 × *g*; 14,000 rpm) for 1 min to elute the RNA.

3.2.6. Isolation of Total RNA, Including the LMW RNA Fraction from Serum or Plasma using the miRNeasy Mini Kit

This protocol has been adapted from the miRNeasy Mini Kit (QIAGEN) and is intended as a guideline for the purification of total RNA, including small RNAs, from serum and plasma using the miRNeasy Mini Kit (Fig. 1c).

1. Prepare serum or plasma or thaw frozen samples.
2. Add 5 volumes of QIAzol Lysis Reagent. Mix thoroughly by vortexing or pipetting. Incomplete homogenization leads to significantly reduced RNA yields and can cause clogging of the RNeasy® Mini spin column. Homogenized lysates can be stored at −70°C for several months.
3. Place the tube containing the homogenate on the benchtop at room temperature (15–25°C) for 5 min. This step promotes dissociation of nucleoprotein complexes.
4. Add 1 volume chloroform to the tube containing the homogenate and cap it securely. Shake the tube vigorously for 15 s. Thorough mixing is important for subsequent phase separation.
5. Place the tube containing the homogenate on the benchtop at room temperature for 2–3 min.

6. Centrifuge for 15 min at 12,000 × *g* at 4°C. After centrifugation, heat the centrifuge up to room temperature (15–25°C) if the same centrifuge will be used for the next centrifugation steps. After centrifugation, the sample separates into 3 phases: an upper, colorless, aqueous phase containing RNA, a white interphase, and a lower, red, organic phase. Note: If you want to purify a separate miRNA-enriched fraction, follow the protocol described in Subheading 3.2.2.1 after performing this step.
7. Transfer the upper aqueous phase to a new collection tube (supplied). Add 1.5 volumes of 100% ethanol and mix thoroughly by pipetting up and down several times. Do not centrifuge. Continue without delay with step 8. A precipitate may form after addition of ethanol, but this will not affect the procedure.
8. Pipette up to 700 μL of the sample, including any precipitate that may have formed, into an RNeasy Mini spin column in a 2-mL collection tube (supplied). Close the lid gently and centrifuge at ≥8,000 × *g* (≥10,000 rpm) for 15 s at room temperature (15–25°C). Discard the flowthrough. Reuse the collection tube in step 9.
9. Repeat step 8 using the remainder of the sample. Discard the flowthrough. Reuse the collection tube in step 10.
10. Add 700 μL Buffer RWT to the RNeasy Mini spin column. Close the lid gently and centrifuge for 15 s at ≥8,000 × *g* (≥10,000 rpm) to wash the column. Discard the flowthrough. Reuse the collection tube in step 11.
11. Pipette 500 μL Buffer RPE onto the RNeasy Mini spin column. Close the lid gently and centrifuge for 15 s at ≥8,000 × *g* (≥10,000 rpm) to wash the column. Discard the flowthrough. Reuse the collection tube in step 12.
12. Add another 500 μL Buffer RPE to the RNeasy Mini spin column. Close the lid gently and centrifuge for 2 min at ≥8,000 × *g* (≥10,000 rpm) to dry the RNeasy Mini spin column membrane. The long centrifugation dries the spin column membrane, ensuring that no ethanol is carried over during RNA elution. Residual ethanol may interfere with downstream reactions. Note: Following centrifugation, remove the RNeasy Mini spin column from the collection tube carefully so the column does not contact the flowthrough. Otherwise, carryover of ethanol will occur.
13. Optional: Place the RNeasy Mini spin column into a new 2-mL collection tube (not supplied), and discard the old collection tube with the flowthrough. Centrifuge in a microcentrifuge at full speed for 1 min. Perform this step to eliminate any possible carryover of Buffer RPE or if residual flowthrough remains on the outside of the RNeasy Mini spin column after step 12.

14. Transfer the RNeasy Mini spin column to a new 1.5-mL collection tube (supplied). Pipette 30–50 μL RNase-free water directly onto the RNeasy Mini spin column membrane. Close the lid gently and centrifuge for 1 min at ≥8,000 × *g* (≥10,000 rpm) to elute the RNA.
15. If the expected RNA yield is >30 μg, repeat step 14 with a second volume of 30–50 μL RNase-free water. Elute into the same collection tube. To obtain a higher total RNA concentration, this second elution step may be performed by using the first eluate (from step 14). The yield will be 15–30% less than the yield obtained using a second volume of RNase-free water, but the final concentration will be higher.

3.3. Quantification and Quality Control

3.3.1. Quantification and Determination of Quality of total RNA Including Small RNAs

1. The concentration of RNA should be determined by measuring the absorbance at 260 nm (A260) using a photometer e.g., a Nanodrop™ instrument (Thermo Scientific) for yield and sample quality. The ratio of the readings at 260 nm and 280 nm (A260/A280) provides an estimate of the purity of RNA with respect to contaminants that absorb UV light, such as protein. Pure RNA has an A260/A280 ratio of 1.8–2.2 in 10 mM Tris–HCl, pH 7.5. As photometric measurements are strongly pH dependent the use of a buffered system is essential for accurate measurements. Always calibrate the photometer with the same dilution buffer–elution buffer ratio and always use the same elution buffer that was used for the RNA preparation, since some elution buffers contain small quantities of salts that could result in high readings at low UV wavelengths.
2. The integrity and size distribution of total RNA including small RNAs can be checked by denaturing agarose gel electrophoresis (1% w/v) and ethidium bromide staining. The respective large ribosomal bands should appear sharp on the stained gel. 28S ribosomal RNA (5 kb) bands should be present with an intensity approximately twice that of the 18 S RNA (1.9 kb) band. Beside this the fraction of small RNA species should be clearly visible as seen in Fig. 3a. If the large ribosomal bands in a given lane are not sharp, but appear as a smear of smaller sized RNAs, it is likely that the RNA sample suffered major degradation during preparation. Alternatively, capillary electrophoresis instruments such as the Bioanalyzer 2100 (Agilent Technologies), the Experion (BioRad) or the QIAexcel (QIAGEN) can be used to analyze the quality of the extracted RNA. If using the Agilent system in combination with common Pico or Nano LabChips (Fig. 3b), you have to keep in mind that high amounts of small RNAs purified with the methods described above can have a negative impact on RIN values, since the algorithm behind the RIN calculation assesses these

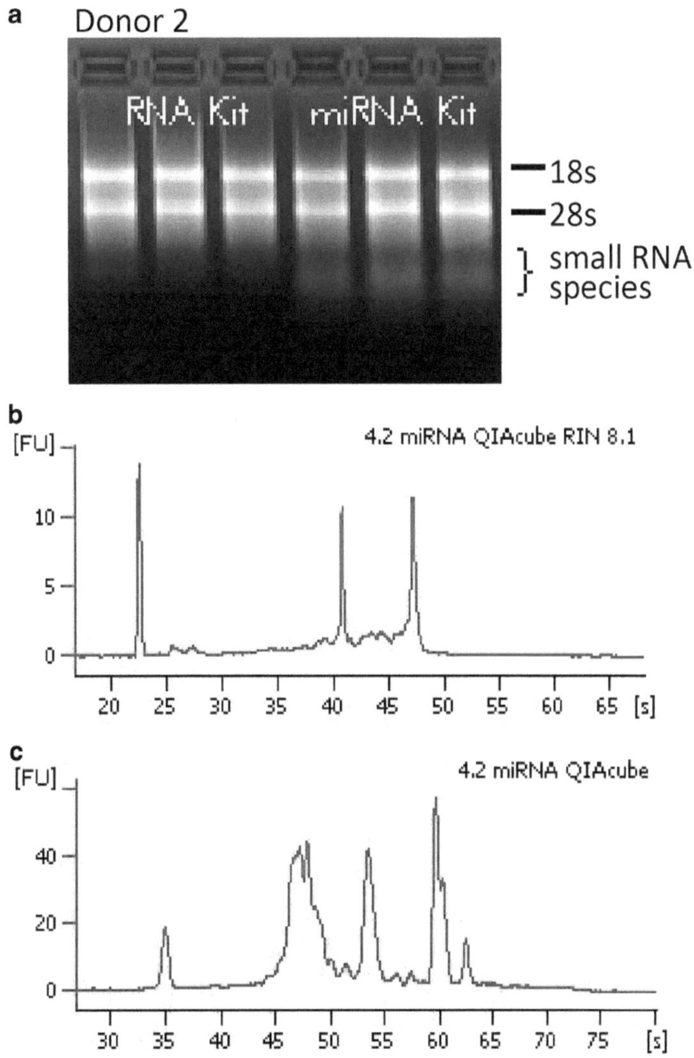

Fig. 3. Determination of quality of total RNA including small RNAs. Analysis of RNA eluates by denaturating agarose gel electrophoresis (**a**) or by using Agilent Bioanalyzer in combination with Agilent Nano LabChip designated for analysis of total RNA (**b**) or Small RNA LabChip designated for analysis of small RNAs in the range of 6–150 nt (**c**). The *y*-axis represents fluorescence units (FU) and the *x*-axis runtime in seconds(s). In (**a**), the enrichment of LMW RNA using the PAXgene Blood miRNA extraction kit in comparison to the PAXgene Blood RNA extraction kit is clearly visible. In (**b**) and (**c**), the total RNA samples from the same donor prepared by the use of the PAXgene Blood miRNA kit was loaded on the two different chips. (**b**) shows the whole RNA spectrum with the small RNAs at 23–28s, whereas (**c**) shows the small RNA region in detail with distinct peaks for 5S rRNA, 5.8S rRNA, and different tRNAs.

small RNAs as degradation products. So, the RIN and similarly of the RQI (RNA Quality Index) of the Experion system are not the method of choice to quality-control RNA eluates containing

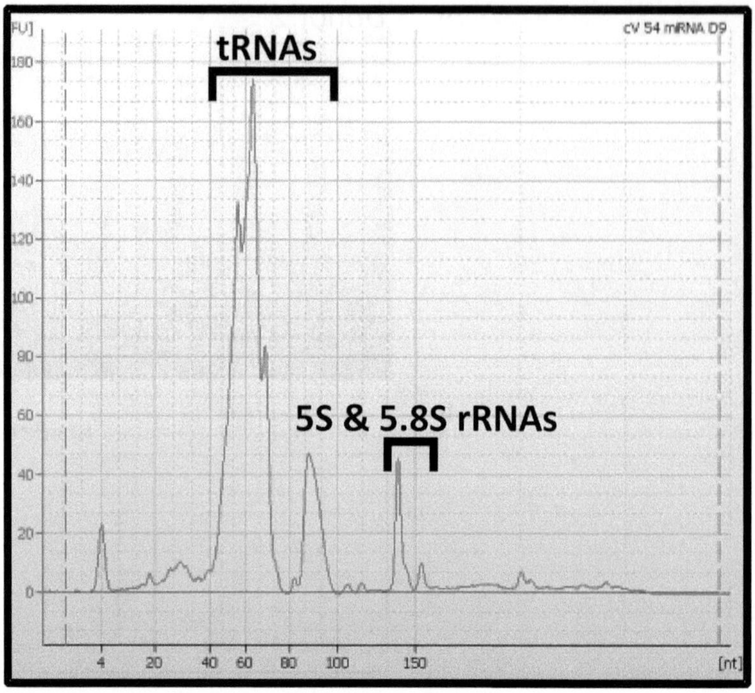

Fig. 4. Determination of quality of LMW RNAs. LMW RNA eluate was analyzed with the Agilent Small RNA LabChip. The *y*-axis represents fluorescence units (FU) and the *x*-axis nucleic acid length in nucleotides (nt).

small RNAs. For the Agilent instrument, a special chip (small RNA LabChip) system is offered that allows a more detailed look at the small RNA species (Fig. 3c).

3.3.2. Quantification and Determination of Quality of LMW RNA

Compared to total RNA preparation procedures eluates generated with methods that are designed to enrich the LMW RNA fraction often contain low RNA concentrations. Therefore results of spectrophotometrical measurements can often not be used for quality control. For small RNA size distribution and quality control, these samples can be analyzed with the Agilent Bioanalyzer small RNA LabChip as well. Using this system, LMW RNA samples of good quality as seen in the electropherogram in Fig. 4 show clearly visible and distinct peaks for 5 and 5.8S rRNA as well as tRNAs.

4. Note

Please carefully follow all special notes which were already included during the description of the protocols if necessary and also always follow manufactures recommendations.

Acknowledgments

Protocols referring to the PAXgene Blood miRNA Kit, PAXgene Blood RNA Kit, QIAamp Criculating Nucleic Acid Kit, miRNeasy Mini Kit, RNeasy Mini kit, RNeasy MinElute Cleanup Kit are established by QIAGEN and were mainly cited literally. We are grateful to receive copyright permission.

References

1. Boldrick, J. C., Alizadeh, A. A., Diehn, M., Dudoit, S., Liu, C. L., Belcher, C. E., et al (2002) Stereotyped and specific gene expression programs in human innate immune responses to bacteria. *Proc Natl Acad Sci USA* **99**, 972–7.
2. Chaussabel, D., Allman, W., Mejias, A., Chung, W., Bennett, L., Ramilo, O., et al (2005) Analysis of significance patterns identifies ubiquitous and disease-specific gene-expression signatures in patient peripheral blood leukocytes. *Annals New York Acad Sci* **1062**, 146–54.
3. Horwitz, P. A., Tsai, E. J., Putt, M. E., Gilmore, J. M., Lepore, J. J., Parmacek, M. S., et al (2004) Detection of cardiac allograft rejection and response to immunosuppressive therapy with peripheral blood gene expression. *Circulation* **110**, 3815–21.
4. Kittleson, M. M., Ye, S. Q., Irizarry, R. A., Minhas, K. M., Edness, G., Conte, J. V., et al (2004) Identification of a gene expression profile that differentiates between ischemic and nonischemic cardiomyopathy. *Circulation* **110**, 3444–51.
5. Chaussabel, D., Quinn, C., Shen, J., Patel, P., Glaser, C., Baldwin, N., et al (2008) A modular analysis framework for blood genomics studies: application to systemic lupus erythematosus. *Immunity* **29**, 150–64.
6. Burczynski, M. E., and Dorner, A. J. (2006) Transcriptional profiling of peripheral blood cells in clinical pharmacogenomic studies. *Pharmacogenomics* 7, 187–202.
7. Lonneborg, A., Aaroe, J., Dumeaux, V., and Borresen-Dale, A. L. (2009) Found in transcription: gene expression and other novel blood biomarkers for the early detection of breast cancer. *Expert Rev Anticancer Ther* **9**, 1115–23.
8. Chaussabel, D., Pascual, V., and Banchereau, J. (2010) Assessing the human immune system through blood transcriptomics. *BMC Biol* **8**, 84.
9. Staratschek-Jox, A., Classen, S., Gaarz, A., Debey-Pascher, S., and Schultze, J. L. (2009) Blood-based transcriptomics: leukemias and beyond. *Expert Rev Mol Diagn* **9**, 271–80.
10. Chitwood, D. H., and Timmermans, M. C. Small RNAs are on the move. *Nature* **467**, 415–9.
11. Ghildiyal, M., and Zamore, P. D. (2009) Small silencing RNAs: an expanding universe. *Nat Rev* **10**, 94–108.
12. Visone, R., and Croce, C. M. (2009) MiRNAs and cancer. *Am J Pathol* **174**, 1131–8.
13. Lagos-Quintana, M., Rauhut, R., Lendeckel, W., and Tuschl, T. (2001) Identification of novel genes coding for small expressed RNAs. *Science* **294**, 853–8.
14. Bartel, D. P. (2004) MicroRNAs: genomics, biogenesis, mechanism, and function. *Cell* **116**, 281–97.
15. Valencia-Sanchez, M. A., Liu, J., Hannon, G. J., and Parker, R. (2006) Control of translation and mRNA degradation by miRNAs and siRNAs. *Genes Dev* **20**, 515–24.
16. Iorio, M. V., Ferracin, M., Liu, C. G., Veronese, A., Spizzo, R., Sabbioni, S., et al (2005) MicroRNA gene expression deregulation in human breast cancer. *Cancer Res* **65**, 7065–70.
17. Subramanian, S., Lui, W. O., Lee, C. H., Espinosa, I., Nielsen, T. O., Heinrich, M. C., et al (2008) MicroRNA expression signature of human sarcomas. *Oncogene* **27**, 2015–26.
18. Calin, G. A., Ferracin, M., Cimmino, A., Di Leva, G., Shimizu, M., Wojcik, S. E., et al (2005) A MicroRNA signature associated with prognosis and progression in chronic lymphocytic leukemia. *N Engl J Med* **353**, 1793–801.
19. Calin, G. A., Liu, C. G., Sevignani, C., Ferracin, M., Felli, N., Dumitru, C. D., et al (2004) MicroRNA profiling reveals distinct signatures in B cell chronic lymphocytic leukemias. *Proc Natl Acad Sci USA* 101, 11755–60.
20. Lawrie, C. H., Soneji, S., Marafioti, T., Cooper, C. D., Palazzo, S., Paterson, J. C., et al (2007) MicroRNA expression distinguishes between germinal center B cell-like and activated B cell-

like subtypes of diffuse large B cell lymphoma. *Int J Cancer* **121**, 1156–61.

21. Mattie, M. D., Benz, C. C., Bowers, J., Sensinger, K., Wong, L., Scott, G. K., et al (2006) Optimized high-throughput microRNA expression profiling provides novel biomarker assessment of clinical prostate and breast cancer biopsies. *Mol Cancer* **5**, 24.
22. Dai, Y., Huang, Y. S., Tang, M., Lv, T. Y., Hu, C. X., Tan, Y. H., et al (2007) Microarray analysis of microRNA expression in peripheral blood cells of systemic lupus erythematosus patients. *Lupus* **16**, 939–46.
23. Poy, M. N., Eliasson, L., Krutzfeldt, J., Kuwajima, S., Ma, X., Macdonald, P. E., et al (2004) A pancreatic islet-specific microRNA regulates insulin secretion. *Nature* **432**, 226–230.
24. Sarnow, P., Jopling, C. L., Norman, K. L., Schutz, S., and Wehner, K. A. (2006) MicroRNAs: expression, avoidance and subversion by vertebrate viruses. *Nat Rev Microbiol* **4**, 651–9.
25. Skog, J., Wurdinger, T., van Rijn, S., Meijer, D. H., Gainche, L., Sena-Esteves, M., et al (2008) Glioblastoma microvesicles transport RNA and proteins that promote tumour growth and provide diagnostic biomarkers. *Nat Cell Biol* **10**, 1470–6.
26. Valadi, H., Ekstrom, K., Bossios, A., Sjostrand, M., Lee, J. J., and Lotvall, J. O. (2007) Exosome-mediated transfer of mRNAs and microRNAs is a novel mechanism of genetic exchange between cells. *Nat Cell Biol* **9**, 654–9.
27. Cortez, M. A., and Calin, G. A. (2009) MicroRNA identification in plasma and serum: a new tool to diagnose and monitor diseases. *Expert Opin Biol Ther* **9**, 703–11.
28. Keller, A., Leidinger, P., Lange, J., Borries, A., Schroers, H., Scheffler, M., et al (2009) Multiple sclerosis: microRNA expression profiles accurately differentiate patients with relapsing-remitting disease from healthy controls. *PLoS One* **4**, e7440.
29. Murata, K., Yoshitomi, H., Tanida, S., Ishikawa, M., Nishitani, K., Ito, H., et al (2010) Plasma and synovial fluid microRNAs as potential biomarkers of rheumatoid arthritis and osteoarthritis. *Arthritis Res Ther* **12**, R86.
30. Tan, K. S., Armugam, A., Sepramaniam, S., Lim, K. Y., Setyowati, K. D., Wang, C. W., et al (2009) Expression profile of MicroRNAs in young stroke patients. *PLoS One* **4**, e7689.
31. Hoekstra, M., van der Lans, C. A., Halvorsen, B., Gullestad, L., Kuiper, J., Aukrust, P., et al (2010) The peripheral blood mononuclear cell microRNA signature of coronary artery disease. *Biochem Biophys Res Commun.* **394**, 792–7.
32. Chen, X., Ba, Y., Ma, L., Cai, X., Yin, Y., Wang, K., et al (2008) Characterization of microRNAs in serum: a novel class of biomarkers for diagnosis of cancer and other diseases. *Cell Res* **18**, 997–1006.
33. Hausler, S. F., Keller, A., Chandran, P. A., Ziegler, K., Zipp, K., Heuer, S., et al (2010) Whole blood-derived miRNA profiles as potential new tools for ovarian cancer screening. *Br J Cancer* **103**, 693–700.
34. Keller, A., Leidinger, P., Borries, A., Wendschlag, A., Wucherpfennig, F., Scheffler, M., et al (2009) miRNAs in lung cancer - studying complex fingerprints in patient's blood cells by microarray experiments. *BMC Cancer* **9**, 353.
35. Leidinger, P., Keller, A., Borries, A., Reichrath, J., Rass, K., Jager, S. U., et al (2010) High-throughput miRNA profiling of human melanoma blood samples. *BMC Cancer* **10**, 262.
36. Resnick, K. E., Alder, H., Hagan, J. P., Richardson, D. L., Croce, C. M., and Cohn, D. E. (2009) The detection of differentially expressed microRNAs from the serum of ovarian cancer patients using a novel real-time PCR platform. *Gynecol Oncol* **112**, 55–9.
37. Chen, J., Lozach, J., Garcia, E. W., Barnes, B., Luo, S., Mikoulitch, I., et al (2008) Highly sensitive and specific microRNA expression profiling using BeadArray technology. *Nucleic Acids Res* **36**, e87.
38. Gaarz, A., Debey-Pascher, S., Classen, S., Eggle, D., Gathof, B., Chen, J., et al (2010) Bead array-based microrna expression profiling of peripheral blood and the impact of different RNA isolation approaches, *J Mol Diagn* **12**, 335–44.
39. Kruhoffer, M., Dyrskjot, L., Voss, T., Lindberg, R. L., Wyrich, R., Thykjaer, T., et al (2007) Isolation of microarray-grade total RNA, microRNA, and DNA from a single PAXgene blood RNA tube. *J Mol Diagn* **9**, 452–8.

INDEX

Jian-Bing Fan (ed.), *Next-Generation MicroRNA Expression Profiling Technology: Methods and Protocols*, Methods in Molecular Biology, vol. 822, DOI 10.1007/978-1-61779-427-8, © Springer Science+Business Media, LLC 2012